The "Peak Oil" Scare and the Coming Oil Flood

The "Peak Oil" Scare and the Coming Oil Flood

Michael C. Lynch

Foreword by Leonardo Maugeri

 PRAEGER™

An Imprint of ABC-CLIO, LLC
Santa Barbara, California • Denver, Colorado

For Nazli, who got me started
For Morry, friend and mentor,
For Kozue and her patience and support
But mostly, my Dad, Robert L. Lynch

Library of Congress Cataloging-in-Publication Data

Names: Lynch, Michael C., author.
Title: The "peak oil" scare and the coming oil flood / Michael C. Lynch ;
    foreword by Leonardo Maugeri.
Description: Santa Barbara, California : Praeger, [2016] | Includes
    bibliographical references and index.
Identifiers: LCCN 2016009907 (print) | LCCN 2016016174 (ebook) |
    ISBN 9781440831867 (hardcopy : alk. paper) | ISBN 9781440831874 (ebook)
Subjects: LCSH: Petroleum products—Prices. | Petroleum industry and trade. |
    Petroleum reserves.
Classification: LCC HD9560.4 .L963 2016 (print) | LCC HD9560.4 (ebook) |
    DDC 333.8/232—dc23
LC record available at https://lccn.loc.gov/2016009907

ISBN: 978–1–4408–3186–7
EISBN: 978–1–4408–3187–4

20 19 18 17 16   1 2 3 4 5

This book is also available as an eBook.

Praeger
An Imprint of ABC-CLIO, LLC

ABC-CLIO, LLC
130 Cremona Drive, P.O. Box 1911
Santa Barbara, California 93116-1911
www.abc-clio.com

This book is printed on acid-free paper ∞

Manufactured in the United States of America

Contents

Figures and Tables

FIGURES

TABLES

Foreword

The oil price collapse of 2014–2015 has caught most experts by surprise, proving once again the apparent unpredictability of the global petroleum market. Much of the reason for that unpredictability lies in persistent analytical errors and decades-old flaws in the approach to the study of oil.

As one of the few who correctly predicted (in 2012) the possibility of the 2014 oil price collapse, and also indicated the right reasons for that collapse (sometimes there are correct predictions based on incorrectly identified causes), and previously anticipated other major oil trends and phenomena by going against the prevailing views, I've always been surprised by observing how poor and superficial were (and remain) the analytical instruments used by most would-be experts to penetrate the apparent mysteries of the oil market.

That's incomprehensible, given the huge importance of petroleum to society.

As one of the world's largest industries and the source of half of the commercial energy consumed, petroleum market fluctuations ripple throughout the global economy, and lower prices hold the potential for a variety of side effects—some benevolent, some benign, and some troublesome. At the same time, uncertainty about whether prices will remain low or recover bedevils not just petroleum producers, but plenty of industries, from transportation to recreation. Inflation rates, stock and bond prices, and government deficits are all affected by the price of oil and accordingly plagued by this uncertainty.

Which explains the attention the oil industry receives from both analysts and the media. Unfortunately, during times of high prices that attention can be deleterious as many novices are attracted to write about a popular subject and their work usually reflects their inexperience.

Not just journalists but bankers, physicists, urban planners and even oceanographers have felt the need to write what they consider to be the definitive analysis about the future of oil supplies, typically repeating truisms that they presume to be valid, often without even minimal fact-checking.

Sadly but unsurprisingly, the field is rife with error and myth. In particular, the past decade has seen much ink spilled on the question of "peak oil," the point at which global oil production ceases to grow and begins to decline, causing havoc with the global economy.

From a 1998 article in the magazine *Scientific American*, the issue has exploded into dozens of books, thousands of articles, and even a number of video documentaries, all repeating a few superficial claims and theories.

Michael Lynch has examined the theories and research of those arguing that peak oil is imminent to demonstrate precisely what errors they have made and how so many were led astray. Beyond that, he shows that many others have seized on their theories for their own purposes, from oil company executives to leaders of producing countries, from renewable energy providers to environmentalist groups, without apparently understanding either the arguments or their weaknesses.

More astonishing is the extent to which so many decision-makers appear to be unaware of the history of resource behavior and especially the 1970s debate, which almost precisely mirrors the arguments of the past decade. And the many neo-Malthusian fears about resource availability more generally all suffer from the same theoretical mistakes as peak oil advocates, which the theorists then ignore while insisting that their arguments are correct, claiming their only error was in the timing.

The result has been huge economic losses, as valuable resources like natural gas were left in the ground to provide for "future generations" while the use of coal, the dirtiest of the fossil fuels, was encouraged globally. Some governments, such as Canada, restrained natural gas exports when prices were high in the early 1980s, then let them soar after prices collapsed.

To this debate, Lynch brings an array of analytical arguments and large amounts of data, as well as a sense of both theory and history, to demonstrate to the reader precisely what we know about oil supply and what we don't. In particular, he elucidates the difference between what is physically possible and what is likely given expected political trends; that chasm between the possible and the likely illuminates the precise challenges facing the industry and their chances of resolution.

Resource nationalism has been a special problem for the petroleum industry, not only for the usual reasons, but because oil prices are so volatile that many governments find it difficult to create fiscal policies that will be optimal in all likely circumstances. And where oil revenues are an important part of government budgets, the volatility creates a wide variety of problems.

Combined with arguments that oil prices must increase sharply over the long term, many politicians are thus encouraged to take an aggressive stance towards possible interested upstream investors and lose money when explorers choose to look elsewhere. This often creates a positive feedback effect, where high prices encourage governments to slow investment, which further tightens supply. But this works in reverse as well, and the current low prices are likely to see countries like Iran, Mexico, and Venezuela reform their fiscal and legal systems to attract foreign oil companies and boost supply.

Now, facing a new era of energy production from shale, many of the same voices are insisting that the technical and political challenges cannot be overcome, and that production from shale oil and gas will prove to be short-lived phenomenon at best, or a source of expensive supply that will put a high floor on the price of oil and gas worldwide. Others insist that ongoing progress will mean ever lower production costs and higher output.

The economic and geopolitical implications are enormous. A partial shift of global petroleum production from the Middle East to North America will change the political dynamics of one of the world's most unstable regions, and could affect U.S. military and foreign policy. Lower energy prices should mean more rapid economic growth in many parts of the world, with reduced inflation and improved incomes for most nations, possibly ushering in an era of stability such as was experienced in the United States in the 1990s.

However, a drop in oil revenues would also entail potentially negative consequences, including economic pressures in oil producing nations that could mean political turmoil and even cause more oil price instability. Higher economic growth and lower energy prices will also mean higher emissions unless action is taken to offset them, while at the same time making many other energy sources less competitive.

This book brilliantly demonstrates the value of intensive research in clarifying complex issues and the danger of relying on superficial analyses and writings, even by apparent experts. The future of energy contains many uncertainties, but this work goes a long way towards reducing them.

Leonardo Maugeri,
Senior Fellow, Belfer Center for Science and International Affairs,
Harvard Kennedy School, Harvard University
Author of *Beyond the Age of Oil: The Myths, Realities, and Future of Fossil Fuels and Their Alternatives*
Rome and Boston, February 2016

ONE

The End of (Oil) Days

According to Professor Kenneth Deffeyes, today (as of this writing) is the peak in world oil (all liquids) production, ... plus or minus three weeks. We hope all our readers have adequately prepared for the occurrence which Colin Campbell has suggested could lead to the extinction of mankind.[1]

At a peak oil conference in Cork, Ireland, in September 2007, James Schlesinger, the former U.S. Secretary of Defense, former head of the U.S. Central Intelligence Agency, PhD economist, and, most important of all, the first U.S. Secretary of Energy, announced that the debate about peak oil had been won. The reaction from the audience was undoubtedly triumphant, with perhaps a mixture of relief at having won a long, contentious debate. Now, the fight for postoil civilization could begin.

Apparently no one in the audience was aware that Schlesinger, upon leaving the office of Secretary of Energy in August 1979, had proclaimed that the production was unlikely to rise from its then level of 65 million barrels per day. Since then, the production has increased by another 25 million barrels per day, mostly at prices lower than when he left office, ignoring his proclamation the same way as the tide ignored King Canute.

So begins the strange saga of how a handful of retired geologists convinced thousands and thousands of people, several oil companies, and numerous investment bankers around the world that a peak in oil production was imminent and unavoidable, leading to "$500 a barrel oil," "$20 a gallon gasoline," "the rationing of oil supplies," "social and political unrest," "food riots," "the end of globalization and pristine wilderness," and "the possible extinction of mankind." Oh, and the ultimate horror: "So forget about that avocado salad in the middle of winter."[2]

Except: the writers arguing for an oil peak are not experts in the subject of forecasting oil supply, their theories and methods have been repeatedly shown to be wrong, and most of their primary points are either vague or irrelevant. And to top it off, their predictions have proven not just erroneous but rashly so. And no mineral has ever experienced such a peak, so they are casually predicting—with great confidence—that something that has never occurred is imminent, and without the world's experts realizing it.

OIL MATTERS

As Daniel Yergin wrote in *The Prize,* "Of all the energy sources, oil has loomed the largest and the most problematic because of its central role, its strategic character, its geographic distribution, the recurrent pattern of crisis in its supply."[3]

The oil industry is beyond a doubt the world's most important industry, and oil is the most important commodity on the globe. Not only does oil power many industries and play a role in providing electricity to others, but nearly every manufactured good, and especially consumer good, is delivered by oil-based transportation, sometimes exclusively. Few, if any of us, have our things delivered to our doorstep by electric-powered freight trains. Everything from the food we eat to the flowers that decorate the table is delivered by oil-based transportation.

(My local "natural" grocery store has sometimes had a sign saying that some carrots for sale had been delivered by bicycle from a local farm. Nearby is a display of mineral water imported from Italy, presumably not by bicycle.)

In the industrialized world (and increasingly, the Third World), most people rely on oil-fired cars for their employment. Hundreds of millions of people not only drive to work every day, but many of them rely on their vehicles to do their jobs. Beyond that, the world petrochemical industry relies on oil and gas to create a variety of products, from fertilizers to medicine to plastics. Indeed, many argue that we are really "eating oil" given the heavy reliance of agriculture on oil-based fertilizers, oil-fired machinery, and oil-powered transportation.

Any number of countries are all but single-commodity economies, from Abu Dhabi to Venezuela, with a number of others, such as Yemen and Mexico, receiving a significant portion of their income from oil revenues. Probably the most important of these is Russia, where oil and gas production make up nearly 20% of its economy and provide most of its export earnings and much of its government budget. Indeed, even the U.S. government gets approximately $20 billion a year just in corporate taxes from the oil industry (beyond gasoline and other sales taxes).[4]

While the world's wealthiest individuals are no longer the owners of oil trusts (like John D. Rockefeller, everyone's whipping boy in the 19th century), three of the ten largest U.S. companies are still oil companies. Of course, Rockefeller provided a reliable service, unlike, say, Bill Gates, whose Microsoft is notorious for buggy software. Indeed, the name of his trust, Standard Oil, was meant to portray that very fact that the kerosene lamp oil he sold was made to reliable quality standards. Still, this fact hardly endeared him (or his successors) to the public at large.

THE MOVEMENT

At present, googling "peak oil" brings up nearly 15 million hits; the Association for the Study of Peak Oil includes 23 national chapters holding numerous meetings around the globe, ranging from an annual international conference to talks to community groups.[5] There are dozens of books on the subject, and the number of articles is beyond count. The press has given enormous attention to this subject, with stories on CNN and NPR, a major article in *Newsweek,* as well as more specialized magazines like the *Oil & Gas Journal* and *Petroleum Economist.*

There have been peak oil specials on the Discovery Channel, the National Geographic Channel, and the History Channel (part of its *Mega Disasters* series), as well as not one but four DVDs explicitly on peak oil, including the award-winning documentary "A Crude Awakening." Okay, the last only won the prize at the Zurich film festival, but a win is a win.

The Internet contains many Websites devoted exclusively to peak oil theories, and a few debunking them. Particularly useful is www.oilcrisis .com, which archives many of the papers written by prominent peak oil theorists.[6] Others such as www.dieoff.com, www.theoildrum.com, and www.peakoil.com include news, commentary, and discussion. God only knows how many political, policy, and investment Websites include at least some material devoted to the subject, occasionally debunking it but usually presenting it uncritically.

There is advice on how to survive peak oil, the effect on food prices, housing, medicine, transportation, the need for urban planning reform, and many socioeconomic factors. There's even a *Post-Petroleum Survival Guide and Cookbook,* as well as a book on prison reform by Michael E. Lynch (no relation) with a chapter on the effect of peak oil on prison policies. And that ultimate arbiter of pop culture, *The Simpsons,* has featured peak oil in an episode on survivalists.

THE DANGER

With oil prices in the past decade reaching two, three, and five times the historic level, economic damage is clearly occurring, with consumers having to shell out more and more to fill their tanks, tourist sites losing business as people cut back on travel, airlines suffering even more than usual (if that's possible), and transportation firms seeing their profits evaporate like gasoline on Texas asphalt. Even food prices have been affected, as the price of fertilizer and trucking costs have both risen sharply, while farm acreage is increasingly devoted to producing biofuels instead of food.

Yet peak oil is like antibiotic-resistant pneumonia compared to the 2008 oil price spike's cold: far worse *and* incurable. Oil production supposedly drops by 3–5% per year instead of increasing by 1.5% per year. Offsetting this "gap" requires enormous amounts of conservation— prices so high that $147/barrel would seem cheap—and creates a massive global recession. The extinction of humankind, as Campbell suggests, is hardly likely but virtually every aspect of life would be far more severely affected.

This is particularly true given the fact that many nations still use limited amounts of petroleum. China has five times the population of the United States, but uses half of the petroleum, a typical case. As these nations develop economically, they will need ever-larger amounts of oil and would not like to hear that it needs to be conserved so that wealthy Westerners can take pleasure cruises in their humongous vehicles.

THE FUTURE IS NOW!

In March 1998, the magazine *Scientific American* published an alarming article, "The End of Cheap Oil," which predicted that global oil production would soon peak and begin to decline and that nothing could be done to prevent it. Coming as it did when oil prices were collapsing, it received only a modest degree of attention. However, a mere six years later, one of the authors was featured on the front page of the *Wall Street Journal:*

> Colin Campbell got a phone call that made him shriek with joy. "Holy Mother!" he yelped after he put down the receiver. "The good ol' moment's arrived!" Oil prices had just hit $50 a barrel, confirming (he thought) his theories and years of warnings.[7]

Other peak oil advocates insist that the point has come and gone. *Playboy* magazine also found itself warned by the oilman T. Boone Pickens in its January 2007 issue.

PLAYBOY: But are we capped out? What about untapped reserves?
PICKENS: What untapped reserves? We're currently getting 85 million barrels of oil a day worldwide and using it all. We won't be getting more.

The most extreme views of the imminence of peak oil come from the late Matthew Simmons, a Houston banker who served the oil industry for decades; T. Boone Pickens, former CEO of Mesa Petroleum; and Kenneth Deffeyes, a professor emeritus of geology at Princeton University. This august collection of apparent experts believe that we passed the peak in world oil production in May 2005 (or thereabouts).

They base this belief largely on the data from the U.S. Department of Energy for crude and condensate production, that is, excluding natural gas liquids and biofuels. As Figure 1.1 shows, production from this category did in fact decline after May 2005.

Unfortunately for this trio, production has since risen about 5 million barrels per day from this level. In fact, a week after Simmons presented the slide in Figure 1.1, data were released showing that production had surpassed his supposed all-time peak!

And a longer view, going back to, say, 1995 (Figure 1.2), shows that such peaks and declines are not at all uncommon. The fluctuations reflect a combination of factors, including weather events (Katrina), accidents, political disruptions (invasion of Iraq), seasonal factors (summer shut-ins for maintenance), and cutbacks by OPEC to strengthen prices.

For example, in the late 1990s, OPEC cut production in response to low demand and prices, as well as after the 9/11 terrorist attacks in 2001, when the subsequent economic disruption led to another cutback by OPEC. President Chavez of Venezuela, by firing about half the employees of the

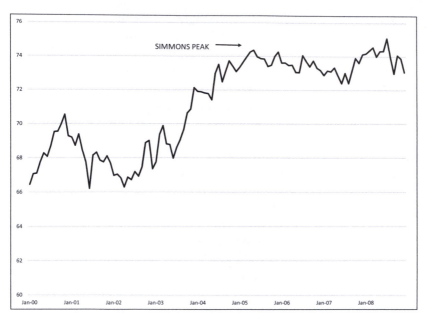

Figure 1.1   Production Data Thought to Show a Peak (Energy Information Administration, US Department of Energy)

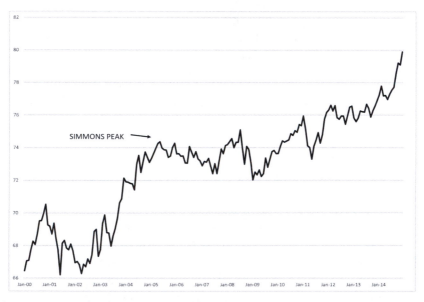

Figure 1.2   Updated Production Data (Energy Information Administration, US Department of Energy)

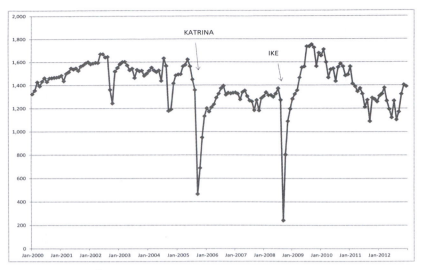

Figure 1.3   Gulf of Mexico Oil Production (Energy Information Administration, US Department of Energy)

state oil company in early 2003, sacrificed about half a million barrels per day of Venezuelan oil production since then (and probably more, but the data are contested). The invasion of Iraq in 2003 caused a loss of roughly 1 million barrels per day for at least five years. Since 2005, Hurricanes Katrina and Ike cut U.S. production by about 300 thousand barrels per day (Figure 1.3), with the Macondo oil disaster in 2010 slowing production further, and the unrest in Nigeria led to a loss of about 600 thousand barrels per day of high-quality oil production for an extended period. And, of course, the price collapse in late 2008 saw OPEC reduce production by about 3 million barrels per day. None of these reflect permanent or geological problems, but that is how they are being interpreted.

Those peak oil advocates who think this is the all-time peak have not explained why this particular peak is any different from earlier ones: in fact, much of the argument is based on the assumption that production, having dropped, can never increase again. This argument is not only counterfactual but obviously so—to everyone but them.

WHAT IS PEAK OIL?

Much of what is written about peak oil either isn't clear or is contradicted by other writings. Indeed, many seem unaware of which theories have been discarded and which have been embraced. To summarize, peak oil theorists maintain two sets of views. The resource-based arguments maintain the following:

- World oil production has peaked or is about to peak, probably within two to four years, but possibly as early as mid-2005.
- Nearly all conventional oil has been discovered, and unconventional oil will not fill the gap for many years.
- The peak is or will be due to resource limitations, and as such nothing can be done to alter it significantly.
- Oil prices will soar, and the economic and social consequences will be severe.
- Unlike earlier warnings of resource exhaustion, the new research is based on reliable data, analyzed by scientists, and, as such, is trustworthy.

Alternatively, some peak oil advocates simply argue that, regardless of the resource base, it has become too difficult to keep increasing production. They say that the "easy" oil is gone, that the industry faces growing challenges as it moves into more difficult political and geographical environments, and that rising depletion makes it "run harder just to stay in place" such that oil production is unlikely to go much higher.

YOU'LL DO IT TIL YOU GET IT RIGHT

Hurry, before this wonderful product is depleted from Nature's laboratory!
                                    Ad for "Kier's Rock Oil" 1855

[T]he peak of [U.S.] production will soon be passed—possibly within three years.
                                    David White, chief geologist, USGS, 1919

[I]t is unsafe to rest in the assurance that plenty of petroleum will be found in the future merely because it has been in the past.
                                    Snider and Brooks, *AAPG Bulletin,* 1936[8]

Why, I'll drink every gallon produced west of the Mississippi!
                                    John Archbold, Standard Oil Executive, 1885.[9]

There you go again
                                    Ronald Reagan

Authors like Yergin, Porter, and Maugeri have noted the many previous predictions of peak oil, since the beginning of the oil industry, and Chapter 2 will describe others.[10] (More general fears of resource scarcity began millennia ago.) But few of the early warnings were based on the serious analysis of global resources until the 1970s and, as such, can hardly be taken very seriously. Certainly, they constitute little more than a curiosity in the history of forecasting.

But one example has stood out: the 1956 prediction by M. King Hubbert, a noted geophysicist, that U.S. oil production would peak somewhere between 1965 and 1970.[11] At the time, his work was derided not only because the industry was quite optimistic about the potential of U.S. supply but also because his method appeared too simplistic to represent such a complex subject. However, the peak did occur in 1971, and, as a result, his approach came to be accepted by many, based on no more evidence than that.

In the mid-1990s, his work began to be revisited by a number of petroleum geologists, most notably Jean Laherrere and Colin Campbell, who published two reports for a leading oil consulting firm, Petroconsultants, analyzing their proprietary database of world oil fields.[12] The work did not receive much attention at the time but was followed up in Campbell's 1997 book, *The Coming Oil Crisis,* and a March 1998 *Scientific American* article by the two authors, "The End of Cheap Oil."

Coming out at a time when oil markets were glutted and prices collapsing, the arguments were not embraced by the industry, nor did they receive much media coverage. However, with the rise in oil prices after 2003, the arguments were judged much more favorably by the media and public. Numerous peak oil books were rushed into publications, academics whipped up articles examining oil production trends, and various governments and international organizations have published reports discussing the issue, including the U.S. and German militaries.

CONVENTIONAL WISDOM

To disagree with peak oil does not necessarily imply a rosy future, however. Even those who believe that peak oil is a scientific dead end are pessimistic about oil's future, especially in terms of price. Nearly every major organization that makes regular forecasts predicts prices remaining high for the foreseeable future, as Figure 1.4 shows, far above levels that were experienced historically.

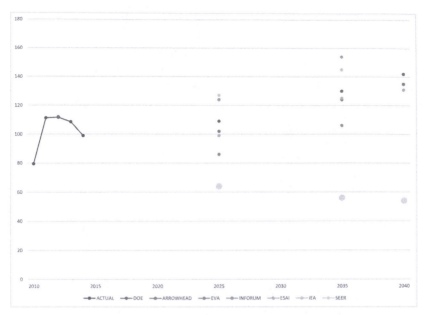

Figure 1.4   Long-Term Forecasts of Oil Prices (Energy Information Administration, US Department of Energy, Annual Energy Outlook, 2014)

Of course, a mere decade ago, those same voices called for prices to stay below $30 a barrel for the next two decades (Figure 1.5), highlighting just how treacherous economic forecasting can be. Old-timers in the oil business are loathe to react to such forecasts, but the younger Wall Street analysts, who don't remember the price collapse in 1998, let alone the one in 1986, pressure them to reinvest to maintain production levels, despite the rising costs.

This shift in expectations reflects the "cheap oil is gone" view that the industry and many analysts have. It explains the near-total consensus that prices will not decline and the increasing interest in expensive sources of energy, including renewables. So, theoretically, this view does not agree with the peak oil advocates, while functionally echoing some of their arguments.

(My own forecast, the lowest by far, has frequently come under ridicule. In June 2012, at the International Energy Seminar hosted in Vienna by OPEC, the moderator suggested that my forecast of a long-term price of $50–60 a barrel was a joke, and when I submitted the forecast in Figure 1.4, a friend at DOE joked that people wondered whether I was drunk.[13])

Two points are typically made: most future oil seems likely to come from OPEC, and the remaining oil is from fields in difficult environments

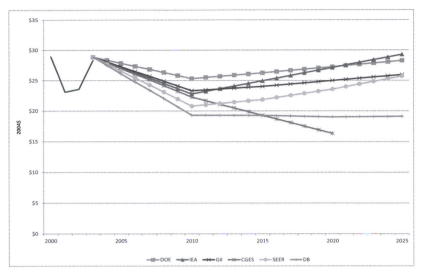

Figure 1.5   Long-Term Oil Price Forecasts from 2004 (Energy Information Administration, US Department of Energy, Annual Energy Outlook, 2005)

that require massive investment. As the previous CEO of the French oil company argued at the 2012 seminar, if price decreases below $100 a barrel, investment will reduce, markets will tighten, and prices will soar much higher. This is roughly what happened in 1998, when the price hit $12 and some industry voices feared it would go down further: drilling was reduced by 35%, the primary factor causing production there to plateau.

BILL CLINTON'S WORST KISS

When he was running for president, Bill Clinton argued that the best strategy to communicate ideas was "*Keep It Simple, Stupid,*" or KISS. Given the media's focus on soundbites, or short, clever comments, this argument proved pretty successful (he served two terms as president and his wife came relatively close to achieving the Democratic nomination eight years later, which certainly meets my definition of success).

To some extent, peak oil advocates follow this strategy, making certain clear-cut points that are obvious to everyone: oil is finite, oil is getting much harder to produce, oil discoveries aren't keeping pace with production, and so forth. All of these have been repeated endlessly without much more thought given to the issue, or even the relevance of these facts.

And, while there are many who disagree that the evidence supports an imminent peak in world oil production, few have analyzed the peak oil writings in any depth. Most scientists, even social scientists, prefer to do original work, not review that of others, and industry executives are far too focused on operational matters to devote much attention to something so hypothetical.

Enter your humble narrator—someone who has not only analyzed and forecasted oil and gas supply for decades but has the patience to work through most of the peak oil research, test the arguments, and compare them to other more conventional ideas.

The next section will examine the various approaches, methods, theories, and arguments (and myths) separately and in detail. By doing this, it shall be shown that the peak oil advocates are variously uninformed, unsophisticated researchers, and/or ignoring the pertinent facts to reach their conclusions.

The next part of the book will address the real issues the oil industry faces, including heightened political risks, the higher cost of production, and the rising decline rates in existing fields. Then, we shall turn to the future of oil supply, seeing not only that conventional oil supply is far from a peak but also that the revolution in production from shale using hydraulic fracturing will herald a new era of abundant supply. Finally, the way prices change and the impact of lower oil prices on both the global economy and the move toward new clean-energy technologies will be discussed in some detail.

All in all, it appears that the world is approaching a new oil age, where supplies will be abundant and cheap, with many benefits to consumers, challenges to producers (and competitors), and a potentially significant impact on the global environment.

TWO

Déjà vu All Over Again

The oil and natural gas we rely on for 75 percent of our energy are running out.
    President Jimmy Carter, televised speech on April 18, 1977

We learn from history that we never learn from history.
    George Bernard Shaw

Many will object that it is all but impossible for so many experts to be so wrong (ignoring the question of who the experts are and how expert they are), and the experience of the 1970s provides an excellent rejoinder. Most are aware that the expectations in the 1970s were seriously flawed, but even in the oil industry, few recall exactly how and why that occurred.[1] A careful examination of the record will provide a very illuminating comparison to the current debate.

DISCO DAYS

Prior to the 1970s, relatively little effort was made to forecast long-term oil markets, and most assumed that oil prices would be flat or perhaps declining slightly over the long term.[2] Constraints on oil supply were not considered significant, and in 1956 M. King Hubbert's prediction of a

U.S. oil production peak was actually rather exceptional (and, though ultimately prescient, controversial at the time).The rapid growth in oil
demand in the 1960s did put stress on both the upstream and downstream
segments of the industry, but at that time most oil-exporting nations were
happy to see their production expand, and the industry was generally able
to add enough capacity in areas where the effort was minimal, notably the
Middle East.

References to the "energy crisis" before 1973 were focused on product
shortages (gasoline and diesel fuel) and electricity blackouts, as booming
demand was hard for the industry to meet. (Regulated markets, which are
by nature prone to shortage, bear a large share of blame as well.) In fact,
upstream oil prices fell during most of the 1960s (relative to inflation).
In the early 1960s, M. A. Adelman had commented that prices would
likely decline, and they did, as inflation eroded the value of the nominal
price.[3]

The shift in power from oil companies to exporting governments
began when Libya's Gaddafi successfully pressured Occidental to
increase its per barrel tax payments in 1970, which inaugurated several
years of back-and-forth increases between Iran (the Gulf producers,
more generally) and Libya, albeit still relatively small compared to
what was to come. With the second Arab oil embargo in October 1973,[4]
and the attendant production cutbacks, a group of exporting nations
managed to impose much larger increases, ultimately taking prices
to $12 a barrel in 1974, more than tripling in a year ($35 per barrel in
2010 dollars).

But the first oil price shock in 1973 opened a great debate that had been
foreshadowed by the publication of the 1972 Club of Rome book *The Limits to Growth* and the Ford Foundation's Energy Policy Project, both of
which were concerned about the sustainability of exponential growth in
energy demand and its impact on the environment.

Now, with an actual "crisis" occurring, two schools of thought appeared
to explain the nature of the oil price increases and their durability. The primary discrepancy in interpretation concerned whether the price increases
could be explained as a result of OPEC behavior or the underlying fundamentals of the market. In this chapter, two threads will be followed to their
rather embarrassing conclusion: belief that prices had to rise exponentially
as a natural law of economics and expectations of continued scarcity—
both being completely wrong. (Sound familiar?)

PRICE EXPECTATIONS: FOOLS RUSH IN WHERE DEAD MEN HAVE PREVIOUSLY TROD

> Practical men, who believe themselves to be quite exempt from any intellectual influence, are usually the slaves of some defunct economist.
>
> John Maynard Keynes[5]

The annual meeting of the American Economic Association, the premier gathering of the nation's—and world's—great economic thinkers, is so large it is spread out over a variety of hotels (and a great many more bars). December 1973 was no different, except that the nation was then in the grip of an oil crisis, engendered by the decision of OAPEC (not OPEC) to declare an embargo against many consuming nations and with a massive production cutback to accompany it.

(The embargo proved nearly worthless, as the major oil companies managed to redirect supplies so that the supply shortage was roughly equally shared by importing nations, although foreign policy experts to this day continue to focus on the threat of embargo rather than the economic damage caused by higher prices, the real danger. But that fallacy is the subject of another book, although the reader would do well by referring to the book *The Genie out of the Bottle*.[6])

Naturally, the then-occurring spike in prices was addressed by the members of the profession, primarily two economists from MIT: Robert Solow, a leading macroeconomist who went on to win a Nobel Prize in Economics, and M. A. Adelman, a resource economist who had just published a book on the microeconomics of the petroleum industry. The former argued that an economic theory (described later in this chapter) proved that oil prices should rise exponentially, whereas the latter argued that the production cutbacks were causing higher prices.[7]

(Personal note: I consider Adelman my mentor, and, although I took one class with Solow—intermediate macroeconomics—my primary recollection is of his impersonation of an unemployed West Virginia mountaineer refusing to take a job at the local mill, "not for no eight dollars an hour, I won't." Despite being descended from a long line of West Virginians, I took no offense at the impersonation; he was, and is, a scholar and a gentleman—which is completely irrelevant to the validity of his interpretation of oil markets.)

History has clearly shown that Adelman's view was correct, hardly surprising since he was intimately familiar with the subject, but the arguments of the former soon became holy writ to many in the profession, who believed *that rising oil prices could be explained by natural economic law.* Specifically, a large number of economists adopted the argument that the resource economist Harold Hotelling had demonstrated in a 1931 article that resource prices had to rise exponentially, as profits had to rise at the real rate of interest.[8] Since oil production costs in the Middle East were said to be near zero, prices equaled profits, and so the market price of oil had to rise at the rate of interest, about 3% per year above inflation. Other minerals would rise exponentially, though not at such a high rate.

In fact, Hotelling's work referred to investment in and the profits from a specific deposit of known size, not to prices for a total resource, and a number of resource economists have argued this but in vain.[9] What is perhaps more astonishing is the contradiction between the theory and reality: oil prices have not risen exponentially at any point in history, nor have any other mineral prices shown such a path (Figures 2.1 and 2.2).

> An economist is someone who sees something work in practice and says, "Yes, but is the theory sound?"
>
> Ronald Reagan

Yet the fact that the so-called Hotelling theory is contradicted by the historical data proved no impediment to its being embraced by numerous

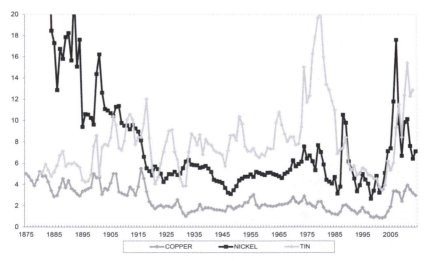

Figure 2.1   Mineral Prices (Log Scale) (US Geological Survey)

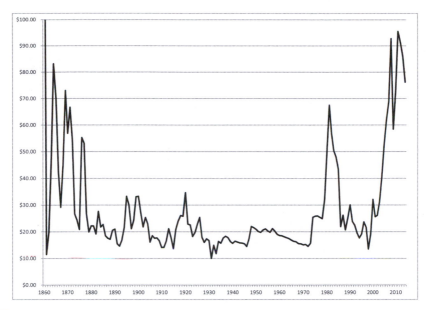

Figure 2.2   Long-Term Oil Prices (US Department of Energy)

economists over the decades and used to justify predictions of ever-rising prices. Indeed, any number of subsequent articles and books cited Solow's 1973 article about the Hotelling theory, while ignoring the arguments of Adelman, the resource economist, made at the same conference.[10]

To this day, it is easy to find economists trying to explain that: "The oft-cited fact that the Hotelling model is frequently rejected by the data ... must be interpreted with caution."[11] In other words, the theory still doesn't fit the facts. And the theoretical contortions made to "fix" the Hotelling theory so that it will approximate the actual behavior have been many—none with success. But too many economists tend to ignore the arguments, made by resource economists, that they are simply misinterpreting the theory. Indeed, it is surprising to see review articles such as that by Slade and Thille that don't even reference these pieces.[12]

THE EXPERTS CASH IN

With the rise in prices, the oil industry became a hot topic and economists (and others) begin to do research in the area. Initial studies after 1973 argued that higher prices could not be sustained, due to the presumed normal response of supply and demand.[13] These included short, relatively simplistic pieces by economists and also two major works, one by the

OECD in Paris, which projected energy markets out to 1985, and the other done as part of Project Independence, the Nixon administration's attempt to make the United States energy independent.[14]

Massive amounts of research were conducted as part of Project Independence, but the major policy finding was that energy independence was not only too expensive but also not very useful politically, since U.S. allies would still be dependent on oil imports, proving once again that (a) economists can't be trusted to do as they're told and (b) politicians never learn from history—not that anyone else does.

Sadly, the main lesson was that simplistic models produced simplistic results (something many have yet to learn). The models did not take into account depletion and, more important, the inflationary effect on drilling costs that a huge increase in drilling would generate. Or price controls in many countries, restricted access for exploration, rising taxes, nationalizations, and so on—all of which had the effect of making investment harder and raising apparent costs to companies, meaning higher prices didn't translate directly into higher investment levels and thus supply.

There was also a rush into the field by nonexperts. Energy economist Dr. A. Alhajji, at a 2011 conference in Boulder, remarked that in 1973 there was only one petroleum economist in the United States, but by the end of 1974, nine academic articles had been published on the subject by others who, with little knowledge, set the discourse in the literature and help draft energy policies.

Thus, many of the books on the 1970s energy crisis(es) were about related or tangential subjects, assuming as a starting point that resource scarcity was the cause of higher prices and that they would continue to grow.

GREEN DAWN: PARADIGM SHIFT

The gas lines and rapid increases in oil prices during the first half of 1979 are but symptoms of the underlying oil supply problem—that is, the world can no longer count on increases in oil production to meet its energy needs.[15]

Central Intelligence Agency, 1979

The second oil crisis, the Iranian Revolution of 1979, saw a major change in the beliefs of the analysts. Those who had argued that oil was abundant and likely-to-moderate prices were completely discredited (especially to those with the opposite point of view, as well as the media and policymakers). The theory that oil prices should rise inevitably and inexorably due to natural economic law was somehow considered validated by the political disruption of 6 million barrels a day of Iranian oil production.[16]

The tight oil markets that the Iranian Revolution created were interpreted as a sign of geological scarcity rather than a transient political event, and afterward nearly all forecasters became much more pessimistic about the amount of oil likely to be produced, especially from outside the Middle East. This, combined with the aforementioned belief in the so-called Hotelling theory, led to expectations of much higher prices, as Figure 2.3 shows far beyond reality.

Still this did not stop many from incorporating this theory into their forecasts. An excellent example is the trend in oil price forecasts by the U.S. Department of Energy, which regularly showed a rising price trend even as prices declined again and again. It is obvious that the *error* is to the *slope of the curve,* that is, the trend, but the correction was repeatedly made to the initial point, dropping it, but still showing the same rate of increase in the future (Figure 2.4).

This focus on the record of the Department of Energy is not because its work was somehow deficient: rather, it makes better targets than most other forecasters for the simple reason that it is required, by law, to publish a detailed annual forecast. Many others, particularly in the industry, slunk away in embarrassment after their earlier forecasts proved so devastatingly wrong and kept their expectations hidden from the public afterward.

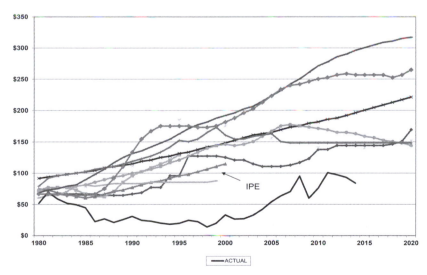

Figure 2.3   Computer Generated Oil Price Forecasts (1980) (Energy Modeling Forum 6 1982; This Author Worked on the IPE Model)

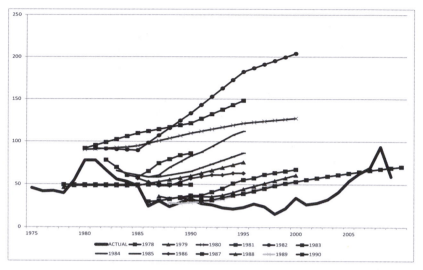

Figure 2.4 Evolution of U.S. Department of Energy Oil Price Forecasts during the 1980s (US Department of Energy, International Energy Outlook, various years)

PLAYING WITH THE TURK: VON MAELZEL'S GHOST

Perhaps no exhibition of the kind has ever elicited so general attention as the Chess-Player of Maelzel. ... But in reality ... the wonderful mechanical genius of Baron Kempelen could invent the necessary means for shutting a door or slipping aside a panel with a human agent too at his service.[17]

In the 1970s, personal computers did not exist and all modeling was done on mainframes requiring specialized knowledge of various computer programs. This excluded the general public from both creating models and examining the modeling done by others. In fact, at the time I personally owned a rubber stamp that said, "This came from a computer and is not to be questioned or disbelieved," a satiric—yet not inaccurate—view of the awe in which many held computer models in days of yore.

And while it might be asked how supposedly scientific, objective computer models could be so badly wrong, the reality is that the models of the oil market were typically Maelzelian machines, an automaton that was actually driven by its human operator. The price was "predicted" as a function of OPEC capacity utilization, according to the curve shown in Figure 2.5.[18] The only problem is that all of the models using this formula

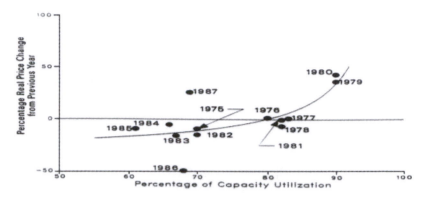

Figure 2.5   Price Formulation in Computer Oil Models (US Department of Energy, International Energy Outlook, 1988)

relied on *assumptions* of OPEC capacity, since upstream investment in OPEC responded primarily to government policies. (The IPE model was the only exception, of which I am aware.) In other words, the modeler could generate any price path desired by simply altering the assumed OPEC capacity—thus, the "man" in the "machine." Since the prevailing theory—Hotelling Principle—said prices should rise at 3% per year, the models were tweaked to produce exactly that result.

Thus, for nearly a decade, forecasts of ever-rising prices were driven to a large degree by expectations that modelers inserted into their models—expectations driven by misinterpretation of a long-dead economist's work. The price forecasts could be easily duplicated by merely taking the current price and assuming a growth rate of 3% per year above inflation, suggesting not just a brave new world of resource scarcity but price increases far above the historical experience. Figure 2.6 compares the forecasts from the mid-1980s, including those after the 1986 oil price collapse, and shows how unrealistic they appear in the historical context.

THE SHIFT TO RESOURCE PESSIMISM

Although, as peak oil advocate Ron Patterson pointed out to me, the 1973/74 Arab oil embargo and subsequent price increases were widely blamed on OPEC by the media, resource pessimism came to be the major factor behind the late-1970s' belief that prices would rise inevitably and sharply, although it was not as clearly defined or as severe as the present arguments made by peak oil theorists. A number of estimates of the

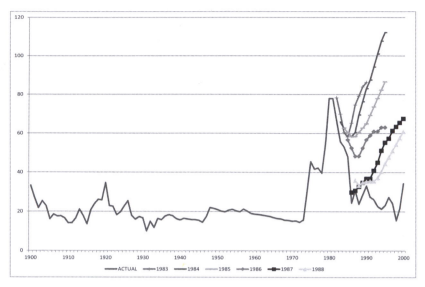

Figure 2.6   DOE 1980s Price Forecasts in Historical Context (US Department of Energy, International Energy Outlook, various years)

world's recoverable resource of petroleum suggested that the total amount was about 2 trillion barrels, implying that about 25% of the world's oil had already been consumed. (See Chapter 15) And given the expectations of continued high demand growth, despite the recent increase in prices, some argued that the peak could be only a couple of decades off.

Limits to Imagination

In 1972, a group of MIT computer modelers published the results of their study of world resource limitations, known as *The Limits to Growth,* which was sponsored by the Club of Rome, an interest group/think tank that has now become synonymous with neo-Malthusianism, or resource pessimism. This study, done by experts in modeling, not resources, assumed exponential growth in demand for various minerals and fuels and projected their depletion, given an assumed resource base. (Other factors, such as pollution and population growth, were also analyzed.)

The results were alarming and received widespread publicity, to the point that even today, the name *Club of Rome* is synonymous with concerns about exponential growth and resource scarcity. Critics have pointed out that the fear of resource exhaustion proved overblown: Ron Bailey notes they predicted gold would be exhausted by 1981; mercury by

1985; tin by 1987; zinc by 1990; petroleum by 1992; and copper, lead, and natural gas by 1993. None of these is now even in short supply after more than four decades of rising consumption.[19]

What went wrong? There were two faults, which are instructive. First, while they tested the effects of a variety of consumption growth rates, they did not have any economic feedbacks; that is, demand followed an exponential path whose rate of growth remained constant, even if the resource was on the verge of total depletion. And while the ranges of growth they predicated for petroleum, 2.9%, 3.9%, and 4.9%, was all below historical growth rates and thus seemed quite reasonable, the actual growth for the subsequent two decades proved to be even lower: 2.4%.

Dead Frog Ponds

A French riddle for children illustrates another aspect of exponential growth—the apparent suddenness with which it approaches a fixed limit. Suppose you own a pond on which a water lily is growing. The lily plant doubles in size each day. If the lily were allowed to grow unchecked, it would completely cover the pond in 30 days, choking off the other forms of life in the water. For a long time the lily plant seems small, and so you decide not to worry about cutting it back until it covers half the pond. On what day will that be? On the twenty-ninth day, of course. You have one day to save your pond.

The Limits to Growth, 29

This lack of feedback is clear from the book's reliance on a French riddle to explain why exponential growth is so dangerous. The implication is that, in this case, the lily grows so fast that you would be unable to stop it from killing the pond. However, this requires that you consciously wait until the last minute (the 29th day) and that the doubling time is faster than the time it takes to clear the pond. Who is to say that the farmer would need more than an hour to chop up the lily and has thus actually benefited from the availability of biomass? (I don't know any French farmers, but the French energy economists I know are all much too intelligent to stand by idly while their ponds die.)

Needless to say, there are very few dead ponds, metaphorical or otherwise, plaguing us in this fashion. Those who fear exponential growth are generally making these assumptions, namely, that we won't recognize that the growth is stressing the resources or that we won't be able to act fast enough. Many continue to argue that such is the case now, but they are basing that on a variety of assumptions, rather than real-world cases. (Or at least, relevant real-world cases; small islands like Easter Island hardly serve as surrogates for the global economy.)[20]

THE ZEITGEIST TIPS[21]

As mentioned earlier, there was very little pessimism about oil supply prior to the 1973 oil crisis, in part because companies seemed to be finding huge supplies in many places, especially the Middle East and North Africa, and a number of governments, such as Iran and Iraq, were putting pressure on the operating companies (mostly the Anglo-American majors) to increase production so as to raise their revenue.

But after the first oil crisis, the question began to be studied more urgently. Early work, such as Project Independence and the OECD's *Energy: Prospects to 1985,* assumed that there would be a supply response proportional to the price increase and that the tripling of prices in 1973/74 would bring on a significant increase in U.S. and other (non-OPEC) oil production.[22] (This is probably one reason that geologists sometimes argue that economists think higher oil prices "create" supplies.)

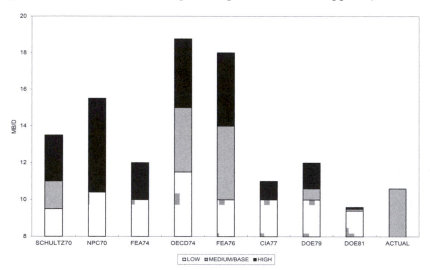

Figure 2.7   Forecasts of 1985 U.S. Oil Production (Lynch, Michael C., "Bias and Theoretical Error in Long-Term Oil Market Forecasting," in *Advances in the Economics of Energy and Natural Resources,* John R. Moroney, ed., JAI Press, 1994)

By 1979, when the Iranian oil crisis landed on markets like a ton of oil, there had been only a small supply response to the first oil price spike, and so most analysts became much more pessimistic. Figure 2.7 shows the evolution of forecasts for U.S. oil production through the 1970s. They had increased quite a bit after oil prices rose in the early 1970s, with expectations that 15 million barrels per day or more might be produced by 1985—not illogically, although it proved to be completely wrong. Beginning with the CIA's projection in 1977 (and following no notable growth in supply), expectations of 1985 supply became much more moderate.[23]

But the Iranian Revolution caused a sea change in the zeitgeist, or prevailing opinion. Between 1979 and 1981, the U.S. Department of Energy, for example, lowered its forecast of U.S. oil production in 1985 by over 1 million barrels per day even though higher prices resulted in an explosion in drilling, soaring from just over 10,000 oil wells in 1973 to over 44,000 in 1981. (Oddly, their original forecast was almost exactly correct; see Figure 2.7. The revisions increased the error.) Most other forecasters followed suit, to where it became the norm to project that nearly every non-OPEC nation would see declining production.

In part, this was a reaction to the excessive optimism about oil supply in the mid-1970s. Oil production failed to respond as expected in most parts of the world by 1979, and costs were seen to be rising sharply, which, combined with the 1979 oil price spike, discredited the optimists and convinced many that resource scarcity was the primary problem, with depletion driving up costs. In fact, costs rose primarily for cyclical reasons, as the drilling boom stressed the limited equipment and personnel resource, raising their cost (see Figure 2.8).

THE NEW PARADIGM: HIGHER PRICES, RESTRICTED SUPPLY

This new conventional wisdom is probably the best exemplified by the IEA's *World Energy Outlook* (1982), which concluded that only the Middle East and Mexico would be able to expand production into the future.[24] Since then, the IEA has retained a similar position, usually very pessimistic about both OECD production (Figure 2.9) and non-OPEC Third World production (Figure 2.10), until the last decade, when forecasts became more optimistic.

Occasionally peak oil theorists describe the IEA as congenitally optimistic about world oil production, a demonstration of their ignorance of the history of oil supply forecasts. Dieoff.org quotes Richard Duncan as saying, "According to Richard Duncan, this represents a significant

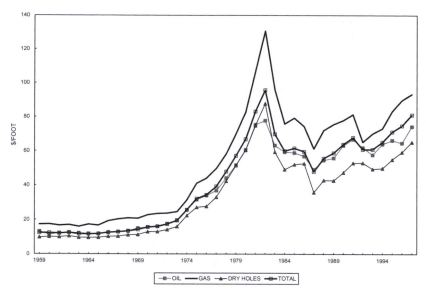

Figure 2.8   Escalation of U.S. Drilling Costs in the 1970s/1980s (Energy Information Administration, US Department of Energy)

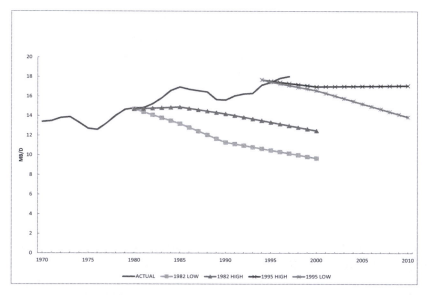

Figure 2.9   IEA Forecasts of OECD Oil Production (World Energy Outlook, various years. Paris: International Energy Agency)

reversal of the IEA position: 'This is a real stand-down for them because until recently they were in the Julian Simon no-limits camp.' "[25]

The same problem is seen in the U.S. Department of Energy forecasts, where the non-Persian Gulf producers are described as "resource constrained" even as some of them, such as Venezuela, expanded production rapidly in the 1990s. They have also been too pessimistic about non-OPEC supply, with non-OPEC Third World oil production providing an excellent example of supply forecasting, in general: a near-term peak and decline is consistently projected and has been consistently corrected as non-OPEC production rises[26] (Figure 2.11).

They were, in fact, reflecting the consensus among the experts. We've already seen that the Secretary of Energy thought production had peaked in 1979, but there were many other such predictions. A USGS study predicted a peak in 1993, while the CEO of ARCO said that "world output is at or near its peak. This year or next could represent the highest level to be achieved."[27] The CEO of Mobil argued that oil companies needed to diversify away from the mature oil-and-gas sectors "or go the way of the buggy-whip makers."[28] (The astute reader will notice that neither ARCO nor Mobil exists anymore, but oil production continues to rise and the industry prospers.)

The Worldwatch Institute published a paper "The Future of the Automobile in an Oil-Short World," by noted environmentalists, at least one of whom, Lester Brown, remains a prominent resource pessimist.

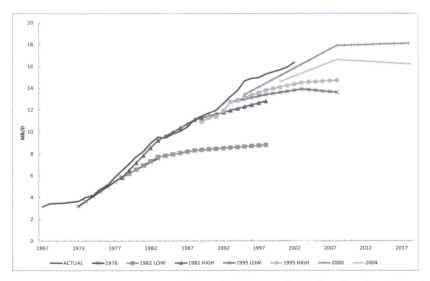

Figure 2.10   IEA Forecasts of Non-OPEC LDC Oil Production (World Energy Outlook, various years. Paris: International Energy Agency)

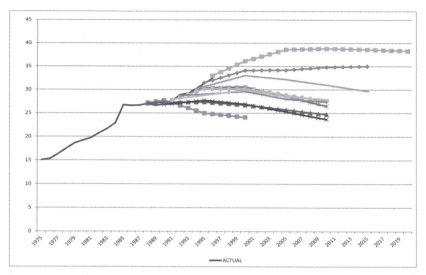

Figure 2.11   Evolution of DOE Forecasts of Non-OPEC LDC Oil Production (US Department of Energy, International Energy Outlook, various years)

And perhaps the most influential study, the Harvard Business School Energy Project, said, "Over the next decade, Mexico and possibly the People's Republic of China represent the only likely sources of significant new supplies for the world oil market outside of OPEC."[29] This was the overwhelming consensus among oil market analysts, although my own work began to challenge that in the next few years.[30]

What did happen? Well, of the major increases in production from 1980 to 1990, China and Mexico were important, accounting for a total of 1.5 million barrels per day in new production, a very impressive showing. But there was growth scattered around the world that occurred largely without fanfare: the 1.8 million barrels per day in additional production from Australia, Canada, especially Norway, and the United Kingdom was more than the two "obvious" sources of new supply. More important, "small" Third World countries like Angola, Brazil, Colombia, Egypt, India, Malaysia, Oman, and Syria added a total of 3 million barrels per day to world oil supplies. This supply increase occurred despite the price collapse in 1986 to levels said to be unsustainably low.

1989: IF YOU HAVE TO ASK, YOU CAN'T AFFORD IT

After the 1986 price collapse, instead of disappearing, the view that prices needed to be higher remained prevalent, and there were recurring bouts of bullishness among analysts. The year 1989 saw an attack on the optimistic point of view from two directions. First, an OPEC researcher examined the published capital investment plans of the member countries and found that they were intending to add 5 million barrels per day of capacity over five years at a cost of $60 billion. This was then trumpeted by some to suggest that oil prices were so low that the members couldn't afford the necessary investment to meet demand, meaning that markets would soon tighten and prices would rise significantly.

Many took this seriously, including Bob Horton, CEO of BP, who repeated it often and argued that the higher prices in the future would mean increased cash flow for BP.[31] In response, he increased the company's debt levels, anticipating that new production and higher prices would allow them to repay their debts easily. (It didn't work, and he is long gone from his post, though not necessarily for that reason.)

At the same time, many voices were heard from the Soviet Union (whose collapse was not then considered imminent) to the effect that rapid cost inflation in the Siberian oil fields meant that production could not continue to increase. Some even argued that it was no longer attractive to invest in that area, implying a tightening of supplies.

Neither of these arguments was solid. The required investment for OPEC, even if the amounts were not exaggerated, represented *at most* 10% of their annual oil revenues and as such could easily be covered by cash flow. And in the case of Siberian costs, most articles simply reported the industry's complaints about inadequate financial resources, without actually estimating costs (which were usually reported in rubles per capacity, not dollars per barrel).

The same misinterpretation has happened recently, when Pemex trumpeted warnings about its inadequate budget leading to falling supply, and most peak oil advocates seized upon it as evidence of a permanent peak in Mexican supply, as well as evidence of an imminent global peak. In fact, more money has meant that production has bottomed out and is likely to increase.[32]

FRAIDY CATS

A cat that jumps on a hot stove will never jump on a hot stove again, but it will never jump on a cold stove either.

Mark Twain

Rather like cats on stoves, many oil market analysts learned not that bad models generated bad results, but to avoid all models. Companies generally eschewed both forecasts and any public announcements of their price expectations and only those who had to—government organizations and some investment banks—continued to produce long-term oil market forecasts.

Still, the market consensus did not change until at least the late 1990s, with most holding the following beliefs:

- A strong consensus existed that prices had to rise more or less steadily (and this consensus was proof that it was valid).
- Physical constraints were the root cause, particularly resource scarcity.
- Non-OPEC supplies had peaked, as had production in many OPEC nations.
- Massive supplies of unconventional fuel would be needed within a decade.
- Markets were too myopic to recognize this change, and prices remained too low, despite their increases, to reflect future supply–demand balances.
- The oil-and-gas industry were mature and would decline.
- The South (commodity-producing nations in the southern hemisphere) would be economically and politically ascendant.

Every one of these proved to be wrong.

And while it could be argued that conservatism is appropriate for policy makers, these mistakes were not without costs, as will be discussed in the next chapter. More important, there are actual *lessons that should have been learned.*

- Consensus does not prove accuracy.
- Experts can badly misinterpret events, including believing theories contradicted by reality (And observers can mistakenly identify novices as experts).
- Nonexperts often join a debate, assuming the consensus is correct and thus reinforcing that consensus by sheer weight of numbers.
- Short-term, transient events can be misinterpreted as changes in underlying trends.
- Policy changes can be misinterpreted as representing physical constraints.
- Most important, *casual study of a subject can be very dangerous.* (A little knowledge is indeed a dangerous thing.)

RESURRECTION: YOGI BERRA MEETS KARL MARX

It's déjà vu all over again! Yogi Berra

> Hegel notes that history always repeats itself. He fails to mention
> that the first time is tragedy, the second time farce.
>
> Karl Marx, 18th Brumaire

But instead of learning these lessons, the old myths have been revived. In 1996, several articles appeared once again warning about the scarcity of petroleum resources, globally and outside the Middle East, in particular, with titles such as "Middle East Oil Forever."[33] At that time, as Figure 2.12 shows, nearly every major forecast called for a sharp increase in the market share of OPEC, which had already risen substantially since the 1986 price collapse. However, even a casual perusal of the figure shows that OPEC market share has not risen since that time.

And roughly at that time, a number of reports and articles began appearing describing "peak oil" and (allegedly) scientific estimates of resources that supposedly proved that industrial society was facing the end of its cycle and "hydrocarbon man" would be on the way out. Largely ignored at first,

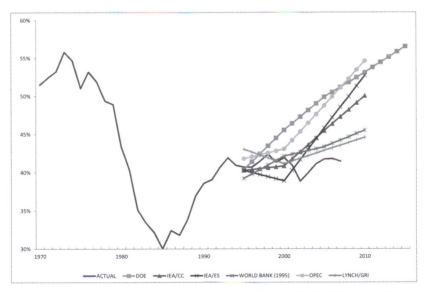

Figure 2.12   Projections of OPEC Market Share (1996 Vintage) (Lynch, Michael C., "The Wolf at the Door or Crying Wolf: Fears about the Next Oil Crisis," in El Mallkah, Dorothea, ed., *Energy Watchers IX, ICEED*, Boulder, CO, 1997)

the price increases after 2002 have given them great credibility—except with those who remember what happened in the 1970s and fear that, as Marx warned, when history repeats itself, the first time is tragedy, the second, farce.

Fooled me Once, Shame on You, Fooled me Twice, Shame on Me—Gomer Pyle

Many will shake their heads in disbelief at the idea that so many can be fooled repeatedly, but the reality is that, first, most peak oil advocates are only vaguely aware of the thinking during the 1970s. Few of them had begun thinking about future oil supplies until very recently, and most appear to be unwilling to delve into writings that predate the Internet.

And governments are hardly better, sometimes falling into the same trap over and over, again, due to the short attention spans of politicians and even of policymakers. Greenhill describes three Cuban refugee crises in three decades, each of which caught the U.S. government by surprise despite clear warnings, and in each of which the Cuban government behavior was nearly identical, yet met with fumbling responses from the other side. Indeed, during the Mariel boatlift, one administration official said that it was only after a week of meetings on the crisis that he heard the mention of the earlier crisis, a mere decade and a half before.[34]

Nor is this shortsightedness a new phenomenon, as a study of the classics reveals. In 66 BC, while in the pursuit of Mithradates, the so-called Poison King, a thousand Roman legionnaires were felled by poison honey by the very same tribe that used the same trick described in Xenophon's Anabasis—a book studied by every classical military official.[35]

It is, however, noteworthy that most (not all) governments and oil companies have disregarded the current alarms over peak oil, in part as they recall earlier experiences and have become wary of apocalyptic alarms. Still, it remains to be seen how the current set of warnings differs from the earlier ones, and why they should be disregarded.

THREE

Going Wrong with Confidence

The widespread misunderstanding of oil market events in the 1970s serves not only as a warning that forecasting can be done wrong, and of the consequences, but also as a cautionary tale for those who argue that it is better to be conservative or "prudent" and guard against the worst-case scenario, given the potentially high costs of being wrong. Past experience shows just how high the cost can be for being too conservative.

Forgotten by many is that the misinterpretation of oil market developments was not only made by academics or environmentalists, but by nearly everyone, most amazingly by executives in the oil companies themselves, who should have known better.

Because of the neo-Malthusian interpretation of the 1970s oil crises as reflecting resource scarcity rather than short-term disruptions of production, governments and the industry both made a variety of moves that proved to be useless or even disastrous. These moves can be separated into the categories of coping with the crisis and responding to long-term scarcity. As later chapters will show, the industry learned its lesson (mostly) and has not repeated these mistakes, whereas many governments are following similar strategies to those that failed in the 1970s.

CRISIS MANAGEMENT

Governments rushed to develop energy policies to cope with the first major oil crisis in the late 1973, but, as Stagliano noted, they largely ignored existing policy proposals from expert communities and instead rushed to respond to voter concerns, particularly gasoline lines in many countries.[1] Later, they sought to improve diplomatic relations with oil exporters and then to invest in foreign upstream operations themselves as ways of assuring "access" to oil supplies.

Gasoline Lines

In both oil crises of 1974 and 1979, gasoline lines appeared in a number of countries, more or less simultaneously. There have been numerous explanations, but two factors seem to bear most of the blame: panicked motorists rushed out to fill their tanks, creating lines and temporary shortages, and governments tried to direct supplies themselves rather than letting the markets cope.

This is not to say that there haven't been numerous other theories, mostly conspiratorial in nature, including the supposed tanker trucks loaded with gasoline hidden in the desert, barges or tankers floating offshore waiting for higher prices, and so forth.

Given that there were price controls on gasoline at the time, it was very hard to engage in profiteering, although the industry certainly benefited from higher prices in markets without price controls, especially for overseas crude oil production.

Blaming panic and gas hoarding by motorists may seem simplistic, but it doesn't take much to realize that, with the news dominated by stories of lost oil supply and government scrambling for any barrels they could locate, the public would be inclined to fear that shortages would reach the retail level. Any news or rumor suggesting empty gas station tanks would send scores of motorists out to fill their tanks before it was too late.

Given roughly 150 million cars in the United States at the time, with an average gas tank capacity of 15 gallons, if most motorists normally filled their tanks when there was 2 gallons left, then the average tank would hold 8.5 gallons. If they panicked and decided to fill up as soon as they were half empty (7.5 gallons), the average tank would then hold 11.25 gallons, meaning a transfer from the oil industry of 600 million gallons or 15 million barrels. And while gasoline inventories had 200 million barrels at that time, more than 80% would be unusable, that is, filling pipelines or in tank bottoms where it couldn't be accessed. Therefore, given the "usable" inventory of 40 million barrels, consumer hoarding could absorb nearly half of it in a few days.

This shortage is referred to as the Thomas effect and the proof that it works came when Johnny Carson joked on *The Tonight Show* that there might be a toilet paper shortage and store shelves were quickly emptied of toilet paper, creating a three-week shortage.[2]

John Sawhill related the story of the shortages on the New Jersey Turnpike, which had led to fights and even a shooting. He said his boss, William Simon, the Energy Czar, had received the unanimous opinion of all his advisors (mostly economists) that moving supplies to New Jersey would simply create or worsen shortages elsewhere, but he decided it was politically imperative that he act. The move was announced on a Friday afternoon, just in time to make the evening news shows (remember, this was pre-Internet) that supplies would be shipped to stations on the Turnpike as soon as possible. The next morning, the gasoline lines were gone–before any supplies were actually arranged.

In today's political environment, it is hard to remember that the political economy was very different in the United States in the 1970s. People seriously debated forcing a breakup of the major oil companies, creating a national oil company, rationing gasoline supplies, and even price controls. In fact, this last policy was implemented by a Republican president, Richard Nixon, who was advised by a very conservative economic team. Today, none of these policies would have received serious consideration.

Price controls certainly contributed to the gasoline lines. Any economist will tell you that freeing up prices will end any shortage–and, eventually, at an unpredictable price. The problem is that few politicians want to let prices rise phenomenally, even if that would wipe out the shortage. The government's allocations of gasoline supplies during the two 1970s oil crises certainly provided the illusion of effective action and enabled politicians to tell voters they were taking measures, but they mostly just worsened the situation.

Surrender, Dorothy

In both crises, many nations tried diplomacy to obtain supplies, but the efforts were fruitless, although not because of a poor understanding of resource economics, rather because of a failure to understand microeconomics, especially of the oil industry. In 1973/74, one of the demands of the embargoing OAPEC nations was that their customers recognize the Palestinian Liberation Organization as the official representative of the

Palestinian people. Most major countries hastened to do so and were removed from the list of embargoed nations. The United States and the Netherlands were the only major importers to refuse.

However, in a lesson that many still have not learned, the oil market proved fungible. Primarily due to the actions of the Seven Sisters,[3] who controlled the vast majority of overseas oil trade, countries that lost more oil when the nations supplying them reduced production received additional supplies from non-OAPEC exporters. This appears to have represented the industry's efforts to share supplies (or conversely, the shortage) equally, but it might also be argued that the free market caused supply and demand to equilibrate.[4]

The situation was much different in the 1979 oil crisis, because those responsible, antigovernment protesters in Iran, were in neither the position nor the mood to make agreements with importers. Worse, given widespread nationalizations, the major oil companies no longer produced most of the oil in OPEC, and, as the crisis went on, they found that the contracts they had to buy oil from those nations (usually where they had earlier held concessions) were cancelled or severely reduced.

Again, governments jumped into the fray, and smaller oil companies and even some large buyers, like utilities, also stepped in. When some buyers lost their supplies from Iran, they went to other oil exporters and offered premiums to get contracts. Some exporters even began cancelling sales contracts at official OPEC prices to resell the oil to new buyers offering more. Again, however, supplies sloshed around, this time more the result of market forces, and no importing country seems to have prospered significantly from arranging new supplies.

(This is a simplified version of events that will be addressed in more detail in a later book.)

You've Got a Friend in Me

At a conference in France in the early 1990s, a leading French official explained that French diplomats were scouring the globe in search of oil supplies. I remarked to my neighbor that we just bought the stuff. But this highlights the difference between statist and free-market approaches to oil supplies, namely, the attempt to gain "access" to "secure" supplies by governments.

Two specific methods have been used: establish good relations so as to ensure a steady flow of supplies, and invest in oil production overseas to guarantee control over the oil being produced. Both have repeatedly been found wanting, but both are pursued because of the Anglo-American oil embargo of 1941 against Japan, one of only two effective oil embargoes historically. (The embargo against Iraq was the other, but it was carried out by consumers against an exporter.)

Because a lot of oil has been produced in areas where governments have significant control over not just investment but operations in the oil fields (unlike the United States), the notion that diplomatic relations have an influence on oil supplies would seem to make sense. Thus, some importing governments have pursued various approaches to keep their suppliers happy. In the 1973/74 oil crisis, this approach meant adopting a pro-Palestinian diplomatic policy (to an extent), but later smaller, more mundane measures were typical.

Most common, it would be argued, is the decision to overlook human rights violations, real or perceived, in many exporting countries so as not to anger the governments. Few refused to buy oil from the murderous regime of Saddam Hussein, and Muammar Gaddafi was actively courted by a number of countries looking for oil as well as some of his country's financial investments. Indeed, it was Jimmy Carter, the first U.S. president to consider human rights as an important area of diplomacy, who went to Teheran to celebrate 1978's New Year's Day.

This makes an important object lesson: the United States was quick to jettison support for its long-time ally, the Shah, when he couldn't deliver the oil due to unrest. Also, on his departure, the United States not only lost a friendly oil supplier; it gained a bitter enemy. Our "friendly" relations could not guarantee supplies during an oil crisis, the one time when they were most needed.

Not all friendship measures are as dramatic as presidential visits. A number of importing governments offer foreign aid specifically directed at oil-exporting nations, such as training missions or investment programs. These packages are usually, by their nature, valuable for both sides, but they do not provide any guarantee of access to supplies, particularly in time of dire need.

Ownership

The other approach some importers have used, controlling upstream investment overseas, has its origin in Winston Churchill's 1913 decision to buy into Burmah Oil, which had just discovered a supergiant field in Persia. He was concerned about converting the British fleet to oil that had many physical advantages but whose production was largely controlled by

American companies. Again, this approach provides a clear object lesson, as the British navy never ran out of oil, but rather found itself strained to protect oil imports during two world wars: the challenge was never *ownership* of the oil, but the *delivery* of it.

Again, the Anglo-American oil embargo against Japan made many countries feel that they might be vulnerable to such action in the future and led to the creation of national oil companies in nations like Italy. (The French had done so after World War I.) Other importing nations, like Korea and Japan, have national oil companies that are focused on overseas investment with the thinking that oil production under the control of domestic corporations would be safer than purchasing it from others.

An argument can be made for creating such companies as "national champions," that is, to be large profitable enterprises that will generate cash and employment, and it is often difficult to separate those objectives from the energy security mission. Countries like Brazil and India have relied on national oil companies to maximize domestic control of resources, although recent years have seen major reforms in both nations. But alternatively, inefficiencies in state mineral enterprises could be said to offset any economic gains.

The move to create national oil corporations was accelerated during the Iranian oil crisis, as exporters began cancelling sales contracts with the major oil companies and seeking their own deals with customers. Some of the more statist oil-exporting governments made it clear that they preferred government-to-government deals, and so some importers set up national oil companies specifically to deal with this concern.[5]

So, while both approaches—diplomacy and investment—had benefits of their own, neither was particularly effective in improving importing nations' energy security, and the costs were often significant. Striking oil workers in Iran (1978/79) and Venezuela (2002/03) were probably unaware of relations between their government and its customers and certainly didn't care. Friendship with countries like Saddam Hussein's Iraq and the Shah's Iran proved to be detrimental, if anything, after new governments came to power.

EXPECTATIONS OF SCARCITY

Both governments and the industry leapt wholeheartedly into strategies to deal with the perceived long-term scarcity of petroleum resources and ever-higher prices, which resulted in significant losses and what can only be considered some outright blunders. The private sector, which should never have bought into the "ever-rising price" paradigm, was also influenced by other factors, while most governments simply took the neo-Malthusian paradigm and ran with it. Combined with a perceived vulnerability to reliance on Middle Eastern oil, this paradigm meant that governments, first, avoided that energy source and, second, sought to reduce the consumption of "scarce" fuels in favor of those perceived to be more abundant.

ABMEO (Anything but Middle Eastern Oil)

The common element in all the major supply disruptions since the Russian Revolution in 1917 and the 1937 Mexican/Bolivian nationalizations was that they originated in Middle Eastern politics, primarily in the Arabian/Persian Gulf. After the second Arab oil embargo in 1973/74, many oil-importing governments decided that they needed to substitute anything for oil from the Middle East.

Unfortunately, this was, to a degree, accompanied by an indifference to cost and an obsession with volume: find enough supply for consumers, no matter what. Aid packages were offered to oil-exporting nations in an effort to ensure "favorable access" of oil supplies, which usually meant paying market price.

In addition, fuel substitution was much sought after, as when European countries purchased natural gas supplies from Algeria, Libya, and the Soviet Union. Sadly, the supplies were priced at the equivalent to oil prices, so the economic savings when oil prices spiked were nonexistent.

And nuclear power, which had already been seen as the energy of the future, was even more heavily promoted in France and Japan, while other countries, like Germany and England, ultimately faced rising public opposition, which proved too much to overcome, and ceased or reduced investment in the sector.

The U.S. federal government has promoted a number of large-scale energy projects intended to solve the energy crisis or at least improve the U.S. economy. Some of these projects predate the 1970s' energy crises, and others were more clearly driven by the self-interest of specific regions and their politicians, but nearly all relied on the perception of an age of energy scarcity for moral support.

FORWARD TO THE FUTURE

One editorial cartoon in the 1970s traced the evolution of man's use of energy from coal to oil to gas to nuclear power to ... coal. After the oil boom in the post–World War II era, when cheap supplies from the Middle

East undercut coal use in many areas (but hardly wiped out the industry), coal seemed to be the fuel of the past. Higher prices of oil and the misperception that natural gas was scarce led many to move back toward coal; with the Carter administration avidly promoting its use, the situation was reversed.

The promotion of nuclear power as the new, space-age technology, which would soon make electricity "too cheap to meter" as one promoter put it (but certainly didn't mean to be taken seriously), also seemed to presage the end of coal in the power sector, its primary customer worldwide. The idea that a shipment of uranium, weighing about 100 tons, would replace a year's worth of coal was tremendously appealing. One story, possibly apocryphal, described how a utility executive, after hearing about a nuclear power plant described, replied, "You mean I can tell the railroad that delivers my coal to go to hell?"

Given the increase in pre-1973 electricity demand of 7% per year (and more), the possibility of a boom in nuclear power was quite welcomed. Of course, by 1973, experience with the technology on a commercial scale was still minimal; the U.S. government had poured money into research to convert nuclear reactor designs for submarines to serve as power stations, but only few had been built. Expectations of massive investment were widespread, even as some questions began to arise about their design.

This optimism was so great, in fact, that some began to fear that there wouldn't be enough uranium to power all the reactors. At that point, only 3.5 billion tonnes of uranium had been discovered in the West, and many feared a scarcity of the resource.[6] It is amusing to recall that the origin of Hubbert's 1956 article on resources was an effort to estimate how much uranium would be required by the nuclear power sector. (He didn't provide an estimate of the resource, but in 1965, the first survey was done by the Nuclear Energy Association, which put the number at 1.7 billion tonnes of uranium.[7])

The solution was apparently designing a new type of nuclear reactor, known as the fast breeder reactor. This reactor, by bombarding nuclear material, turned nonfissile atoms into fissile material, actually creating more nuclear fuel than it consumed. The United States, Japan, and a consortium of European countries each began to construct prototypes, planning them as a second-generation design to gradually replace the initial light-water reactors.

The story has a happy ending—unless you were building a fast breeder reactor (or paying for it). First, the energy crisis of the 1970s led to much slower increase in the demand for electricity, reducing the need for huge numbers of nuclear reactors and, thus, uranium. In addition, the fast breeder reactor's costs soared to the point where its output was no longer competitive with conventional power or even light-water reactors.

Most important for our story, uranium resources proved to be far more abundant than had been imagined, with reserves doubling from 1975 even as consumption increased.[8] The problem had been not one of geological scarcity, but minimal exploration due to low prices and almost no demand before the 1970s. Once it appeared that the utility industry would be a reliable customer, and prices rose in the 1970s, geologists fanned out and found huge new discoveries in Canada, Australia, and various other places. Even old mine tailings in South Africa, waste piles from the production of gold, were processed to extract, among other things, the uranium contained therein.

At the same time, many nations turned back to coal, especially to serve as power plant fuel. The abundance of cheap coal made it desirable despite the dangers of mining and the pollution from its consumption. Oddly enough, coal consumption soared despite the downturn in coal mining in a number of countries, including Britain and Japan, where deposits had been exploited to the point that they were no longer competitive with seaborne imports. Consumption grew faster in the 1980s, slowed in the 1990s with the economic collapse of the former Soviet economies, and is now soaring along with Chinese and Indian economic growth (Figure 3.1).

As mentioned earlier, utilities were discouraged from using the "valuable" natural gas to make power, given the availability of nasty old coal.

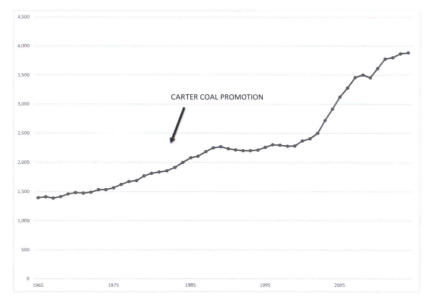

Figure 3.1   World Coal Consumption (BP Statistical Review of World Energy 2015)

In the United States, billions of dollars were spent on clean coal research, prompted by politicians from coal-producing states who were hoping to keep their industry—and its jobs—a little longer.

Natural Gas: Tres Chic

In the early 1980s, my colleague (okay, boss) Loren Cox referred to the European attitude toward natural gas as it being a boutique fuel, one that you bought in small amounts and carried off in fancy little shopping bags. This was a very clever way of describing the widespread attitude that natural gas was a "superior" fuel, of high value, which should be reserved for only the best people, I mean, uses, such as conversion to petrochemicals.

(Actually, another colleague, Henry Jacoby, for his part once said that incoming students to energy economics normally assumed that natural gas was more valuable than oil, and it was the job of the class to help them understand why that wasn't true. Apparently most of his students weren't European.)

Producers were happy to encourage this mistaken belief, since it implied they could get higher prices for their output. (However, U.S. oil companies later regretted having discouraged its broader acceptance in the power sector.) As mentioned earlier, many governments, in part due to the pressure from the Carter administration, discouraged the use of natural gas to make electricity, while seeking new supplies with alacrity.

Again, the Carter administration was at the forefront, pushing aggressively for the development of a 4,700 mile, $30–45 billion (2010$) natural gas pipeline from the north slope of Alaska to the continental United States, where the cost at the citygate was $7.50–8.50/Mcf (2010$), nearly twice the current citygate price.[9] This price was justified, the government argued, because natural gas prices would rise 2% per year above inflation, and so the gas would be competitive by the time the project was finished.

Diplomacy also came into play, as the European countries sought foreign supplies, often putting forth the added inducement of financial aid. The French were quite pleased to seek commercial deals with Algeria, even offering to pay a premium above the price of oil for their purchases. The British were paying exorbitant prices to Norway for natural gas imports—while maintaining prices to domestic producers at as little as one tenth that amount.

Once again, reality trumped the research. True, Carter had inherited a shortage of natural gas, which had resulted in periodic closings of schools and factories due to insufficient supply in the winter. Despite economists' insistence that this was due to wellhead price controls, which had been

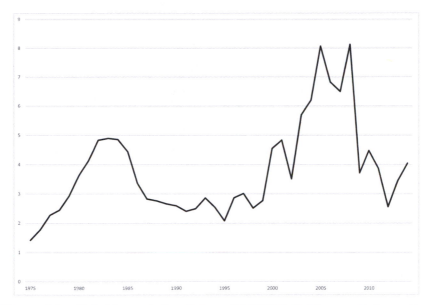

Figure 3.2   U.S. Natural Gas Prices (2010$/Mcf) (Energy Information Administration, US Department of Energy)

imposed in the 1950s, the neo-Malthusian ambience, together with the distrust of the oil industry, led to that explanation being rejected. Carter's solution was to impose a Rube Goldberg–like schedule of prices for old gas, newly discovered gas, tight gas, deep gas, and so forth, attempting to encourage new drilling, especially for unconventional supplies, while keeping average prices—and so-called windfall profits—low.

By the time natural gas prices (and markets) in the United States were fully deregulated, a supply glut had been created, which was originally referred to as a "bubble" but persisted so long it became known as a "sausage." As Figure 3.2 shows, prices were depressed for roughly a decade and a half, from 1986 to 2000, during which time the industry constantly complained that prices were too low for them to make a profit—and they kept drilling.

Canada

The politically liberal government of Pierre Trudeau in Canada, from 1980 to 1984, not surprisingly embraced this view of resource scarcity as well as a posture of resource nationalism, directed primarily at the U.S. oil companies that had a large role in Canada's industry. For natural gas, he took two important, but flawed, steps. First, he heavily subsidized drilling in northern Canada, such as the Mackenzie delta, believing

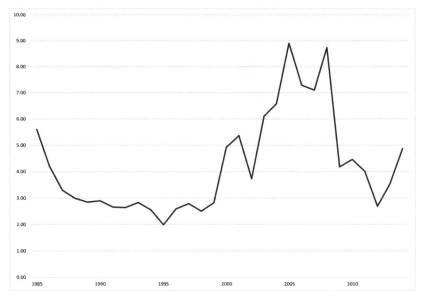

Figure 3.3   Price of U.S. Natural Gas Imports ($2010/Mcf) (DOE/EIA. Pre 1985 Prices Total Imports (mostly pipeline); Subsequent Prices are for Pipeline Imports from Canada)

(emulating Carter) that the resources would soon be needed critically and that markets were so myopic that they didn't price natural gas high enough to make it economic to drill there.

Second, fearing that natural gas exports to the United States would deplete Canadian resources and leave future generations short of the supply needed to fuel their industry, he insisted that oil companies could not export any gas until they proved they had set aside a 25-year supply for domestic use, that is, discovered reserves would be 25 times that of annual production. In contrast, in the United States, the unregulated industry generally maintained a 10-year supply (or inventory).

More amazingly, the United States was prepared to pay a price for imports from Canada equal to $14/Mcf (in 2010$) for supplies, which, as Figure 3.3 shows, was far above what was ever achieved subsequently. By hindering sales at that price, the Trudeau government cost the industry—and the country—tens of billions of dollars.

THE OIL INDUSTRY REGROUPS

In response to the above-mentioned problems and expectations, the oil industry undertook three primary strategic changes. First, it invested heavily in petroleum exploration and development in the politically "safe" United States. Second, it tried to diversify away from petroleum and into other energy sources, seen as dominating the future economic picture. Third, it moved out of energy and into a variety of other unrelated industries. All were mistakes.

Go Home, Young Man!

The private oil companies found themselves in a difficult situation, even before the Iranian Revolution. Most of their foreign holdings, particularly in the Middle East, had been nationalized, and many other countries became more hostile to outside investment, especially from U.S. oil companies. Most politicians, then as now, were not very sophisticated in their understanding of microeconomics and did not understand that the prices companies were willing to pay during an extremely tight market, as in 1973/74, did not represent a "fair" price, or a market-clearing price over the long term. Instead, they concluded that the oil companies had used their power to suppress prices prior to 1973, which only increased their hostility.

This certainly reinforced the political leanings of many governments, particularly but not exclusively outside the OECD, where leftist or socialist policies were being actively implemented as superior approaches to market-based capitalism, or simply preferred by the reigning politicos, many of whom had risen to prominence resisting imperialism/colonialism. Resource nationalism, as it's now called, dominated the political thinking in most countries, including the West.

The funny thing was that higher prices resulted in much more cash flow for the oil industries, even as they had less places to explore for oil. In many ways, their situation resembled that of the second-generation rich, who inherited a business but didn't know what how to proceed themselves. Shareholders were not pleased with rising cash hoards, which provided small returns on investment, and there was pressure for them to do something with the money—or give it back to shareholders.

One result was a boom in the domestic oil industry, as, from 1976 to 1980, the number of rigs operating doubled from 1,600 to 3,200. Unfortunately, costs also rose, absorbing much of the additional investment funds without physical results, and productivity dropped, as less experienced workers were pressed into service. Some quipped that high school football in Texas suffered a crisis, as all the players were in the oil fields.

Beyond Petroleum: Let the Sun Shine In

Mr. Nader said President Carter and Energy Secretary James Schlesinger have deceived the American public by failing to inform them that energy generated from solar, wind, geothermal and wood waste sources "could easily replace nuclear power," and characterized these energy alternatives as safe, competitively priced, and decentralized."[10]

Given these three elements—the belief that oil and gas were "mature" and "declining" industries, the lack of upstream investment opportunities, and the desire to diversify their revenue streams—the industry began to move into new fields. Some made investments in nonpetroleum energy, like coal, nuclear, and solar, but others were completely unrelated to the energy business itself.

Given the consensus that nonpetroleum energy would experience a boom, seemingly justified by the difference in prices, if nothing else, a number of companies looked for alternative investments. Exxon developed a massive coal mine in Colombia and set up its own nuclear division; in 1979, Gulf Oil opened the largest uranium mine in the United States in New Mexico, although it closed by 1989 due to depressed prices.[11] Phillips Petroleum actually invested in nuclear fusion research "as a hedge against depleting oil supplies," based on an inventor's idea for an approach cheaper than large-scale magnetic containment.[12]

Perhaps a more telling trend was the decision to invest in solar power, principally photovoltaics, which formed an amusing footnote. At the time, consumer and environmental activists denounced this move, saying the petroleum industry was trying to "monopolize the sun." Obviously, that was ridiculous, but reflective of a bias against both the oil industry and large corporations among supporters of solar power. Now, of course, crusaders like Ralph Nader criticize oil companies for *not* investing in the sector.

Most of the major oil companies invested in photovoltaics in the 1970s, and that was cited by many environmentalists as evidence of the impending success of solar power. Few companies argued that they were transforming themselves into renewable energy producers, but there was a sense that solar was a "sunshine" industry (sorry) and would soon be prosperous. Thirty years later, photovoltaic electricity still costs about twice conventional power in the United States, and in countries like Spain, where subsidies were withdrawn, the sector has collapsed.[13]

In what appeared to be the largest single private investment ever made in solar energy, the Atlantic Richfield Company agreed yesterday to pay Energy Conversion Devices of Troy, Mich., $25 million to accelerate development of substances that, among other things, convert sunlight into energy.[14]

'RCA's laboratories in Princeton and Energy Conversion Devices in Troy, Mich., are leaders in the race to produce solar cells from amorphous silicon. ... If perfected, it could turn out solar cells like newsprint, providing electricity for about 20 cents per kilowatt hour.[15]

As part of a companywide restructuring in 1989, Arco sold its pioneering Arco Solar Inc. unit, headquartered in Camarillo, to Siemens Solar Industries for $35.9 million.[16]

[A]morphous (non-crystalline) silicon was used in the first thin film to be developed, suffers from low efficiency levels and high production costs.[17]

Arco's decision to invest in solar energy was clearly a mistake, and it is worth noting that Arco executives were among those loudly proclaiming the end of the oil age. (Oil is still here, but Arco is gone.) Stanford Ovshinsky, who founded Energy Conversion Devices, was a darling in the energy industry, as he appeared on the cutting edge of many aspects of photovoltaic and electricity storage technology. Unfortunately, he proved to be more of an idea man than an implementer, and his once-promising solar innovations did not yield fruit, despite rampant enthusiasm from promoters.

Enter the Theorists

Just at this time, a new academic approach to asset valuation was becoming popular: the capital asset pricing model or CAPM. The basic idea was that some assets moved together in price, others diverged, and investors should seek balance or diversification. For example, trucking industries would prosper with low oil prices but suffer with high oil prices. To be safe then, you would weight your portfolio with stocks of trucking companies and oil companies: whichever way oil prices moved, one part of your portfolio would prosper. Untold wealth was generated by means of this model—at least for the consultants.

The industry responded to this model, as well as to the constrained opportunities for upstream petroleum investment, by buying into other types of businesses in a multibillion dollar–spending spree.

Beyond Energy

The oil business has come to maturity, and with this maturity comes a new set of challenges ... oil companies have no other choice. They must diversify or go the way of the buggy-whip makers.[18]

Mobil CEO

Exxon bought Reliance, a Cleveland-based manufacturer of motors, electrical communications and weighing equipment, for $1.24 billion in 1979.

The company hoped to profit from technological developments contained in an energy-saving device for electric motors, for which it held commercial rights. But the technology proved commercially unfeasible.[19]

Probably the most symptomatic of the oil industry's move away from energy was Exxon's investment in Reliance Electric (not to be confused with Reliance Industries in India). At the time, the company heralded this step as groundbreaking, based on Reliance's development of a variable-speed electric motor, because the company thought this represented a breakthrough.

But this product was not novel or unique, except to Exxon. Other manufacturers had similar motors, and the investment community quickly realized that Exxon had grossly overpaid for Reliance—it was to resell it later for about one-fifth of what it paid—simply because it was not familiar with the electric motor business and technology[20]—an expensive price for ignorance, but a lesson well-learned, as will be seen later.

When I was young, owning an IBM was everyone's dream—not a computer, but the Selectric Typewriter. The company, International Business Machines—has been transformed several times, from a leading maker of business machines like typewriters to the world's primary computer manufacturer to now, a consulting firm. In the 1990s, nothing could compare with either the IBM Selectric typewriter or the various mainframe computers IBM supplied worldwide. Other systems were poor seconds to the point that the company was accused of monopolistic behavior based on their market dominance.

In the fall of 1984, after 10 years and at least a US$500 million investment, the world's largest company concluded that oil and office products do not mix. Exxon says it pulled the plug on its Exxon Office Systems ... "maybe we recognized earlier (than most) that the industry was not as good as people thought it was going to be," says Lawrence G Rawl, Exxon Corp.'s president.[21]

Thus, Exxon moved into (or bought into) the office equipment business, thinking it could compete with minicomputer companies like Wang, which itself later declined sharply from competition from the personal computer. The synergies between oil production and office equipment are few if any, and it appears that Exxon got in—and out—almost whimsically.

But this was the type of business that seemed attractive even to the oil industry, as the Mobil CEO Warner best expressed it. The oil industry was simply "cashing in" its assets, petroleum reserves, producing the oil for cash without replacing the reserves. Instead, it believed it should put the cash flow to better use by moving into other areas. This led Mobil to buy Montgomery Ward department stores, consider purchasing Ringling Bros. Circus, and, ultimately, go the way of buggywhip makers.

Not that Mobil completely abandoned the oil industry. It agreed to build an expensive petrochemical plant in Saudi Arabia in return for guaranteed long-term supplies of crude oil—at official prices. This reflected a belief among many companies and countries that future supplies would be tight, as analyzed in Raymond Vernon's *Two Hungry Giants* (which was actually not alarmist despite the title) or, more recently, Kent Calder's *Pacific Defense* (which warned of competition in Asia for energy resources).[22]

What did the company get for its investment? The right to buy oil at the market price, in other words, getting what everyone else was getting for free. The market weakened shortly after the agreement was signed, and "access" proved of no value. (In fact, Mobil lost significant amounts of money because the Saudis kept their "official" prices stable in the early 1980s while spot prices declined.)

WHIPLASH FROM WALL STREET

Louis V. Gerstner Jr. caused a stir two summers ago when he declared that the last thing IBM needed was to proclaim a grand vision.[23]

Twenty years ago, the CEO of IBM learned about the power of Wall Street when he denounced the fad for "vision" and his stock price immediately dropped. The oil industry can certainly sympathize, having been whipped about from fad to fad over time, and especially in the 1980s.

There is probably not a single corporate executive who likes being dictated to by Wall Street, and the oil industry in the 1980s helps explain why. Consider this quote from Standard & Poor's in August 1980:

Diversification into alternative energy fields should offer promising new opportunities for increasing profitability.[24]

Followed only four years later by a complete reversal:

Diversification out of the oil business has been disastrous for most of the majors. . ..[25]

Things began to change in the early 1980s. For one thing, T. Boone Pickens began to use what's known as "greenmail" to pressure the large oil companies to stop making investments that had poor returns. This included the upstream in the United States, which, though having a very low political risk, also faced very high costs, which soared cyclically with investment (see Chapter 2).

But there was an increasing recognition that the original diversification strategy had several flaws. Most notable was that oil prices were weakening, not rising. Also, the diversification strategy relied on the assumption that returns would not change for the acquired company after it was taken over. This proved false, or, as an Exxon executive wryly said to me, "we thought an office electronics company would prosper with the same management style that a large natural resource producer used."

Wall Street also embraced the shift in strategy away from diversification under the new rubric of "focus on your core business." If the public is generally unhappy with Wall Street, imagine the feeling among corporate executives who are pressured to follow a given strategy, see their stock price punished if they don't, and then find that they are pressured to revert to their original focus *by the same institutions.*

GOVERNMENT KNOWS BEST

Government strategies to cope with the perceived resource scarcity fared, if anything, worse, although it was taxpayers and industry that paid the price. The oil price increases had been a surprise to most and seemed to confirm the analysis of both the resource pessimists and those in favor of what is now known as "industrial policy." The latter believe the government should guide a country's industries down the correct path, which they are incapable of perceiving for themselves and for which markets are improperly signaling.

Bear in mind, too, that it was just after the 1979 Iranian oil crisis that the Japanese economic bubble began, which led to the publication of numerous books about the superiority of Japanese economic policies and, in particular, Chalmers Johnsons's *MITI and the Japanese Miracle*, which argued that the Japanese government had successfully instructed its industries in the appropriate investment paths to yield the phenomenal success seen in automotive, consumer electronic, shipping, and other industries.[26] This theory became a template for many liberal economists and is still the slogan of "cleantech" supporters. (Now, China fills the same role as role model that Japan did in the 1980s.)

The subsequent bursting of the Japanese bubble has tarnished industrial policy to many, as did the work of academics like Richard Samuels and his students, such as David Friedman, who noted in some detail that the Japanese government's assistance was much more general than that claimed by Johnson, and more in the form of cheap capital than tactical advice.[27]

The two primary industrial policy mistakes in the energy sector were the Carter administration's creation of the Synthetic Fuel Corporation, whose primary mission was to commercialize oil from kerogen and natural gas from coal, and the Trudeau government's decision to heavily subsidize "frontier" exploration in the Arctic and East Coast offshore. The basic concept was that the resources would not be developed by the time they were needed because the market was not sophisticated enough to understand the kinds of analysis informing governments as to the best strategy.

But reliance on these beliefs led to investment in technologies and fuels that were not ready, overconsumption of coal and underdevelopment of natural gas, and the wasting of billions of dollars on energy resources that are, for the most part, still not economically viable 35 years later.

Synfuels: I can't believe it's not Oil!

"Synthetic" fuels can be said to have a very long history—if you define them that way. Hannukah, the Hebrew festival of lights, celebrated the miracle of long-lasting oil, presumably linseed or flax. Pretty minor stuff for an all-powerful God, but it demonstrates the long-term use of what would now pass for biodiesel.

In modern times, the exploitation was more industrial in nature, involving chemical conversion: the Fischer–Tropsch process was used in World

War II by the Germans to make synthetic oil products, coal had been gasi-
fied to supply early "town gas" companies since the 19th century, and
shale (kerogen) was long recognized as a rock that would burn. But each
of these was supplanted by cheaper conventional oil and gas in nearly all
markets during the 20th century.

But in the 1970s, the belief that petroleum resources were "running out"
as Jimmy Carter said and that prices for conventional oil and gas would
always rise meant that the time had come for some of these to become
competitive in the marketplace. Again, it was believed that markets were
too short-sighted to recognize the needs, and the government had to accelerate the transition—nearly identical to the current "peak oil" argument
that postpeak transitions would be fraught with difficulty.[28]

Carter tried to deal with the problem by setting up the Synthetic Fuels
Corporation, which was to provide funding and loan guarantees to enable
large projects to start up that would generate unconventional hydrocarbons. One was the Great Plains Coal Gasification project, which was
intended to turn coal into natural gas, supplementing the nation's supply
of conventional gas.

The second that nearly proceeded was the Old Colony Shale Oil project.
Exxon's CEO, enthusiastic about ever-higher oil prices and the urgent
need for new supplies, proposed a $12 billion project that included a strip
mine in Colorado, employing 7,000 workers to produce 48 thousand
barrels a day of synthetic crude from kerogen. In the end, the project was
abandoned long before significant amounts of money were spent, and I have
heard that Exxon actually cleared a profit because the housing it built for
workers was sold at a profit. (That could be rationalization, of course.)

BACKSTOP TECHNOLOGIES AND REALITY

Many studies in the 1970s assumed that the world would transition from
conventional energy to what economists often called "backstop technologies." Oil from kerogen was considered a prime candidate, and the oil
price was perceived as rising from the then-low price to the higher price
necessary to make kerogen economically viable over the period it was
thought necessary to achieve the changeover.

But this was an extremely simplistic view, in essence assuming a two-
step supply curve when the reality was that there were a myriad of sources
of energy at costs between the two levels. The real world supply is much
more continuous than the leap from "easy" oil to "hard" oil or from conventional electricity sources to renewables.

Thus, the biggest elements contributing to avoidance of an ever-worsening energy crisis were actually quite mundane. Higher prices led to conservation, just as economic theory predicted. Energy use per GDP in the United States actually increased in the years before the first oil crisis in 1973, but from 1980 to 1990, it declined by about 2.5% per year. Government might have contributed with policies such as automobile efficiency regulations, but the broader gains in the economy were due to rational consumer choices.

The second element could be considered "synthetic fuel" in a sense, but much more humdrum than giant, multibillion dollar plants producing liquid fuels. Substituting coal and natural gas in stationary uses such as power plants freed up 5 million barrels per day of low-quality oil in the OECD countries, which was then upgraded by advanced refineries to high-quality products like gasoline and diesel fuel—for a fraction of what coal liquefaction or kerogen fractionation would have cost.

FOUR

Abundance and Its Enemies

The strongest witness is the vast population of the earth to which we are a burden and she scarcely can provide for our needs; as our demands grow greater, our complaints against nature's inadequacy are heard by all.

Tertullan, 3rd century[1]

The term *Malthusian* has come to refer to pessimism about either over-population or resource scarcity and originated with the English demographer Thomas Malthus, who estimated in 1798 that population growth was faster than that of food production. His projection that this growth would lead to famine is often cited as an example of both an early warning and a badly failed prediction. In fact, his was a ceteris paribus work, arguing that agricultural reform was necessary to overcome population growth, and in later editions, with more data, he became much less pessimistic.[2]

But Malthus was hardly the first to be concerned about the issue. Perhaps economics is not the oldest profession, but sometimes it must seem as if demography is. Ancient societies, always facing fluctuations in weather conditions that determined the availability of food, must have often feared an inability to feed themselves. And the migrations of peoples, usually because of population pressure, were frequently the cause of the fall of kingdoms and empires, from Assyria to Rome.

Barbara Tuchman, for example, refers to the ancient Greeks as having thought that the Trojan War was initiated by the gods to reduce an excessive population and the earlier comment from Tertullan shows that such thoughts also occurred among the Romans, despite their having centuries of experience overcoming insufficient resources, importing grain from Sicily, Egypt, and later the Black Sea region.[3] Of course, they also faced frequent disruptions of supply due to war, piracy, and weather.

MALTHUS AND HIS WOULD-BE DISCIPLES

Famine regularly visited the more primitive societies, as inferior transportation infrastructure meant a local crop failure that could not be offset by supplies from distant sources. Wolmar points out that a Chinese famine before the railroads resulted in 25 times as many dead as a subsequent one, although there are obviously many other variables.[4] The Little Ice Age from the 16th to 19th centuries saw widespread hunger in Europe, but, more commonly, drought or heavy rains frequently caused crop failures and, if proceeded by a period of good weather (and peace), a population increase was likely to mean that subsequent bad harvests would find them doubly stressed.

But it was Englishman Thomas Malthus who first performed serious empirical research that allowed him analyze the relative growth of agriculture and population, warning that the former was falling behind the latter. Indeed, he proved correct in the sense that England has long been a major food importer—but largely because the population became urbanized.

The opening up of new agricultural lands in the United States, South America, and Australia, combined with railroads and steam power, enabled the transportation of food—grains and livestock—to great distances, and depression of agricultural prices seemed more of a threat from the late 19th century than widespread famine.

The 20th century saw both a population boom and massive hunger and starvation; however, the two were largely unconnected. The Ukrainian famine was deliberately inflicted by Stalin in the 1930s as a means of suppressing his enemies, including successful farmers (who were the first to suffer), but more generally a "nation" that had been somewhat rebellious, attempting to express its cultural and political identity separate from the Russians after the fall of Tsar Nicholas in 1917.

Mao's 1956 Great Leap Forward was the inverse, where many suffered starvation due to government attempts to promote industrialization, and agricultural policies were determined by those in the central government who were largely ignorant about farming. Their biggest mistake was to think that revolutionary fervor could triumph over nature, yielding the desired out-comes. This was arguably the largest mass famine in history.[5] Modern ana-lysts like Amartya Sen have argued that modern famine is due primarily to poor management, and so democratic societies will not be prone to them.[6]

THE BOMB FIZZLES

The population boom, as a result of 20th-century industrialization improving medicine and sanitation, brought revived fears that it would be physically impossible to provide sufficient food to feed the global pop-ulation. Most famous of these was *The Population Bomb,* published in 1968 by the biologist Paul Ehrlich, who believed that the population was about to reach the planet's ability to support and that mass global starva-tion was not only imminent but also unavoidable.[7]

Published at a time of rising environmental consciousness, this book struck a nerve with many people, and Paul Ehrlich not only sold huge numbers of books but also became idolized, appeared frequently on talk shows (he was one of the few scientists to appear on *The Tonight Show* with Johnny Carson), spoke and wrote frequently, and received numerous awards. He himself, in a typical show of immodesty, went so far as to claim that "Since *The Population Bomb* was published, millions of Americans have made responsible personal decisions to limit the size of their families to at most two children."[8]

Since then it has become obvious that his work was flawed (to put it mildly) and that his predictions proved to be as wrong as the many apoca-lyptic predictions from some in the religious community (or UFOlogists). However, because so many are biased to believe his view, it has continued to be influential, with many defending his work as sound and granting him many awards. (Few other than advocacy groups have done so.) One of his early coauthors, physicist John Holdren, served as President Obama's first science advisor, which presumably explains the Obama administration's talk of the finite nature of resources.

THE LIMITS TO GROWTH

If something can't go on forever it probably won't.

Herbert Stein[9]

In the 1970s, "computer" meant a mainframe from IBM or Honeywell, or possibly a "mini-computer" from DEC, and there were only a few

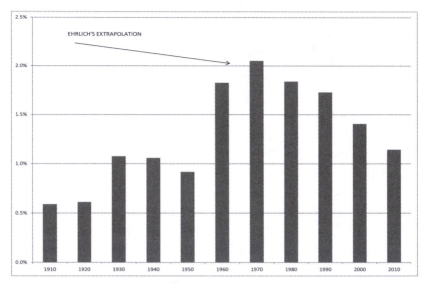

Figure 4.1   World Population Growth Rates (http://www.census.gov/population/international/data/worldpop/table_population.php)

people who were capable of working with them (except perhaps for data entry, which was basically typing). To create a graph from data using a computer, for example, required the writing of a specialized program, and performing statistical analysis usually required finding a grad student who knew how to use the relevant software. In fact, two companies prospered because they had computer models of the U.S. economy—and there was little competition, given the difficulty of the task. (Note: neither still exists, having been merged, renamed Global Insight, and acquired by IHS.)

MIT's Jay Forrester developed Systems Dynamics, whose principles were used to create a computer modeling software called Dynamo, which allowed researchers to make computer models with less effort than earlier languages. The difficult part, then as now, lies in knowing what the numbers were to be input. Some of his colleagues, working for the Club of Rome, created a model of "the world," which was used to test the feasibility of continued exponential growth.

The results, published in the 1970 book *The Limits to Growth,* used estimate of resources, combined with population and economic growth, to find when exhaustion would occur, as well as estimating pollution levels.[10] Their conclusion that existing trends were unsustainable was, again, seized on by many to promote deindustrialization, zero economic growth, and a number of similar ideologies. It certainly helped to influence the Carter administration's views on resources, as Chapter 2 demonstrated.

AN ENORMOUS BUT

Unfortunately, this entire area is based on mistaken principles. *The Limits to Growth* made the very basic mistake of assuming no change in growth rates, when in fact, the post–World War II economic trends were elevated by recovery from war-time damage, just as Chinese and Indian growth rates are now elevated by the movement of the workforce from agriculture to industry and services. Already, global population growth rates have dropped sharply, and a peak is in sight (although demographic trends do change!), and, more important for commodities, they ignore the mechanism by which growth rates change: scarcity yields higher prices, which reduces consumption; causes switching to substitutes; and leads to new supplies being developed. The early models assumed no feedback effects.

That mistake is compounded by the use of resource estimates that are much too conservative. *The Limits to Growth* team attempted to avoid this problem by using estimates that were multiples of then-official numbers; however, they didn't realize that the data they were relying on were not actual estimates of global resources, but known developed reserves, a small subset of the resource base (see Chapter 15).

Petroleum is the clearest example of how these numbers were far too conservative. The authors thought that they would avoid such criticism by testing the results of a resource base that was five times more than known reserves of 455 billion barrels, or 2,275 billion barrels.[11] Since then, the world has used 1,100 billion barrels of oil and still has about 1,700 billion barrels of reserves. Even excluding heavy oil and the so-called spurious reserve revisions, the assumption in *The Limits to Growth* would mean no more oil remained to be discovered.

MY FAIR MALTHUSIAN

"Just you wait, 'Enry 'iggins!" Eliza Doolittle, "My Fair Lady"

The amazing thing is that, like the Millenarians, the Malthusians never seem to be concerned when their predictions fail. Indeed, many of them look back and find that their expectations were (somehow) confirmed. *The Limits to Growth* authors are quite enthusiastic about the fact that growing mineral reserves (and production) had confirmed their warnings, since it implied that we are "closer to the end."

There seem to be three approaches to dealing with the problem of model and/or theoretical failure: claim you missed on the timing, ignore much of what you said earlier, or insist you were right all along. All have been used by various neo-Malthusians, including the peak oil advocates.

Gardner discusses a number of neo-Malthusian analysts who proved not just wrong but woefully so, and yet none admits to it. This includes not only Ehrlich but also Heilbroner and Kunstler, the latter now a peak oil advocate. Gardner explains that "the first line of defense of the failed prophet is ... the clock is still ticking."[12]

Two recent appraisals of the original *Limits To Growth* work carry this even further. Validating the model's forecasts, both Turner and Alexander and Hall and Day argue that they were quite accurate.[13] In reality, the model's forecast of continued growth was the accurate element, since the impending collapse remains in the future. Thus, the socioeconomic trends of the past four decades in no way confirm the model and especially its principal warning that resource scarcity would have a major impact in the 21st century.

This is reminiscent of the London astrologers, described by Charles MacKay, who predicted that River Thames would catastrophically flood London on February 1, 1524. When it failed to do so, and the citizenry considered dunking said astrologers in the river, they explained that "By an error (a very slight one) of a little figure, they had fixed the date of this awful inundation a whole century too early. The stars were right after all, and they, erring mortals, were wrong."[14] In other words, the model was correct; a bad piece of data was to blame.

Peak oil advocates emulate this approach, with Campbell stiffly quoting Keynes, "When I have new information, I change my conclusions. What do you do? Sir."[15] Of course, he ignores the fact that his most important claim was to have robust data that were not revised (see Chapter 7). He also doesn't mention his many earlier predictions of a peak in production, rather showing the "depletion midpoint," which he says changed only by five years over a decade.

Robert Hirsch lists predictions of peak oil, but only the most recent ones at the time of the report, ignoring that a number of other dates have already passed. Few discuss forecasts for individual countries, and there appears to be no recognition that Simmons's warning of the impending collapse of Saudi oil production failed abysmally (see Chapter 8).

The same thing occurs with neo-Malthusians, the classic example being *The Limits to Growth* work, which continues to be updated and, to the uninitiated, appears sound, while those sympathetic to its views continue to trumpet its quality. Simmons, for example, said, "The most amazing aspect of the book is how accurate many of the basic trend extrapolation worries . . . still are, some 30 years later, . . . there was nothing that I could find in the book which has so far been even vaguely invalidated."[16]

He goes beyond this to say, "There was not one sentence or even a single word written about an oil shortage, or limit to any specific resource, by the year 2000."[17] Why he thinks the year 2000 was referenced is not clear, but the authors clearly did describe one scenario as finding, "Eventual depletion of nonrenewables brings a sudden collapse of the industry system."[18]

The authors themselves, in the first update, note that energy reserves have been rising, even relative to consumption. Does this mean their model was wrong? They change this question into, does this mean that there was more fuel in 1990 than in 1970? "No, of course not. There were 450 fewer billion barrels of oil, 90 billion fewer tons of coal, and 1100 trillion few cubic meters of natural gas."[19]

Changing the subject is also popular. Thus, Simmons rebuts criticism of *The Limits to Growth* by noting that population and economic growth have, in fact, continued, and use of resources has soared.[20] Similarly, Patterson argues that Ehrlich was right because the population has grown as he had predicted.[21] Of course, these were not the points at issue, something that goes unmentioned.

FINITE RESOURCES: THINK INSIDE THE BOX

> Oil is a finite resource that we have been producing at ever increasing rates to fuel our expanding economies. World oil production will soon reach a maximum rate of production, if it hasn't by the time you read this book.
>
> Hirsch et al.[22]

One of the major tenets of peak oilers is the fact that oil resources are finite, and therefore we are running out, and thus they must peak. (Implied is that we might not foresee the peak and that it will somehow be catastrophic.) Repeatedly, I have had hostile questioners ask me to admit that oil is finite.

This is so prevalent that it often becomes a straw man for peak oil advocates, who accuse "optimists" like me of believing in "infinite resources." Campbell goes so far as to turn a comment by M. A. Adelman and me that resources are so large as to be *effectively* infinite into *actually* infinite.[23] And others such as Bartlett have used the term *flat-earth economist* to refer to someone who doesn't seem to recognize that the Earth is a sphere, and, as such, any given resource is finite. This is a simplistic error, which we shall return to later.[24]

Indeed, one reason that many support renewables it that they are "sustainable." The wind doesn't run out and crops are reborn every year. Our great-great-great grandchildren could still be getting wind from nearby mountain passes or offshore sites, and (ignoring climate change and soil erosion and degradation) without fearing that the resource will deplete, while the oil industry must constantly struggle to replace wells and fields whose resource has been exhausted.

Problem 1

The first problem is that the finite nature of oil is not strictly speaking true. Unlike mineral resources, petroleum resources are created in an ongoing manner from biomass. Admittedly, it takes millions of years and the right combination of heat and pressure to turn organic (mostly plant and algae) material into coal, oil, and gas. (And actually mineral resources are born in supernovae explosions, but replenishment of Earth's minerals from a supernova anytime soon seems pretty unlikely.) The length of time is not really important, though, since the process started many millions of years ago; what matters is the rate and accumulated amount.

And there, as Hamlet would say, is the rub. Only about 2 million barrels of petroleum are created in a year, according to best estimates, which represents approximately 0.0064% of what we are currently consuming. In other words, the fact that oil is actually renewable is not important because the rate is too small to matter, which brings us to the point that the (allegedly) finite nature of petroleum is not important either, because only the numbers matter. Pointing to oil's finite nature is valuable only as evidence of either one's ignorance of resources or poor math skills.

Greek Fire

Optimists often remark that "The Stone Age didn't end because we ran out of stones," but more relevantly, neither did the Bronze or Iron Ages end because those more rare and scarce nonrenewable resources were depleted. Indeed, it was prior to the Golden Age of Greece when the primary soldiers were hoplites, that is, landed gentry who had enough wealth to afford the bronze armor that they wore.

Now, roughly five millennia after the Bronze Age began, copper is no longer the province of the wealthy but the stuff of pennies! Granted, copper pipes are valuable enough to attract the attention of the avaricious (thieves, not Congress), but it remains readily available, and every boom in prices has been followed by a bust.

So Much for Solar Power

An astronomer was giving a luncheon talk to a local group, and happened to mention that the sun would burn out in five billion years. An elderly woman in the front row gasped, and asked him to repeat it. When he did, she heaved a sigh of relief, "Thank goodness, I thought you said million!" (old joke)

This brings us back to the point that *numbers matter*. Just like petroleum, the sun's fuel is finite—but don't expect a wave of solar power cancellations when that realization goes around. Numbers tell us that we don't need to worry about the "finite" nature of the sun's fuel.

And while the petroleum resource is hardly comparable to the life of the sun's fuel, the real issue is: *how much is that resource,* not whether it is finite. Given their estimates of only 2–2.5 trillion barrels of recoverable oil, half of which has been consumed, an argument of peak oil advocates that scarcity is imminent seems credible. And this recalls the neo-Malthusian's error mentioned earlier; assume a static resource and use a conservative number.

DEAD FROGS: THE BUGABOO OF EXPONENTIAL GROWTH

Thomas Malthus was one of many people who has observed that population growth seemed to be outstripping the growth in agricultural production, but being an economist, he didn't assume that this was somehow driven by immutable natural laws and instead argued for agricultural reform to increase food production.

More recently, there has been a clamor about "peak everything" based on the idea that, well, everything is finite and we're using it up, so it is "running out." Or at least, production must peak. Or, as one physicist points out, eventually human energy production will generate as much heat as the sun does—eventually being 1400 years.

Flat Earth

Colin Campbell, in the famed (well, famous in the IEA's offices) debate at the IEA in 1997, compared resource optimists to the conservative Spanish court that opposed the visionary, Columbus, and has since referred to those, like Adelman and me, who disagreed with him as "flat-earth economists." Albert Bartlett later explained that the term actually meant that economists thought the earth had two dimensions and thus was infinite, containing equivalently infinite resources.

But this description ignores two important variables: capital and knowledge. Additional investment can often increase the production of renewables like agricultural products and nonrenewables like minerals and oil in the same amount of space, as can better technology. Neo-Malthusians tend to ignore this factor and argue that the rate of technological advance (and greater scientific knowledge) has diminished or disappeared, as described in Chapter 7.

The argument is somewhat specious and relies in part the question of the finiteness of resources, discussed earlier—or a static measure of resources and dynamic view of consumption, as in *The Limits to Growth*.

HOW LONG?

Perhaps the most important factor that raises skepticism is the fact that at least some exponential alarmists fear the distant future. Any number of pundits have looked at long-term forecasts of economic and/or technological development and characterized them as foolish. We have no flying cars, nuclear power is not too cheap to meter, and no one is eating Soylent Green.

On the other hand, most of these were not serious forecasting efforts, but rather off-the-cuff remarks (or the equivalent), and those making them were not particularly serious about achieving them within a specific time frame. And we do eat Soylent Green already; only we call it tofu and vegemite. (Read the book, it wasn't people.)

NEWTON'S FIRST LAW

The biggest mistakes have come from an apparent source: extrapolation of a trend endlessly, as if there were no feedback or other variables involved. Jay Forrester, the inventor of Systems Dynamics, which was used in *The Limits to Growth* model (and which I have used), reportedly once said that feedback effects tend to overwhelm the initial stimuli, which is probably true in many cases. Yet, many neo-Malthusians and especially peak oil advocates tend to extrapolate a given trend endlessly, assuming no feedback effect whatsoever.

Indeed, the first wave of peak oil advocates explicitly argued that no feedback effect would occur: prices didn't affect production or consumption levels. Technological advances were either unimportant or had ceased and so could not increase the resource base.

An important element of the fear of exponential growth is the analysts' choice of particularly high growth rates. As Figure 4.1 showed, Ehrlich chose the highest observed growth in the 20th century for his calculations, even though it represented the post–World War II baby boom and should have been considered an exception, not the norm. Similarly, Bartlett, writing in 1998, talks about the growth in oil demand from the 1950s and 1960s at 7% a year, which causes a doubling of use every decade,[25] which sounds alarming, given the arguments about the difficulty of making a speedy energy transition, until you realize that consumption growth dropped to 3% per year in the 1970s (a doubling time of 24 years), and under 1% per year in the 1980s (a doubling period of 75 years), before recovering to 1.5% in the seven years before his talk (48 years).

This emphasizes the lack of feedback mechanism used in these simplistic models and how important they are in the real world.

REAL SCARCITY

Indeed, the subtext of the fear of resource scarcity is that renewable resources have repeatedly been the source of problems. In Tainter's *The Collapse of Complex Societies,* he talks about resources as causing the fall of a number of (mostly) ancient civilizations; nearly all suffered from problems like lengthy droughts and salt buildup in irrigated farmland.[26]

And similar problems continue today, especially if you consider endangered species, from rhinos to tuna. In all cases, these are *renewable* resources, the very ones that are NOT finite, that are sustainable, that we can rely on for all eternity—in theory. No lasting shortage of nonrenewable resources— minerals and energy—has occurred since the advent of the global economy.

There is a very appropriate analogy in the oil industry itself: oil to the ancients meant plant oil. The Phoenicians relied on linseed oil, for example, the Romans, olive oil. The Jews celebrating Hanukah (or Chanukah, as some of my childhood friends spelled it, the biggest religious controversy I knew before college) were talking about lamp oil, which was flax oil.

But the industry made a "transition" to whale oil, which was found to be cheaper and superior to other oils, and was a major part of the global economy in the 19th century. Indeed, countries sought to develop whaling because oil was a "strategic" industry even then, with the British chasing the French fleet during the Napoleonic wars and the U.S. Confederate raiders attacking the Union's whaling fleet during the Civil War (or the War between the States, as my childhood teachers in Richmond, Virginia, insisted we call it).[27]

But aside from the fact that whale oil is "green," probably carbon neutral, and theoretically "renewable," it became scarce because the world was using it up faster than it could be replaced and ran up close to the absolute limits. Instead, oil use was changed to that nonrenewable resource, fossil-based petroleum, which, even today, a century and a half later, is not being used at a rate that brings us anywhere near the limits of the resource.

THE PAST AS PROLOGUE

Why, then, is there so much focus on the "finite" nature of petroleum among peak oil advocates? Well, in fact, this rationale has appeared on a number of other occasions, such as when Will Rogers said, "Buy land. They ain't making any more of it." Of course, during the Roaring Twenties, agriculture was booming along with stock markets, so land was naturally valuable then —as in the 1970s, for example. And farmers generally bought land, thinking that it was finite and therefore must rise in value as the population and food demand increased. Similarly, during the Japanese bubble in the 1980s, when land values soared to preposterous levels, it was repeatedly noted that land in Japan was finite and scarce—which didn't stop prices from collapsing.

This rationale brings up another related argument, namely, that "people have to eat." I couldn't track down the original version of this statement— and it may not be possible, given the universal nature of the opinion—but when googling it, got 9.5 million hits. A quick perusal suggested that many of them were, you guessed it, brokers talking about agricultural prices or the stock of companies in the food processing industry.

CYCLES NOT TRENDS

In a Calvin and Hobbes cartoon, Calvin notes that the weather is getting colder and concludes that the Earth has spun out of orbit and is receding from the sun. Although the cartoon is one of my favorite, the reality is that children, by the time they notice changing seasons (and certainly understand the role of the Earth's orbit, as Calvin did), usually have sufficient memories of seasonal changes that they are rarely suspicious that the growing or shrinking temperatures are part of a long-term trend. For adults, interpretation of short-term events or trends is more complicated, and something humans have struggled with for a long time. Even Aristotle is said to have remarked something to the effect that a single robin does not a spring make.

Having observed numerous seasonal changes, most adults also notice the deviations from the norm—hot summers, cold winters, and so forth—and thus tend toward confusion about what represents an unexplained deviation and what a new trend. This is true not only for weather patterns but also for social and economic trends, although these tend to be more complicated because of the role of humans (and psychology).

Anyone doubting this should watch one of the cable channels devoted to business and watch the dueling talking heads debate whether that days' stock market moves are an anomaly or consistent with the pattern they believe prevails. Not every down day on the market brings out the bears, but almost any bullish trend seems to have those who defend the likely permanence, ignoring the old mantra "Buy low, sell high." This has been true at least since the Dutch in the 17th century convinced themselves that tulip prices would not decline.

Even those who deny the existence of global climate change are having trouble dismissing the evidence of the last year.

Newsweek 2011[28]

The rains came steadily in the spring in those years, 1926 through 1929, and with wet years, everyone forgot about the dry ones and said the weather had changed—permanently—for the better.

The Worst Hard Time, 52[29]

We are certainly seeing the same phenomenon with regard to weather events and global warming, where Al Gore was embarrassed to seize on the severe hurricane season of 2005 as evidence of a change in the climate, only to see 2006 completely reverse the trend and produce an exceptionally mild season. Similarly, all too many have called the unusually large number of tornadoes in 2011 in the United States as evidence of a long-term trend, not realizing that there have been other similarly violent years in the past, including 1936 and 1953.[30]

Buddy, Can You Paradigm?

In economic terms, there have been any number of cycles or bubbles that have been touted as "new paradigms." The oil price increases in the 1970s were touted as being permanent (see Chapter 2), and the 1986 price collapse back toward the long-term norm was considered exceptional, and unlikely to last. Many still treat the 1986–2002 period of prices as both "low" and an interlude, rather than the norm.

In part, this reflects what Shermer refers to as "agenticity" or the idea that something is deliberately causing a change.[31] If high prices were only due to volatility and random effects such as normal weather fluctuations, then they could be expected to come back down. But when there is thought to be some particular effect causing them, such as Chinese economic demand pressuring commodity prices, then it is easier to believe that the situation will continue.

This problem has been particularly true of Malthusian prognosticators, who sometimes seem more like psychics trying to justify the value of their visions. Gardner describes the classic case of Robert Heilbroner, who, in 1973, predicted a horrific future for humankind, with increasing starvation, a retreat to "iron governments," and the poor threatening nuclear blackmail of the rich. Later, in reviewing the earlier predictions, he noted with satisfaction that "several serious famines" had occurred since his book was published.[32] Of course he was just trying to rationalize his poor prediction, but we see this so many times because agricultural commodities are inherently cyclical: bad weather regularly reduces production and sends prices soaring, seeming to confirm the Malthusian argument. And naturally, higher prices bring on more production and a price collapse.

In one telling instance, Ehrlich actually criticizes resource optimists for pointing to recent abundant grain harvests to support their arguments, then does exactly the same himself, noting the two recent poor harvests.[33] Apparently he couldn't see the contradiction and didn't have an astute editor.

Paradigms Everywhere

Even aside from anthropogenic climate change, the Earth's climate is constantly changing, and droughts can last a season or a century, in the most extreme case. Scientists still can't predict their duration, beyond seeing a brief way into the future, perhaps several months. This is why farmers in what became known as the Dust Bowl in the 1930s were able to convince themselves that a few good years represented the new norm, not a deviation. And who can forgot the warnings of an impending Ice Age in the 1970s, when the global climate was unusually cool?

This is especially true when there is a plausible theory that seeks to explain the anomaly. In the case of the Ice Age fears, they were hardly unreasonable given that the planet seems to be overdue for one based on the historical record; it would not surprise climatologists if it were to commence soon, and some even argue that global warming is already being offset partly as a result. For the Dust Bowl farmers, they were told that "rain follows the plow," which was partly an advertisement to entice them into dry areas, but also represented a belief that local climate could be altered by, for example, farming.

> The surest sign of an imminent recession is when economists argue we've beaten the business cycle.
>
> Old economists' joke (and joke told by old economists)

In terms of economic trends, the problem includes not only the complexity of the system but also an enhanced self-interest on the part of many of those involved. Huge numbers of analysts (including me!) are paid to predict the economy, the stock market, commodity prices, and so forth, and they can hardly respond as financier J. P. Morgan allegedly did when he was asked to predict the stock market. "It will fluctuate."

Still, the stock market not only fluctuates but has a general long-term trend (up) and a variety of cycles, which many try to use to predict future behavior. In hindsight, it is perfectly clear what was a cycle and what part of the underlying trend, but, at the time, it is never so obvious.

Recall during the dot.com bubble, any number of analysts were available to talk about how the Internet economy was a "new paradigm" and the old measures of stock valuation were not relevant. Similarly, in the 1920s bull market, some argued that new, "scientific" management techniques meant that companies were better managed and therefore price/earnings ratios should be higher than before. (Not only geologists can claim to be scientific.) The 19th-century advent of the railroad saw a similar stock market boom, which was followed by a bust, but not until after many experts had explained that this industrial revolution was different and justified much higher stock prices than traditional measures.

And of course the 16th-century Dutch truly believed that the tulip industry was special and that a new tulip really could be worth as much as a fine mansion. While some modern economists have attempted to argue that these were valid valuations, most are relying on the theory that any price is "valid" if people are willing to pay it.

How Is the Business Different from All Other Businesses?

What is most amazing is not that people have done a poor job of valuing new businesses or industries, but that those that are regularly cyclical are still constantly presumed to be moving into new areas. Real estate prices fluctuate and undergo boom-and-bust cycles, but there are still many who will buy at the peak.

Indeed, Gardner describes a number of prominent economists who insisted that real estate prices were not cyclically high in the last decade, giving various rationalizations, and all ignoring the fact that home ownership levels were soaring along with prices.[34]

Agricultural commodities are notoriously cyclical, but, every decade or so, when a drought or flood is sufficient to raise prices, many farmers leap at the opportunity to borrow more money to acquire land and equipment, thinking that now prices are "normal" rather than elevated. The willingness of governments, not least the United States, to then bail them out when the cycle deflates only worsens the problem.

Similarly, peak oil advocates look at the invasion of Iraq and see higher prices and assume that this is a new, permanent condition or see the drop in global production after May 2005 and believe it to be the all-time peak, without noting, first, the way Hurricane Katrina reduced U.S. oil production, and then the upheaval in Nigeria, which together took over 1 million barrels per day off the market. And of course, OPEC cut its quotas by 1.5 million barrels per day (and actual production by nearly 2.5 million barrels per day). Some peak oil advocates seem to think that the last represents an astonishing coincidence of technical problems, forcing a drop in production just as the organization announces a quota cut to support prices.

And when oil prices first declined in 2008, any number of analysts referred to the need to "recover" to $100 a barrel, as if that were the norm. Even prices in the $70–80 range, more than twice the historical level, were thought to be depressed. Now, obviously, $50 is being treated as an aberration, and the only uncertainty to many is how long it will take for prices to recover. Few think that $50 could be sustainable.

Hindsight

Distinguishing between long-term trends and cyclical changes is extremely easy—after the fact. This is especially true given the cacophony of voices that will provide all possible interpretations, up to and including astrological influences. But there are a few guidelines that can be applied.

The most valuable clue is the rate of change: if a commodity price suddenly shows rapid growth or decline, that behavior is unlikely to represent a trend that will continue. Slowly rising housing prices, say 3% a year, are much more sustainable than a 20% per year increase: the former could be explained by a strong economy and growing population in an area, better use of land, and so forth. The latter almost certainly reflects short-term conditions, probably reinforced at some point by speculative investment and momentum trading, and at the least will probably end, and conceivably be largely reversed.

Self-Reinforcing Errors and the Devil Take the Hindmost

The first tenet of investing is "buy low, sell high," yet there are many who carry out momentum trading: buy something that's rising, because it will probably keep rising. While this sounds foolish, the reality is that many traders and investors react to what other traders and investors are thinking, and they know that many others buy into rising markets, so they can buy in and ride them up, hoping to get out.

This works frequently for high-volume, high-speed traders where they are able to spot a downturn and bail out quickly. The big winners in cases like the dot.com and recent real estate bubbles were not those who never got in and avoided losses, but got in early and *got out early*.

An old Wall Street aphorism is "The bulls make money, the bears make money, but the pigs never do." This was complemented by J. P. Morgan's explanation of his riches when he said, "I always sold too soon." It is those who held on, either not recognizing a market top or thinking that they were not experiencing a cycle—the pigs who got greedy—who wound up losing money.

POLITICAL AND SOCIAL CYCLES

The problem is hardly limited to economists and investment bankers, as social commentators often seize on short-term developments or trends and extrapolate them ever onward. It is not clear if this is true of the more apocalyptic visionaries, as witness publication of the book *Dow 40,000,* which was, at best, premature. But like science-fiction writers who seize some change in culture and imagine its dominance in the distant future, many pundits seem incapable of distinguishing between fundamental, long-term trends and those that are only temporary.

The mother of all such social bubbles was the Japanese bubble in the 1980s. Some thoughtful commentators had seen the rise of Japanese manufacturing from a source of cheap, unsophisticated goods to those of the highest quality, which remains true today.[35] However, the 1980s saw an enormous financial bubble in Japan with escalating stock market values and real estate prices going to absurd heights. Perhaps the most telling example of misinterpreting the boom is the way many American pundits saw the high price of real estate in Japan as a sign of a strong economy rather than a financial bubble. The real estate on which the Imperial palace was located, for instance, was said to be worth more than all of that in California.

Any number of commentators then predicted the coming dominance of Japan (or perhaps Asia) on the global scene. The United States would be replaced as the world's superpower, Asia would be ascendant, and "the end of empire" was at hand. Many of these were Japan specialists, but occasionally other observers would fall into the trap, especially those who became "instant experts" on the subject of Japan.

Some of this work was quite soberly done, including *The Rise and the Fall of the Great Powers,* by Paul Kennedy, who warned of imperial over-reach as a cause of imperial decline.[36] But many others simply took for granted that the Japanese economy would continue to boom, the stock market would soar, and the "Rising Sun" would dominate the horizon,[37] in other words, blind extrapolation of both a long-term trend of economic development and the most recent boom years, which proved to be a bubble.

Instead, the United States proved to be going through a weak economic patch from which it recovered, making it the "hyper power," albeit due as much to the decline of its main competitor—the Soviet Union. And the Japanese economy ultimately deflated, leaving it a major economy but stagnant rather than rising.

The Chinese boom is similar to and different from the Japanese boom, and China will undoubtedly become a major economic and political power in the future. It has a much larger population than any other country, with an entrepreneurial spirit, and a government that has embraced economic (and income) growth.

Yet much of the recent improvement has been due to the same things that saw other countries boom, namely, the movement of labor from low-productivity agricultural sector to the higher-productivity industrial sector. This appears to be nearing a finish, which means that economic growth (and oil demand) will slow in the next decade or two. Better education of the rural population might alleviate this, but the point is that nothing last forever, including the breakneck growth China has had in the past two decades.

CYCLING ON THE BIAS: CHERRYPICKING

The problem of cycles feeds into the entire question of resources and Malthusian bias, since many who are concerned about the "finite" nature of nonrenewable resources (minerals and fossil fuels) also worry that the finite (and decreasing) amount of arable land, combined with rising population, leads to a corresponding problem with food scarcity. This was most prominent in the 20th century (and continuing) in regard primarily to commodities, agricultural, as well as mineral and energy.

Before that, Lester Brown looked at higher food prices in China and argued that it was the beginning of a long-term boom. Specifically, he cited prices in February and March 1994 as indication that the nation had reached its productive limits.[38]

Richard Heinberg has published a book called *Peak Everything: Waking Up to the Century of Declines,* and two of the original authors of *The Limits to Growth* produced a 30-year update in 2004, emphasizing primarily the environment and climate change, but not admitting that their earlier concerns about resources had been, at the least, greatly exaggerated.[39]

STAIRWAY TO HEAVEN

This is not to argue that some factors do not move continually upward, such as economic growth, or experience patterns that cannot be truly called cyclical. In modern times, population growth has shown a tendency to rise sharply with better nutrition and medical care and then slow and decline as economic progress leads to people having better things to do than procreate.

Similarly, although economies experience boom-and-bust cycle, there is a tendency in the modernized countries to undergo long-term growth, especially in per-capita income. Again, some interpret every step backward during recession as the beginning of eternal decline, and every economic boom as a new golden age. With every recession—and recovery —economists have to remind the public that employment is very much a lagging indicator, and its slow decline is not proof of a mild recession, nor is its slow recovery evidence that the recession will never end.

FIVE

The Hubbert Curve Ball

If the supply pessimists have a guiding spirit, it surely is the ghost of King Hubbert.[1]

The guru for many peak oil advocates is M. King Hubbert, a geophysicist who worked in the petroleum business for many years and, among many other contributions, developed the theory that oil production must follow a bell curve, from his original 1956 publication. This idea has been seized on by the geologists who started the current round of concerns, Colin Campbell and Jean Laherrere, and appeared in their 1998 *Scientific American* article, "The End of Cheap Oil." This was not their first word on the subject (see Chapter 2), but they brought it to the attention of a general audience.

The curve became famous not when Hubbert first produced it in 1956, but later, in the early 1970s when it appeared to have successfully predicted the peak in U.S. oil production. Given this success, and perhaps because it was created by a geologist, not an economist, it was adopted as a standard means to predict oil production, at least among peak oil advocates, beginning with Campbell in his 1989 article, but described more explicitly in the 1998 *Scientific American* article, where Campbell and Laherrere insisted that

Adding the output of fields of various sizes and ages ... usually yields a bell-shaped production curve for the region as a whole. M. King

Hubbert, a geologist with Shell Oil, exploited this fact in 1956 to predict correctly that oil from the lower 48 American states would peak around 1969.

Other peak oil advocates have adopted the Hubbert curve wholeheartedly, including Roger Bentley and Ugo Bardi, and scientists in Kuwait announced that they had "adopted a newer approach by including many Hubbert production cycles or bell-shaped curves showing the rise and fall of a nonrecyclable resource. Earlier models typically assumed just one production cycle, despite the fact that most oil-producing nations have historically experienced more of a rollercoaster ride in production."[2]

TRUMPING THE ECONOMISTS

A convincing *theoretical* explanation in support of the lognormal or any other particular distributional shape grounded in geologic or geochemical laws is still absent.

Kaufmann, 1993

Recalling the Reagan joke about economists being less concerned with reality than theory, this is an amazing case where *neither theory nor reality* supports the argument. As Hubbert himself initially acknowledged, "We could draw a family of possible production curves, all of which exhibit the common property of beginning and ending at zero, and encompassing an area equal to or less than the initial quantity" (1956, pp. 10, 11). In other words, *the creator of the Hubbert curve admitted that there was no underlying reason to use a bell curve.*

He noted that not only did actual production follow a variety of readily observable paths but there was no reason to believe it had to follow one given path. All that was necessary, he said, was that there be a beginning, end, and peak somewhere in between. This is, of course, self-obvious, but tells us *only* that once production ends, we will know how much oil has been produced. This is a completely self-obvious statement, of no utility whatsoever, rather like saying a race starts at the beginning, finishes at the end, and the halfway point is in the middle.

Peak oil advocates have, however, jumped to the back of the book and ignored the fact that that there is absolutely no theoretical explanation for production following a bell curve. Campbell and Laherrere attempt to alleviate this shortcoming by showing some individual field production curves and asserting that "Adding the output of fields of various sizes and ages (green curves at right) usually yields a bell-shaped production curve for the region as a whole."

EXCEPT THAT IT DOESN'T

Careful examination of their figure shows that not only does there appear to be no connection between the individual curves and the aggregate one portrayed, but the field curves are not realistic presentations of actual field behavior. Real fields follow many different production curves, but there is a tendency toward asymmetry, that is, an early peak, followed by a long decline, not the more symmetrical curve with a flat peak. This is usually done for economic reasons, namely, to get oil out quickly to increase the return on investment.[3]

Explanations of why production from numerous fields adds up to a bell curve have been noteworthy for their absence, the only rough explanation being that the Central Limit Theorem requires that the addition of many numbers results in a bell curve.[4] Others point out that the curve *seems* to work and believe that this implies that it must have theoretical foundations, even if they are not understood. This is rather like stock market forecasters who think they can predict the stock market because they perceive certain patterns.

Peak oil advocates often overlook the fact that the pattern doesn't always work. Bentley, addressing an energy economists meeting, pointed to several national production curves as evidence for the theory; when pressed as to why he was using data for small producers like France and Germany, he admitted it was because other countries did not provide such a close match to the supposed "scientific" formulation. Similarly, his 2007 article showed the United Kingdom, Germany, and Egypt as its only examples—all three of which approximate a bell curve.[5]

This highlights the fact that peak oil advocates are repeating the error of the 1970s' economists, described in Chapter 2, who promoted a theory about price behavior that was at odds with historical reality. A simple examination of production curves for nations finds that, while perhaps one-fifth roughly resemble a bell curve, most do not. Mexico, for example, shows a long gradual increase after the 1920s peak, due not to geology, but to ruinous taxation of the foreign oil companies. The recent decrease there, which peak oil advocates naturally ascribe to geological factors, appears to be in the process of being reversed.

Somewhat astonishingly, peak oil advocates attempt to apply Hubbert curves to countries like Saudi Arabia or Venezuela, where OPEC quotas—especially in the weak markets of the 1980s and 1990s—drove

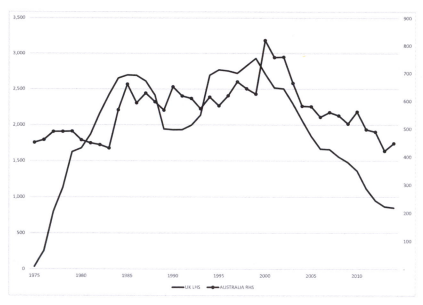

Figure 5.1   Production Curve for Australia and the United Kingdom (BP Statistical Review of World Energy 2015)

production trends. But even in the case of countries like the United Kingdom and Australia, with free market environments and no adherence to quotas, there are multiple peaks and only the former roughly resembles a bell curve (Figure 5.1).

Indeed, apparently due to prodding from certain quarters, Campbell and Laherrere did acknowledge that the Hubbert curve could often have multiple peaks, meaning that the peak in production did not indicate the halfway point of resource exploitation and that the curve itself was not predictive.[6] Laherrere managed to rationalize this by opining that "The important message from Hubbert's work, which is often forgotten by economists, is that oil has to be found before it can be produced."[7] Oddly, the two later defended their work from Leonardo Magueri's criticism in *Scientific American,* by arguing that "Decline typically commences at about midpoint of depletion, as already exemplified in more than 50 countries."[8]

This has been a difficult truth to convey peak oil advocates, although some have quietly abandoned the argument. Still some, like Bentley, Jeff Brown, and Ugo Bardi, *continued* to make references to the curve, and others (especially the mathematicians) debate whether the Hubbert curve is a proper bell curve or a logistic or Gaussian curve, all the while ignoring the fact that they are merely curve-fitting without theory.[9]

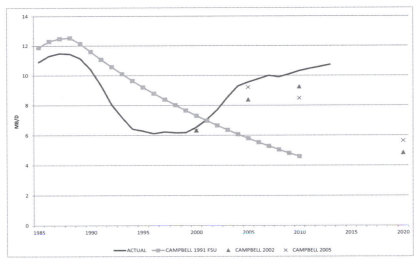

Figure 5.2   Campbell Extrapolates Discovery Trends (but only the decline) (Campbell [2003])

Again, we find the nature of physical science to be a hindrance, not a benefit. In physics, extrapolation is a good strategy, as the laws of physics don't change as you proceed. (Assuming you have included all variables, such as the gravity of the planet, your spaceship is going past.) When studying physical sciences, patterns reflect natural laws: the curve of a planet's orbit allows you to measure gravitational pull without worrying that it will change, and measuring the position of the Earth's moon from different latitudes can be used to estimate its distance with confidence that it won't show up somewhere else in a few days.

But this is because physical laws are supposed to be immutable and unalterable (and usually are). When Alexis Bouvard noted irregularities in the orbit of Uranus, he didn't think that Newton's laws of physics were wrong, nor did he assume that Uranus had changed its mind about its trajectory. Instead, he presumed another gravitational force was acting on it and used that to locate the planet Neptune.

But peak oil advocates assume that discovery and production of oil are solely the result of physical forces: the geology and chemistry of the deposits. Thus, it becomes simplicity itself to observe discovery and production trends, as Campbell does for global discoveries in Figure 5.2— and extrapolate them. Since geology doesn't change, these trends shouldn't either. And to address the argument that technology will affect future trends, peak oil advocates generally argue that it either has ceased or cannot do so (as we will see in Chapter 7).

The astute observer will notice, however, that Campbell *has ignored the most recent trend* of rising discoveries and instead assumed they will start a new decline. Nor do past trends provide much information about future behavior: looking at different periods gives us very different viewpoints, as the added lines in Figure 5.2 show.

Given that the only information he is using to make his forecast trend is the historical trend, this would seem to be a rather egregious error. Clearly, he believes that discoveries must decline, even though they weren't doing so recently, and so extrapolates the opposite of the observed trend. Perhaps he believes that geology, like a fickle woman, had changed its mind twice in the past decade, from decline to growth (observed), but now to resumed—and assumed—decline. Indeed, sophisticated statistical analysis would provide a more precise answer, but, as M. A. Adelman always said, "Don't confuse precision for accuracy." The weakness of the data, as discussed in Chapter 6, prevents any such analysis from having great utility.

THE SIX DEGREES OF HUBBERT CURVES

> Interestingly, galaxies, urban populations and other natural agglomerations also seem to fall along such parabolas.[10]

Amazingly, peak oil advocates think that the existence of bell curves in other phenomena somehow proves that the theory is a valid geological law or perhaps some overarching, universal natural law. For example, Jean Laherrere notes temperature cycles in Siberia and cod landings,[11] while Ugo Bardi comments on obesity following a similar trend. This, they apparently believe, implies something about the universality of the bell curve in forecasting. (I admit that my weight has been declining recently, and I would like to think that it probably follows a bell curve, but I doubt either that it will follow such a path toward zero or that it *must* do so.)

Many other examples can be found just from perusing everyday life. Indeed, there was a book titled *The Bell Curve* about the distribution of IQs among population groups, and in a recent article on nuclear weapons, we see that in both the United States and the Soviet Union, the nuclear arsenals roughly resemble bell curves,[12] as does the recent H1N1 flu outbreak in the United States.[13]

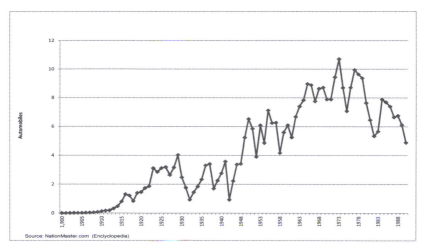

Figure 5.3   U.S. Automobile Production (http://www.nationmaster.com/country-info/stats/Industry/Car/Production)

For something a little closer to oil production behavior, examine the historical pattern of U.S. automobile production, which certainly does resemble a bell curve (Figure 5.3). And while the U.S. automobile industry is certainly suffering, a look outside your window should demonstrate that we are hardly running out of cars. Indeed, rather than blame scarcity, in this case, the switch from production of automobiles to light trucks and SUVs and pickups is the explanation. In other words, the decline is entirely a matter of semantics.

BELL CURVES, THE REALITY

What is the connection between all of these things? Why would cod catches resemble oil production, or even the size of galaxies? There are two primary explanations, neither of which has anything to do with geology or peak oil.

First, a bell curve represents exponential growth and decline, which is common when there is inertia in a system. For example, a government won't suddenly destroy huge numbers of nuclear weapons; it will break them down gradually. Farmers can't suddenly double their equipment purchases; the manufacturing sector doesn't have the capacity. Instead, changes are incremental, a few percent a year. In the oil business, it is impossible to raise production too quickly in areas that are already producing large amounts, and even if all investment ceases, production will decline only slowly as existing fields and equipment wear out: no one is going to tear out operating wells and pumps.

(Contrast this with other variables, such as the population of a prairie dog colony. Disease can wipe out a large percentage very quickly. In the case of agriculture, if the scale is small enough, such as a particular county, production can shift rapidly due to weather or market conditions. On a global level, such is much less likely.)

In effect, peak oil advocates have simply identified the fact that some areas experience growth and then decline, and because of imbedded capital, the change is gradual. This often yields something like a bell curve, but it has little to do with geology, and it is certainly not determined by some type of geological or physical law, except to the extent that fluid dynamics restrict production to slow changes.

DO MONKEYS RING A BELL CURVE?

> The completely implausible theorem … asks us to believe that by taking an infinite number of monkeys and sitting them down at typewriters, eventually one of them will type a known work of literature, such as William Shakespeare's Hamlet.[14]

But the other factor is that certain patterns recur all the time, whether in socioeconomic data or physical data, the difference being that in the latter, scientific laws can be construed with a degree of certainty. In the former, this is not necessarily so. The distribution of galaxy sizes might easily follow a natural physical law, but cod landings depend on a variety of factors, including government regulations, which are not directly visible simply from viewing historical data showing the final result of the many inputs.

But even if there is no physical law creating a pattern, by examining enough data, you occasionally find such a pattern. The infinite monkey theorem is not meant to suggest that, given enough monkeys, one will be capable of doing the work of Shakespeare, but that by observing enough random data, a pattern will be found that fits the presumed theory. In this case, "enough data" is close to infinite. That this "pattern-seeking" is occurring with peak oil and the Hubbert/bell curve is clear from the authors' tendency to ignore countries that do not follow a bell curve, as mentioned earlier.

ISLANDS IN THE SKY

There is also a very human tendency to observe patterns that aren't even there. Indeed, the great satirist, Jonathan Swift, in *Gulliver's Travels,* described the hero's visit to the Grand Academy of Lagado, where a professor had a giant frame containing blocks of wood on wires covered with words. "The pupils at his command took each of them hold of an iron handle, ... and giving them a sudden turn, the whole disposition of the words was entirely changed. He then commanded six and thirty of the lads to read the several lines softly as they appeared on the frame; and where they found three or four words together that might make part of a sentence, they dictated to the four remaining boys who were scribes." Volumes of the collected works were to be published "to give the world a complete body of all the arts and sciences." Like a true academic, he sought public funds to expand the number of word frames to 500.[15]

A real-world example is cited in a book on ghosts that describes an effort to explain why some people believe they hear spirits in the noise coming from between radio stations:

> [D]escribed ... an experiment in which a group of people were handed paper and pencil and asked to help transcribe what they were told was a faint, poor-quality recording of a lecture. The subjects offered dozens of phrases and even whole sentences they'd managed to make out—though the tape contained nothing but white noise.[16]

Ken Deffeyes shows us how a hard scientist can be lead astray by his background. In his first book, he discusses Zipf's law,[17] which noted that there were certain recurring patterns in the world, in this instance, the number of times that the most common letter, "e," appeared in English lit. He then goes on to note that this seems to apply to the size of cities in Denmark and oil fields in Mexico, which he uses to suggest that there is an undiscovered field in Mexico, as well as a missing Danish city (Atlantis, perhaps).

The question becomes: how do cities fall into such a pattern? The implication is that there is some outside factor, a natural law, driving people to congregate in places to form cities of certain sizes. Somehow, once a city reaches the appropriate size, people would unconsciously refuse to move in or move out. Or having caught a certain amount of cod, they cease fishing. And why Denmark? Does this mean that other countries don't fit the pattern and Deffeyes is relying on the one case that does, again, demonstrate confirmation bias?

TOO MANY HORATIOS

> There are more things on heaven and earth, Horatio, than are dreamt
> of in your philosophy.
>
> Hamlet

To repeat, in the physical world, where there is no human intervention, objects do obey such laws, although it is not always apparent, since there may be more than one simple factor affecting their behavior. But clearly, the amount of cod being caught depends not only on the biology of the cod but also on government regulations governing fishing, the technology available to the fishermen, and many other factors that are not carved in stone and, in fact, vary quite a bit over time. To suggest that the Hubbert curve "predicts" when governments will limit catches to conserve the resource is absurd and harks back to ancient haruspices, who interpreted the thinking of the gods (not the cods).

Simply, in the case of oil supply, the discovery of oil depends not only on the presence of oil in the ground ("it has to be there to find it" as peak oil advocates say) but also on whether explorationists are allowed to drill freely, whether they have the technology to access that particular geology, and whether the economics are favorable (which is affected by the price of oil, the tax rate, and the local infrastructure, as well as the geology of the oil). Yet many peak oil advocates argue that only geology matters, or as Campbell says, "Oil is ultimately controlled by events in the Jurassic which are immune to politics" (Campbell 2000) and "discovery and depletion are set respectively by what Nature has to offer and the immutable physics of the reservoirs."[18] The peak oil advocates are so many Horatios, who don't see the many aspects that affect oil production in the real world.

NOTHING SUCCEEDS LIKE SUCCESS?

Naturally, the argument that the curve works would seem to be compelling, *except that it doesn't*. Not only do any number of countries and regions fail to follow the "scientifically determined" bell curve shape, but the use of this shape has often led peak oil advocates astray.

The bell curve theorem essentially assumes that all production paths are caused by geology, and since geology is invariant over time, extrapolation is the best forecasting method. In reality, any number of things can cause production trends to change, so that the bell curve assumption has resulted in a variety of failures. Most obvious one came in Campbell's 1991 book, when, for example, the temporary production decline in the United

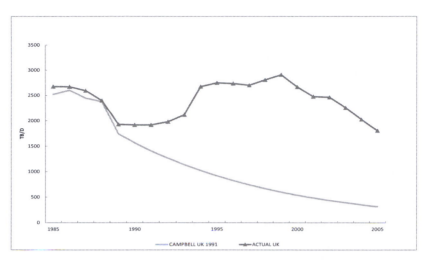

Figure 5.4   Campbell Extrapolates UK Decline Post-Piper Alpha Disaster (Campbell [1991] and BP Statistical Review of World Energy)

Kingdom caused by the Piper Alpha disaster was extrapolated into the future, leading to a massive error in the forecast (Figure 5.4).

Similarly, production in the Soviet Union dropped when the economy collapsed in the early 1990s and falling demand resulted in lower production due to limited export capacity, leading Campbell to project ever-falling production (Figure 5.5). Most other analysts were far more optimistic, and my

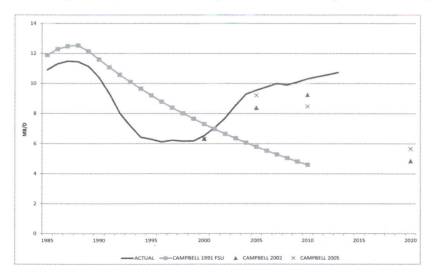

Figure 5.5   Various Forecasts of FSU Production by Campbell (Campbell [1991], [2002], and [2003]. BP Statistical Review of World Energy)

own examination of Soviet/Russian data led me to predict correctly that Russian production would soar.[19]

DEMAND DRIVES SUPPLY

And the role of demand in affecting production is usually overlooked, because of the assumption that production in excess of demand can be exported. However, at the global level, this is obviously not true. The drop in production following the 1970s price hikes resembled the beginning of the downside of a bell curve, but was not caused by geological factors; it was ultimately reversed. (This makes current claims that the recent plateau represents a peak all the more ridiculous.)

U.S. natural gas supply provides a clear-cut case of how this method can mislead analysts. After prices soared from the late 1970s, demand dropped sharply. Since there is no significant export market for natural gas, supply had to be curtailed. However, by looking at the trajectory of natural gas production, M. King Hubbert, among others, came to very pessimistic conclusions about the U.S. natural gas resource. So pessimistic, in fact, that the United States has already produced more natural gas than he thought possible. Figure 5.6 shows exactly where this type of extrapolation leads, and just how wrong it is to, assume that supply trends are driven only by geology. Blind extrapolation is rarely successful.

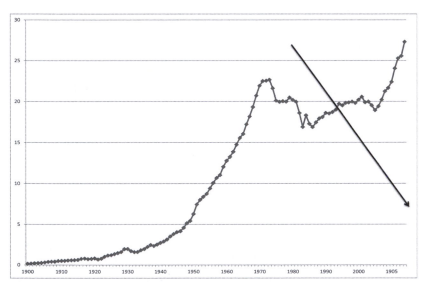

Figure 5.6   U.S. Natural Gas Supply: Extrapolation Gone Wrong (Energy Information Administration, US Department of Energy)

Perhaps more amazing is that none of the peak oil analysts seems to have considered the possibility that other factors might be driving production in these various cases, even in the obvious cases of the economic collapse of the Soviet Union, or the collapse of U.S. gas demand, which was widely described as creating a "bubble" of surplus capacity—and when it was extended beyond the expected length of time a "sausage."

CONCLUSION

The simple, awful truth is that some geologists have noted a pattern in some oil production data and assumed that it represents geological effects—which it doesn't—and then have ignored the fact that most cases doesn't follow the same pattern. Yet they have relied on this pattern to produce spurious forecasts.

Bell curves, being nonlinear, appear much more complex and thus more "scientific" than straight lines, but they represent nothing more than exponential growth and decline. Like sine waves, parabolas, and other shapes, they appear frequently in data for various effects, some physical, others economic or demographic, but they do not represent physical laws *except where no other influences are being felt*—which is not the case for oil discovery and production.

That peak oil advocates would embrace such a theory that has no actual theoretical basis, and fail to note its repeated failures speak volumes as to their powers of observation, the quality of their research, and the reliability of their assertions. As the following chapters will show, however, this is far from a unique failing on their part.

SIX

Get Thee to a Statistics Class

Occasionally, when talking to high school or college students, I'm asked about the extent to which higher math like trigonometry or calculus is used in my work, and I usually laugh and respond that even algebra rarely makes an appearance. It is true that a lot of economics involves higher math (to the point where few laypeople can read most of the articles in the *American Economic Review*), but in the vast majority of work about the oil market, there is almost no usage of even simple equations.

Thus, the statistical work by some peak oil authors stands out as sophisticated to the lay observer, especially if presented as a slide show where the speaker puts up a complex graph, asserts that it makes his point, and quickly moves to the next one. The seeming sophistication of this work, particularly when intricate equations are presented using exponents, integrals and differentials, Greek letters, and arcane symbols (arcane to those who aren't familiar with Greek letters), with no keys to explain them, impresses nearly any audience.

And yet, the vast majority of peak oil work has been done by people who apparently have little experience with statistical analysis, geologists and physicists in particular. Thus it should not be surprising that a number of fairly egregious errors are committed in the course of their analyses such as would earn a failing grade in an undergraduate statistics course. (However, it is surprising that so few, besides myself, have noticed the mistakes—topic for another chapter.)

There are three specific elements of peak oil research that will be shown to violate the most basic principles of statistical analysis, in ways that even an introductory statistics class would warn against. But to those who haven't taken such a class (probably 99% of Americans), the mistakes will not be obvious, rather the work will actually appear startlingly sophisticated instead of amazingly erroneous.

ERROR 1: THE LAW OF LARGE NUMBERS

One of the peak oil advocates who has published large amounts of graphics demonstrating "results" is Jean Laherrere, a retired oil company geologist, who has many fans. (Granted, Googling his name "only" gets 187,000 hits in March 2010.) Because his work is quantitative in nature, because he seems to get extraordinary results, but mostly because he commits a very basic mistake, one of his methods will be described here.

Laherrere often matches discovery and production, in cumulative form, on the same graph, as for U.S. natural gas. His conclusion is that the production curve is merely the discovery curve shifted some years to the right, so that if discovery can be predicted, production can also be predicted. The close match is obvious and seems very compelling: Laherrere appears to have discovered a way to predict production using discovery data.

To those unfamiliar with statistical analysis, the results appear ironclad. Unfortunately, a basic statistics education would show the folly of this approach. One of the first "tests" usually given to the students of beginning statistics is to try to predict something such as U.S. population or GDP. They will typically discover that they can fit a curve with a high degree of correlation (called r-squared, just to confuse the lay population) and congratulate themselves on picking it up so quickly. The point is, of course, that variables such as population do not change much from year to year, so next year's population will be very close to this year's and thus is easy to predict within a couple of percent.

This approach proves fallacious primarily because of scaling and inertia, and a few examples show how this works. In the first place, even major changes can be reduced to invisibility if you use the right graphics. Figure 6.1 shows cumulative U.S. GDP from 1960, and it appears to be a simple exponential growth curve (the kind that makes so many neo-Malthusians clutch their copies of *The Population Bomb* to their breasts). Extrapolating it into the future for the next decade or so would seem to be the easiest way to predict the economy's direction.

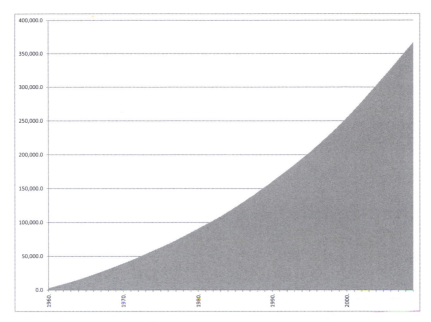

Figure 6.1   Cumulative U.S. GDP, 1960–2014 (Economic Report of the President, Council of Economic Advisers, 2015)

But one might well ask, what happened to the recession of 1980? Or the current slowdown? They seem virtually invisible, *because of the scaling*. In 2008, the "cumulative" GDP for 50 years is $366 trillion, and so the recent recession, with an estimated loss of about $400 million GDP from 2008 to 2009, represents only 0.1% of the cumulative total and is thus imperceptible. An entire recession vanishes!

But if actual, not cumulative, GDP is shown, then the fluctuations are a lot more obvious, as Figure 6.2 demonstrates. Although the recession becomes visible, the level of detail is so poor that this graph doesn't tell us much, and we would hardly use it to try to model future trends (assuming we knew what we were doing).

But our professor might want a little more detail. The slowdowns, such as in 1980/81, are at least visible, but only in an approximate way, due to the combination of scaling—changes are relatively small compared to the total—and inertia. Inertia is important for many socioeconomic activities that change only slowly: the entire U.S. economy is not reborn from scratch every January 1, so that it can be assumed that the level will change only marginally year to year.

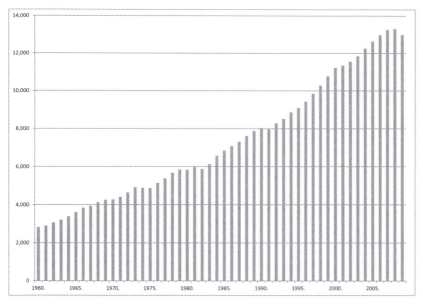

Figure 6.2   Annual U.S. GDP, 1960–2014 (Economic Report of the President, Council of Economic Advisers, 2015)

This is why economists focus on the *rate of change* (hopefully an increase and not a decline) in the economy each year. Forecasting the absolute level of the economy using the graph in Figure 6.2—or annual data—is simple: in any year, predict that the following year will be 3% greater, plus or minus 2%, and you will seemingly achieve a high degree of accuracy. In Figure 6.3, using the growth rate of 3.6% from 1960 to 1980 to create a "forecast" of that period makes it appear as if the forecast is highly accurate, with only small differences between the actual and estimated GDP.

But seemingly is the operative word here. Because the economy has a huge amount of inertia, it is easy to assume that the basic level won't change, only the increment. But that increment is the highly uncertain part and the target of our inquiries. Figure 6.4 shows the actual growth rates during 1960–1980, along with the average, and it is obvious that, while using the average might get us to the same answer at the end of the period, the actual growth rate in any given year often diverges substantially from the average.

Thus, it becomes obvious that using large numbers hides small changes, and cumulating a large string of data almost always leads us to an apparently stable trend, when, in fact, the annual change, and the trend itself, can be rather volatile.

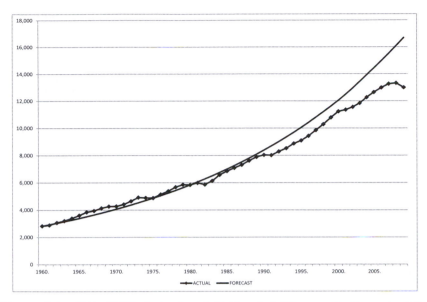

Figure 6.3   U.S. GDP, Actual and Estimated (Economic Report of the President, Council of Economic Advisers, 2015)

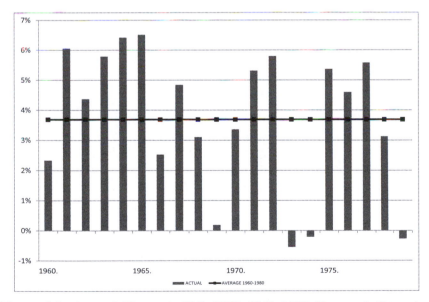

Figure 6.4   Annual Change in U.S. GDP, 1960–1980 (Economic Report of the President, Council of Economic Advisers, 2015)

There is a straightforward example involving oil supply. Consider Figure 6.5a, which shows cumulative British production. Any observer can see what appears to be a nearly straight line, and it appears that this is a highly reliable model of production. Yet examining annual production data (Figure 6.5b) shows that it actually fluctuates enormously. While Figure 6.5a appears to provide a clear insight into British oil production, this is *entirely due to the use of cumulative data* and the scale of the graph, which, at 25 billion barrels, completely obscures the annual changes that, while often enormous, simply don't show up on such a scale. Production dropped 6.3% in 2008, a rather large amount, but represented only 0.01346% of the cumulative production to that date.

In other words, the method whereby peak oil advocates correlate cumulative production with cumulative discoveries provides no meaningful insight and appears valid only because of a trick of scaling. But as a geologist, Laherrere would seem to be a novice at this type of statistical analysis and makes a novice's mistake.

ERROR 2: BAD EQUATIONS

As Newsweek said of Deffeyes work, "[A] simple but elegant piece of mathematics."[1]

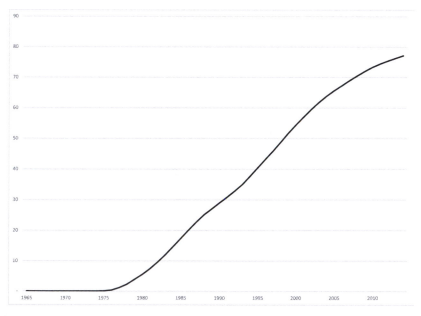

Figure 6.5a   UK Cumulative Production (in billion barrels) (BP Statistical Review of World Energy)

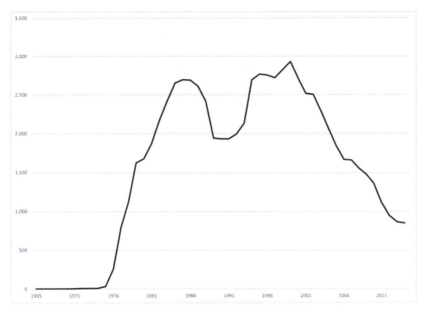

Figure 6.5b   UK Annual Production (thousand barrels per day) (BP Statistical Review of World Energy)

A Math Trick: Number below 10

Step1: Think of a number below 10.

Step2: Double the number you have thought.

Step3: Add 6 with the getting result.

Step4: Half the answer; that is divide it by 2.

Step5: Take away the number you have thought from the answer, that is, subtract the answer from the number you have thought.

Answer: 3^2

The work of Kenneth Deffeyes is on an entirely different level, at least in terms of the apparent math being used. He provides the precise equations used to develop a prediction of peak oil, which he describes as accurate within two to three weeks of Thanksgiving Day, 2005.[3] When he made this prediction to the American Geophysical Union several years ago, it was treated as a joke (he opened with it, which probably didn't help), since the idea that something as complex as the peak in global oil production could be predicted so closely presumably struck the audience as absurd. Given that experienced oil market forecasters have rarely been able to see more than a couple of years into the future, as we saw in Chapter 2, a great deal of skepticism is only natural. (In earlier research, I found that forecasts made of the oil market two or three decades into the future were usually accurate for about three or four years at best.[4])

The funny thing is that speaking to him personally, he will be quite optimistic about new technologies allowing access to additional oil supplies; when we debated at a Wall Street firm, he was quite enthusiastic about a colleague who was redeveloping an old oil field.

This method, which he has labeled "Hubbert linearization," has become quite popular among peak oil advocates, who have applied it to production in Saudi Arabia, the United States, Iran, Romania, and so on.[5] Unfortunately, the results generated by this method are what mathematicians would call "an artifact of the method," that is, resource scarcity has nothing to do with the trend line seemingly pointing toward zero, rather the method always generates the same results regardless of the phenomenon studied. For example, applying it to U.S. population yields a nearly identical curve (Figure 6.6a) and seems to suggest that U.S. population is also heading toward zero, yet this is not true (Figure 6.6b).

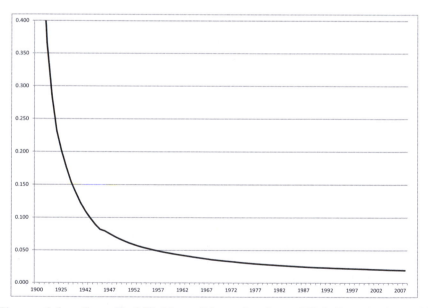

Figure 6.6a   So-called Hubbert Linearization of U.S. Population (US Census Bureau)

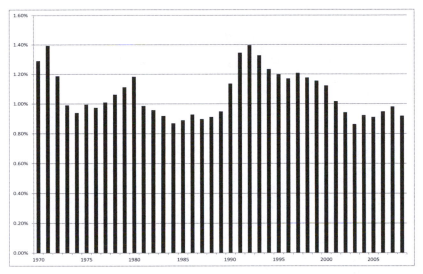

Figure 6.6b   Actual Annual Change in U.S. Population (US Census Bureau)

The first and most obvious question for Deffeyes and others is: Why does the curve work only in recent years, say, after 1980? If it is "scientific" in nature, has the science changed? On the Oil Drum, a number of bloggers comment that this appears to reflect the maturity of a country's production; that is, in the early years, the industry is changing too rapidly for the law to apply.[6]

Yet the obvious answer eludes them. By dividing the current production over total historical production, they are virtually assuring two things. First, the *number inevitably will decline,* because total historical production is getting bigger and bigger, so that the result of dividing current production by total historical production is getting smaller and smaller, unless production is growing very rapidly. For example, in 1900, global production increased by 149 million barrels, which was 9% of the total production till that date. By 2000, annual production was 25 billion barrels, but less than 3% of the historical total. In fact, the increase in 2000 from 1999 was 900 million barrels, far more than the total production in 1900, yet it was only a tiny blip on a graph of cumulative production—about 0.1%! Presto, the big number has been made miniscule!

By 2000, the total historical production (141 years' worth) was so large, at just under 900 billion barrels, that even large changes in annual production *would not be visible* on the graph, similar to what we saw in the case of production in the United Kingdom (Figures 6.5a and 6.5b). As a result, the last several decades show what appears to be a "stable" trend toward the zero point. (In reality, it is actually a curve that would never quite reach zero.) It is also interesting that, despite near-constant increases in global production in the past couple of decades, the curve shows a declining trend, but only because the total historical production keeps growing.

This is almost the same as the math trick highlighted earlier, which always gives the answer 3. In other words, if you apply this method to any data series, from oil production to pickle consumption, once you have enough history under your belt, you will see a stable and declining curve. This is not a proof of a peak but only means that the historical total is very large compared to the current amount. This is the type of error that a high school student shouldn't make, let alone a college statistics student. But again, the peak oil advocates using this method appear to be primarily geologists and are seemingly not familiar enough with this type of analysis to recognize the very basic error being committed.

ERROR 3: THE MISUSE OF CREAMING CURVES (AKA "REAL SCIENCE")

This brings us to an advanced-level case—the use of creaming curves to estimate resources within a region. This method actually has some scientific basis, namely, both in theory and in fact (yes, that sometimes happens). First, resource economics makes the very reasonable argument that explorationists seek to find and develop the "best" deposit available, implying that over time, the quality of the resource declines. In a given basin, at (roughly) the same depth, this translates into shrinking size of new finds.

And such is observed in the real world, whether it's Texas, Western Canada, or the UK North Sea, as shown in Figure 6.7a.

This is a major support for the idea of peak oil, namely, that field sizes are getting smaller and smaller as the resource is used up or depleted. It is not new and is irrefutable. But there is a problem with reality.

REALITY BITES

"Sire, the sultan has come with his hundred wives, each more beautiful than the next."
"Each more beautiful than the next?!"
"Well, only if you line them up that way."

Wizard of Id[7]

But Figure 6.7a is actually constructed by ordering the data by size and does not represent the discoveries as they actually occurred, which was much less well ordered. Figure 6.7b shows the discoveries in the order of discovery, and the decline is visible but the trend is far from smooth. Partly, this reflects the fact that exploration is never a perfect process, moving from the biggest to the smallest, but much more uneven. (The use of cumulative discoveries in the creaming curves of peak oil advocates makes this unevenness largely invisible.)

Some of the dispersion from the trend represents random elements (random usually being an economist's term for "I don't know why"), but there are a number of geological and geographical factors that influence choices by drillers. Explorers (in theory, but not always in reality) will focus not just on size but other physical and economic factors, including the proximity to infrastructure like a pipeline or pumping station. Over time, as pipelines, roads, and other infrastructure are extended, an area might become more attractive to drillers, even though field size hasn't changed, creating lumpiness in the field size trends.

Second, the process suffers from a significant amount of political interference. Governments release areas for exploration piecemeal, and there is no indication that they use the same parameters that explorers do in terms of attractiveness, as opposed to simply parceling up a given region for bidding and then later turning to adjacent regions. Studying any map of

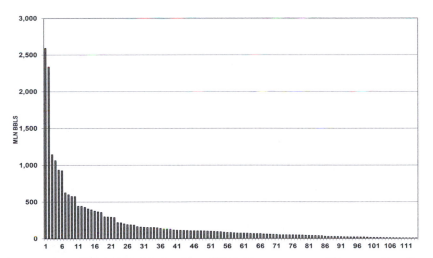

Figure 6.7a   UK Oil Fields by Size (U.K. Department of Energy, Development of the Oil and Gas Resources of the United Kingdom. London: Her Majesty's Stationary Office, Annual [discontinued])

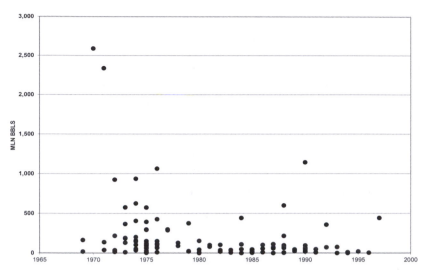

Figure 6.7b   UK Discoveries by Date of Discovery (UK Department of Energy op. cit.)

petroleum basins offered for lease will show that they progress in geometric shapes, not resembling any geological analysis.

Technology also plays a factor, especially offshore where a basin might cover an area with various water depths, which *are thus available for exploration* only as rigs are designed that are capable of operating in those levels. As everyone now knows too well, deepwater often contains some of the largest fields, and as drilling moves into deeper water, field sizes increase. This is not a violation of the rule that "the biggest get found first" but rather reflects the fact that the rule is *constrained to a given basin,* and according to the political and technological availability of the basin for drilling.

REAL SCIENTISTS DO MAKE CREAMING CURVES (BUT KNOW HOW TO INTERPRET THEM)

And yet, despite its shortcomings, this method has been employed by decades by geologists and statisticians (even economists!) to try to predict future discovery sizes as well as resources, not too differently from the manner in which peak oilers have used it, and with a certain amount of success.[8] But geologists have recognized that there are two specific conditions that limit the utility of this approach: uncertain field size data and the discovery of new basins.

Revisions to field sizes are important because they tend to be positive (i.e., increasing) and thus cause the level of the curve to grow with time. This can be seen in a recent study by Ahlbrandt and Klett, where they tested various methods of estimating resources in a mature province in Argentina using historical discovery data.[9] But as part of the research, they updated the field size data and found that revisions to older fields meant that the curve was revised upward significantly, about 700 million barrels (or one-third), in only five years! Not bad for something that isn't supposed to happen at all, according to peak oil advocates.

Not only have peak oil modelers not dealt with this problem, but also they have *denied its existence* by claiming that field growth doesn't occur in the database they used, because it covers proved + probable reserves instead of proved reserves. The evidence in support of this claim is negligible (and actually contradicts it). The company that is the source of the data has itself shown that there are significant revisions to data, and the USGS, studying their data, has come to the same conclusion.[10] In response, Colin Campbell claims that the company no longer produces reliable data, because they are using "consultants" now instead of geologists. Chapter 7 will address this in more detail, but, in fact, Campbell and Laherrere provide ample evidence to refute their own assertions.

Politics Not Geology

The second problem with this approach is that it is traditionally applied to single, homogeneous basins. While the precise delineation of a given basin might be a question, the work by Laherrere (the only one who has used creaming curves in his analysis, to my knowledge) usually focuses on nations and entire regions (such as the Middle East), not geological basins. In fact, the Middle East contains 23 major basins, and the USGS has done its work based on basin-level analysis.[11]

The creaming curve for the Middle East in Laherrere (2003) demonstrates quite clearly the problems involved. First, by using a large geographical region, he (perhaps unintentionally) includes a number of geographical basins, each of which should have its own, independent creaming curve. This helps to explain why the curves provide different answers at different points in time: the exploration of new basins adds new curves with different shapes.

But looking at the entire Middle East means that Laherrere, who often criticizes others for their use of so-called political reserve data, is unintentionally incorporating political decisions into his work. The discovery sequence in a country reflects not just its politics but also its various upstream policies. In mundane terms, this can simply be the manner in which areas are leased for exploration. (Smith also makes this mistake; see Chapter 9.[12])

The Middle East is a classic example of government decisions changing the shape of the creaming curve, as the market collapse in 1980s discouraged drilling in the major oil-producing nations of Iran, Iraq, Kuwait, Saudi Arabia, and the UAE, as did the 1980–1988 Iran–Iraq War. The shift in drilling to countries with smaller fields, namely, Oman, Syria, and Yemen, resulted in a sharp fall in discovery size in the "Middle East" as a whole. Yet Laherrere, by assuming only geology affects the curve, mistakenly interpreted this as that there is *almost no oil left in the Middle East*.

Finally, his inexperience using statistical methods comes to light when considering his use of regional creaming curves. When economists and others try to fit curves to data and get different results at different times, they conclude the model they are using as being at the least incomplete. Laherrere gets four different answers when he uses the creaming curves to estimate regional resources in the Middle East, and he assumes that the last one is the correct one. But if he had performed the work in 1975, he would have thought that the third answer was the final one, and there is no reason to believe that the 2002 estimate will be the final one any more than the three previous, flawed, estimates were. In fact, given the discoveries in Iran and Iraq in the past few years, it seems quite probable that there will be a new, higher estimate when the work is recalibrated.

THE BOURGEOIS GEOLOGIST

> Although most of the current generation of oil and gas geologists and exploration managers have been exposed to more statistical training than their older counterparts, there is a cultural gap that still remains to be bridged.[13]

Just as Moliere's nouveau riche gentleman discovered that he had been speaking prose without knowing it, so too the peak oil geologists appear to have been unaware that what they are doing is statistics, not natural science, as Aleklett claims.[14] And because they are largely older (nearly all retired), they are unfamiliar with most of the modern methods statisticians use or the errors that they have been committing. The results of their research, specifically the estimates of the resource base, should be discarded as completely, irretrievably, flawed.

SEVEN

Reserve Growth: From Small Acorns Grow Mighty Trees

The issue of field size estimates might seem to be minor or irrelevant, but it is not only at the core of the peak oil question but also emblematic of the way observers have been led astray. The original research by Campbell and Laherrere was driven in part by a belief that they had solved the analytical problem of estimating the amount of discovered oil when field sizes were uncertain. This meant that their resource estimates, based on creaming curves, were reliable and stable.

Unfortunately, this turned out to be untrue, and their defense of their methods and assertions are a microcosm of the entire debate. This chapter will walk through the progression of the discussion and review the ways that the industry has expanded the resource.

PREHISTORY

Creaming curves, as discussed in Chapter 6, have long been employed by geologists as a method to get rough estimates of a basin's resource, but, as Drew noted as long ago as 1996, it generally provided good estimates of the number of fields, but not the resource.[1] This was due to a lack of good data on the ultimate recovery of oil from the fields: estimates always reflected the amount thought to be recoverable *at the time of the estimate,* and these numbers changed over time.

But then several studies were published by a company called Petroconsultants in the mid-1990s, made available to the industry for a high price, written by Colin Campbell and Jean Laherrere.[2] These relied on the proprietary database of field sizes developed by the firm Petroconsultants, the only such existing database at that time and one that is still well-regarded by the industry. The authors, in their public work, argued that, unlike the *Oil & Gas Journal* database for reserves by nation, these estimates were made by geologists and thus "scientific."

This had two specific consequences (supposedly). First, they were "technical," not political, and second, they represented P50 reserve, not P90, estimates, meaning that they should have been unbiased and unlikely to show net changes over the long term. (Chapter 15 explains the concept of reserves and resources in more detail.)

The argument about the "technical" nature of the field reserve estimates is completely irrelevant and seems to represent an arrogance on the part of the authors as to the prescience (or omniscience) of their profession and its members. While many of the *Oil & Gas Journal* reserve estimates are not reliable, they are not in every case political, but rather "official," that is, reported by governments. What this means varies from country to country, with some not reporting changes for many years arguably for political reasons, others assigning a junior bureaucrat to handle the chore, but many sending in precise and reliable, albeit often misinterpreted, estimates.

In another example of their novice status, while they discuss early work by, for example, Arps, the peak oil advocates have not been aware that other efforts have been made to create creaming curves, and not just using U.S. data. An MIT group in the 1970s did a number of studies using field data from the United Kingdom to estimate resources there, Gordon Kaufmann described the various methods of petroleum resource estimation and several researchers at MIT have published works in the 1980s and 1990s.[3]

It is true that when Campbell and Laherrere began their careers, the numbers in the *Oil & Gas Journal* were considered fairly trustworthy and were based largely on industry estimates. Until the 1970s, few governments considered oil a strategic material whose reserves should be kept a secret. After that, and especially as the governments nationalized their oil fields, many insisted that only "official" government estimates of reserves be released, at least to publications like the *Oil & Gas Journal*.

Since the initial work by Campbell and Laherrere had used field sizes to estimate resources and predict production trends, the reliability of the data was of utmost importance. In the United States, the petroleum geologist profession long ago adopted the nomenclature "proved reserves" to mean discovered and developed reserves (see Chapter 15). This was intentionally conservative, to improve reliability of the estimates in the early days of the industry and to reduce the possibility that someone might "talk up" the value of a deposit, lease or oil company stock.

But Campbell and Laherrere err in pointing to this as the primary source of what they consider to be the myth of reserve growth: the availability of U.S. field data, which represented "proved reserves" meant that some analysts (read "economists") had mistaken the tendency for estimates to grow in the United States for universal behavior.[4]

Instead, they opined that the use of the data from Petroconsultants avoided this error. It was intentionally based on "proved plus probable" estimates that were thought to represent the amount that had a 50% probability of being correct. Any given field's size might be revised up or down, but there should be no bias, and thus the numbers as a whole should not change. This would allow the creation of reasonably reliable creaming curves—although that wouldn't mean anything other than allowing resource estimates for a given basin.

EVIDENCE—OR NOT

The peak oil argument that the P50 field size estimates made by Petroconsultants don't increase is a supposition, not a fact, and one that a diligent researcher would have tested for validity, given its crucial role in their calculations .Campbell and Laherrere apparently had access to a total of two datasets, created four years apart (1994 and 1998), nowhere near enough information to provide a reasonable test of the stability of the estimates over the long term.

Be that as it may, Laherrere did compare two datasets from 1995 and 2003 and found—surprise—that they did not in fact exhibit the behavior he was assuming. On average, field sizes grew by a net of 20 billion barrels a year, which he explained away by arguing that on abandonment, this amount would probably be revised downward.[5] While that is possible, it would suggest that those making the estimates were not providing scientific, unbiased estimates but overstating field sizes—not in the first estimate, but the second estimates, which should be improved, not degraded.

Campbell, for his part, published a graph from the Norwegian Petroleum Directorate showing reserve growth but claims that there is none—compared to predrill estimates.[6] This is the only reference I have seen to predrill estimates in any of the work on peak oil, and it certainly contradicts their usage of the Petroconsultants database.

Indeed, it is amusing that the successor to Colin Campbell as president of the ASPO has published information showing field growth, directly refuting his predecessor's and mentor's claims. In his 2012 book, he points to research by his colleague Robelius showing reserve growth and then adds, "in the future, we cannot expect the same degree of reserve growth that we saw during the past 20 years."[7] At no point is it mentioned that this vital part of Campbell and Laherrere's "model" has been overturned. And the UK Energy Research Centre's Global Oil Depletion report, whose authors include at least two peak oil advocates, notes, "Reserve growth currently accounts for the majority of reserve additions in most areas of the world and is expected to continue to do so for the future."[8]

Contrary Evidence

What does the evidence actually show? First, the assertion that only U.S. field sizes grow is absurd and easily disproved. Research on Canadian, British, and Norwegian fields—the only ones where size estimates are regularly published—show that growth is quite typical.

Ellerman and Sem analyzed 100 North Sea fields, comparing growth rates by size.[9] In fact, there is enormous dispersion by field among the sizes, and they conclude that "the norm, excepting the medium sized fields, is for oil reserves to appreciate at a statistically significant 2–3% per annum regardless of sector, year, or age." They could not explain some anomalies such as the lack of growth in medium-size fields.

In addition, although the Petroconsultants/IHS Energy database was not set up to allow easy analysis of field size revisions, the USGS, as part of its 2000 revision, slogged through the various datasets and found significant net increases in field size estimates, which they used, in part, to calculate how much oil might come from reserve growth in the future. They were attacked by many in the peak oil community who incorrectly claimed that they had used growth factors calculated from U.S. fields for non-U.S. resources (Chapter 12).

And ultimately, this controversy led to IHS Energy itself speaking up, not only disagreeing with the peak oil theories and arguments of Campbell and Laherrere, but stating clearly that their database did show that the estimated size of fields grew over time. Their findings were that growth ranged from 32% to 44% for most fields.

As showed in the previous chapter, Ahlbrandt and Klett revisited the creaming curve for the Neuquen basin in Argentina, a mature basin that should be expected to show few revisions if Campbell and Laherrere were correct. In actuality, the amount of oil in existing discoveries was increased by 700 million barrels of oil equivalent from the 1996 to 2001estimates, or about 33%. Instead of no net growth, very robust growth was experienced.

It might seem as if this would settle the issue, but both Campbell and Laherrere responded by arguing that Petroconsultants was no longer reliable. In particular, they were accused of changing their position to please their clients in the oil industry, who wanted to keep the world complacent about the looming problem. This contradicted Campbell's many assertions that various companies had "acknowledged" peak oil and left open the question of who the company's clients were earlier when the two produced their reports in the 1990s.

No particular evidence was provided for this argument, leaving one to suspect that it was more of an insult than a substantive comment.

MARCH OF THE ZOMBIE FIELDS

About 2005, the debate took a new turn, as Jean Laherrere developed a method of estimating field sizes that he claimed was superior to those made by firms such as IHS.[10] Specifically, he graphed production data to show that once on a downward trajectory, the trend could be extrapolated to the zero point, indicating the ultimate recovery. The deviation from the trend resulted from a gas injection program in the 1990s, which he says, confirms his argument because, after several years, production returned to the earlier trend.

But this is another iteration of the argument that geology alone determines production paths and they cannot be altered by human decisions, a necessary condition for extrapolating production trends. Petroleum engineers and project managers would no doubt be surprised to discover that their decisions are completely irrelevant to production paths. This method once again falls into the category of spotting an occasional pattern and assuming it is physically determined, rather than largely coincidental.

And the real world clearly refutes the validity of this approach. Laherrere himself shows some fields like East Texas, where a variety of production trends can be observed, although he later stopped publishing the examples that didn't conform to this theory.[11] Other fields, like Forties in the North Sea, have seen new investments cause production to stop declining and sometimes increase. Indeed, only about one-third of UK fields show a similar production pattern to the one Laherrere asserts is "scientific." As for the Forties field, it has been producing at a relatively constant rate of 60 thousand barrels per day for about a decade. This failure to follow the earlier trend not only invalidates the method Laherrere uses, it demonstrates how new investment can add to reserves and extend a field's life.

THE USGS AND ITS CRITICS

The 2000 mega-study by the USGS has come under particular criticism for its estimates of reserve growth, as it suggested that there was perhaps 600 billion barrels of additional recoverable oil that would be available globally due to reserve growth in known fields. Aside from the other complaints by peak oil advocates, many have decried the USGS reliance on U.S. reserve growth factors for non-U.S. fields, which they deem inappropriate.

It is the same with USGS world reserves estimates from 1987 to 2001, when T. Ahlbrandt replaced C. Masters. Masters used inferred reserves considering no reserve growth, when Ahlbrandt uses proved reserves for the US and proven + probable for the rest of the world, and he assumes that proven + probable will have the same reserve growth as proved reserves in the US, which is plain wishful thinking.[12]

Just one problem: they didn't. The article written by Schmoker and Klett is actually quite clear about what they did.[13] First, it repeatedly notes the uncertainty involved and states that the work is preliminary, which seems to escape those like Heinberg.[14] Second, they describe relying, in part, on work analyzing fields in "the Caribbean, Latin America, South America, Western Europe, the Middle East, Africa, non-Communist Asia, and the southwestern Pacific."[15]

They also report using the same database that Campbell and Laherrere used, the Petroconsultants 1996 oil field database, as well as the NRG database for Canada. But they note that growth estimates for fields outside the United States are "not sufficiently reliable and consistent to develop world-level reserve growth functions" (RG-10). In other words, they don't have high confidence in the number used for non-U.S. fields, which is not the same as saying no number should be used.

However, the gist of the peak oil advocates' argument seems to come from the statement "using an analog that incorporates the reserve growth experience of the United States" (RG-11). The words *incorporates* and *analog* have been translated to mean "equal to," which they clearly don't.

And how do the two compare? Well, in the first place, the USGS found that, for a 20-year-old field, a 50% increase in URR could be expected in a 30-year time frame. This led them to calculate that global URR for existing fields would be about 50%. Peak oil advocates didn't do the math, and now this error has become an article of faith, thanks to the circle of citations.

(Of course, the actual equation is more complex, assigning growth rates by age of field, but this is the composite growth over the period.)

THE END OF TECHNOLOGICAL ADVANCES

[T]he technology being used to extract oil today has been in the works since the 1970s.[16]

You can say, he offered, that Lee Harvey Oswald's assassination of President John F. Kennedy depended on the discovery of iron, but so what?[17]

Many peak oil advocates have adopted the rather bizarre argument that the technologies being heralded by the oil industry are not new, Laherrere noting that deepwater wells have been drilled since the 1960s, while Blanchard dismissed the potential for shale oil production in 2011, saying "Although one of your experts gave the impression that the technology is improving daily, fracking technology is quite mature at this point in time."[18]

Simmons made a typically dismissive comment, suggesting that current technology was "in the works" three decades earlier. William Pike, a long-term petroleum geologist who writes for *World Oil,* agrees to a point. In looking over a 1931 book on drilling, he notes, "In addition to fairly sophisticated drilling equipment, our predecessors had most of our current production technology. ... Yes, they had a lot of modern technology back then, and we should be grateful for their lead, but *we have left them in the dust*"[19] [emphasis added].

The idea that there is no significant new technology in the oil fields is constantly being refuted by the industry and would be laughable—if it weren't an important part of the peak oil argument. As such, it provides another example of the kind of assertion often made by peak oil advocates, which has no basis in fact, but is usually not challenged by the media.

FASTER NOT BETTER

You will hear many claims for technology. . . . In Production, it keeps production rate higher for longer, but has little impact on the reserves themselves. . . . Advances in technology have reduced the time lag from peak discovery in 1974 to 27 years. We are getting better at depleting our resources.[20]

An odd argument that arose about a decade ago was that new technologies were not increasing recovery rates, but rather allowing producers to get the same amount of oil out faster. The evidence for this is, again, all but nonexistent, with a couple of examples seeming to support the theory, but the vast amount of evidence refuting it.

Simmons made much out of the production collapse at the Yibal field in Oman, where Shell implemented a program to use horizontal drilling to raise production sharply. Instead, production collapsed, leading some, like Simmons, to describe the methods as "superstraw" technologies, which sucked oil out faster, but damaged the reservoirs.[21]

In his 2005 book, he has a page showing production profile from eight fields, intended to show a particularly sharp decline and confirm his argument.[22] Oddly, those fields were developed before the "superstraw" technologies were available, and both Romashinko and Samotlor, at least, suffered from the application of Soviet engineering methods and would not serve as an example that applies to any non-Soviet fields. It is also worth noting that the fields follow four different production trends, refuting Laherrere's theory described earlier.

And in the next section, numerous examples of technological improvements that have increased the amount of oil that can be recovered in existing fields will be discussed, but the perusal of any petroleum industry journal will find it replete with such cases. I have seen no one in the oil patch, outside of peak oil advocates, argue that technology is contributing little to our oil supply.

RESERVE GROWTH IN THEORY

Geologists, engineers, and economists have long studied reserve growth, but they are often hampered by the lack of good data. Historically, only the United States kept good data on field sizes, done by the industry in what were called the *blue books,* which included annual estimates of the amounts of oil that had been discovered in each year back to 1920.

(Not to be confused with the Brown Books, published by the UK government for years, which had data by actual field.)

Sadly, the Carter administration decided that the government should take over this job, as part of a general mistrust of the oil industry at the time (not much has changed, granted). But they collected data by company, not year, so that the data could no longer be used to estimate field growth.

And it was true that the U.S. fields grew more than foreign fields, and primarily because of the use of "proved" or P90 estimates instead of "proved plus probable" or P50, but this was clear to everyone in the field and only seems to have confused peak oil advocates, again, due to their lack of familiarity with the existing research.

Those considering field growth have made some general assumptions:

- Older fields should grow more than younger fields, as the early estimates were constrained by earlier technology.
- Large fields should grow more than smaller fields, because incremental improvements in recovery are more easily applied to large fields.
- Onshore fields should grow more than offshore fields, as it is easier to drill infill wells onshore, since a new well doesn't require a new platform (although this is changing offshore).
- Heavy oil fields will grow more than light oil fields, because the former start with lower recovery rates due to higher viscosity.

All of these assumptions are based on sound reasoning. The data, however, tell a different story.

RESERVE GROWTH IN PRACTICE

These theories have not been proved out, however. (Shocking, I know.) Various studies have found reserve growth, especially in the United States using proved reserves, but even where other countries have collected data, such as Norway and the United Kingdom. The results are often puzzling, however, including the fact that Norwegian and British fields grow at different rates, that fields grow in different amounts and timing, and that field size isn't always a good predictor of growth. The predicted patterns are largely absent.

The primary work has been done by geologists at the USGS and a number of economists at MIT. In the North Sea, for example, Ellerman and Sem found that small fields and large fields grew the most, while middle-sized fields grew notably less, a rather perverse result. Norwegian fields grew at different rates than British fields. Watkins found that after an average production of nine years, reserves in the United Kingdom appreciated by a factor of 1.23, while Norwegian fields, almost 10 years of production on average, appreciated by a factor of 1.46.[23]

(The late Campbell Watkins was the ultimate in economists, a man who not only worked on real-world issues but understood and used higher math. He was the only person I know who would review an article and carefully double-check all of the math.)

An early Canadian study confirms the often bizarre character of field growth, where, for example, fields found in one year might grow far more than fields in the next year, or far less.[24] The explanation would seem to be that individual fields are very idiosyncratic and that the results for a given year might be dominated by particularly good or bad results for a single, large field.

RESERVE GROWTH AT THE FIELD LEVEL

The entire claim that field sizes don't increase is also strongly contradicted by ongoing developments in the industry, where a flood of reports about expansion of reserves are constantly appearing. While any given anecdote about increased recovery might not be very meaningful, the constant barrage of such reports makes it clear that these are not just promotional in nature, or isolated cases, but symptomatic of industry operations. In the following text, these efforts are broken into the general and the specific.

The Life of a Field

Analysts like Laherrere, Smith, and Aleklett like to point to a traditional pattern of field production, but this is an oversimplification of a very complex process, looking only at the path that actual production takes. Historically, it was common to think of a field with three primary steps: primary, secondary, and enhanced oil recovery. The first involves the flow from the natural pressure in the field and tends to generate 5–10% recovery of the oil in place or in the ground. Secondary recovery involves the injection of water or gas to raise pressure and force more oil out, resulting in 25–30% recovery.

Enhanced oil recovery refers to more complex techniques, such as the injection of surfactant chemicals that allow the remaining oil to flow more easily. It can add another 25–30% to the recovery of the field.[25]

Again, each field is different and the difference is most apparent with different weights, or densities, of oils. Heavy oil tends to be much more viscous and often requires heating to flow, with lower recovery factors than light oil. Infill drilling, that is, adding wells between other wells or aimed at deposits that are above or below the original deposit, has long been done and adds more to oil-in-place than to the recovery factor. For older fields, complete redevelopment with new seismic and more advanced equipment can add to total recovery.

New Investment

Numerous examples exist of fields that, after producing for many years, are redeveloped. This can include seismic studies (and modeling) to find oil left behind and estimate ways to optimize further recovery; drilling new wells; injecting gas, water, steam, or chemicals into the field; or even placing new platforms at offshore fields. In the United Kingdom, Enquest recently chose to redevelop the Alma and Galia fields, adding 29 million barrels (oil equivalent) of proved plus probable reserves.[26]

Shell added a new platform to its Mars B field, allowing it to access an additional 300 million barrels of oil equivalent to the original 700.[27]

Sidetracks: BP drilled sidetracks at its Mad Dog field in the deepwater Gulf of Mexico and found new reserves at Mad Dog North.[28]

Gas injection: Statoil is planning to put a subsea gas compression system in place to boost recovery at Gullfaks in the North Sea. At the Gullfaks C platform, recovery is expected to increase by 22 million barrels, rising from 62% to 74%.[29]

Chemistry

A major factor determining the recovery rate is the viscosity of oil, which varies by type, with the heavier the oil the more viscous as a general rule. That, combined with the rock type, determines how well it flows—or doesn't. Companies are constantly trying new methods to improve the flow rates, and thus recoveries.

Saudi Aramco has developed what it calls SmartWater Flood, which involves fine-tuning factors such as salinity and ionic composition.[30]

Chinese petroleum engineers are addressing poor performance in horizontal wells in the Tarim basin by optimizing drilling fluids, adding fibrous bridging agents, which quintupled per-well production in a test case.[31] One new process is called the water-alternating-gas or WAG process, which Exxon-Mobil is planning to employ at the Guntong field off Malaysia.[32]

Emulsifiers are increasingly employed to allow heavy oil to flow better underground and improve recovery rates. Various chemicals can and have been used, with effectiveness dependent on the type of oil and rock.[33]

Equipment

New sensory equipment is evolving at a rapid pace. Logging while drilling (LWD) is not a new technology, but is continually advancing, in terms of both the capability of the equipment and the software that integrates the information. It is now possible to integrate down hole fluid analysis and sampling with formation pressure while drilling (FPWD), thereby providing environmental, economic, time-saving and data-quality benefits.[34]

Coiled tubing[35] is precisely what it sounds like, relatively thin, flexible tubing that can be kept on a coil and run out or retracted much easier than conventional pipe. Originally used for well interventions, it can be run inside a working well for sand control, cementing, and similar operations.

For shale, the new "hydraulically powered top drive technology and hands-free pipe handling system means that the same high-torque, deep drilling challenges can be met, but with fewer personnel on the ground."[36]

Most impressive, perhaps, are the new drilling rigs that are now being used to exploit shale deposits.[37] Because shale wells are often drilled in close proximity, instead of disassembling a rig and moving it to a new site, which can involve up to 40 trucks, walking rigs are mobile. The latest are also highly automated, reducing labor costs substantially. Not quite independent robots, they are a major step in that direction.

Methods

Fishhook wells have been used to exploit offshore deposits from onshore drilling sites. The well is drilled vertically and then deviated upward toward the pay zone. This has been done in Brunei, for example, to access resources that would have been uneconomic if drilling had to proceed from offshore.[38]

Also, the use of managed pressure drilling (MPD) can address many technical challenges. Hannegan[39] notes that about one-half of offshore conventional oil resources are thought "undrillable with conventional, open-to-atmosphere circulating fluid systems." Also called underbalanced drilling, this involves using lightweight drilling muds and keeping the pressure in the wellbore below that of the reservoir.

Fracturing has received much attention as a result of the shale gas/oil revolution, but it actually has been applied long before that to conventional wells. The use of channels while fracturing is a new development, in which the proppant is added in short pulses. Utilization in the Priobskoye field in Siberia led well productivity to increase by 10–15%.[40]

Toe-to-heel air injection (THAI) involves creating a wall of combusted petroleum in heavy oil deposits, heating the adjacent oil and also upgrading it to a lighter product.[41]

Software

A software model designed to optimize fracturing in a sandstone-conglomerate field in China raised well productivity from 20 barrels a day to 200 barrels a day.[42]

A major element in the increased access to offshore oil was the development of software that allowed for dynamically positioned drilling rigs. By monitoring water movements and controlling engines, a floating rig can be maintained in one place without being attached by heavy, expensive platforms.

3D/4D Seismic

The development of supercomputers made it possible to create 3D images of a geological formation, making identification of potential drilling targets much easier. By adding monitoring equipment to a field, 4D videos can be made showing the flow of fluids within a deposit, enabling operators to maximize recovery.

"Shell had acquired 3D seismic data on these leases, and it was the data from those surveys that informed our decision to pursue leases in the Beaufort Sea in 2005."[43] In an area with discoveries from 20 years earlier that were not developed.

Identifying missed resource: In Australia, Apache used new seismic in the Stag field to see where some oil had been left behind, leading them to realize the "field . . . has plenty of life left in it."[44]

CUTTING EDGE

Not all technologies pan out, naturally, but there are areas where new technologies are being tested that potentially could increase oil recovery rates in the future. For example, microbial enhanced oil recovery (MEOR) methods have not proved successful as of yet but continue to have their advocates as research is ongoing, with Statoil being the first to test it in an offshore field, Norne.[45] A test at the Stirrup field in Kansas raised production from the well tested by 60%, reducing the water cut from 91% to 88% overall.

Glori Energy's scientists are attempting a version of microbial enhanced oil recovery that they refer to as activated environment for recovery of oil (AERO), wherein nutrients are fed into an oil deposit, which enhances the ability of microbes to consume oil, producing a surfactant that reduces viscosity, as well as blocking pores and creating new pathways for oil flow. The company estimates it can recover oil for $10 a barrel with this method, versus $40 a barrel for CO_2 floods.[46]

Nanotechnology has the potential to become a factor in the long run, as it might allow for small-scale modification of geology but, more important, provide real-time monitoring of field performance at a reduced cost. This would improve recovery rates and extend field life as well as allow for the redevelopment of old fields.

EIGHT

The Strange Controversy over Saudi Oil Production

My father rode a camel, I drive a car, my son flies a jet airplane. His son will ride a camel.

Saudi saying[1]

A decade ago, a prominent peak oil advocate named Matthew Simmons began to argue that the Saudi oil situation was not just bleaker than it seemed, but on the verge of catastrophe. After making presentations on the subject for about a year, he published a book called *Twilight in the Desert: The Coming Saudi Oil Shock and the World Economy*. This book was embraced by the peak oil community and even proved to be popular with the general public.

Others echoed his concerns and added to them. Deffeyes noted worryingly that "Early in 2003, Saudi Aramco and the government of Saudi Arabia announced that their production was maxed out at 9.2 million barrels per day" and compared this to the 1970 announcement by the Texas Railroad Commission that the long-standing surplus in Texas had vanished, which presaged the peak in U.S. oil production and, arguably, the oil crises of the 1970s.[2] Newcolonist.org actually had a bumper sticker made stating, "Ghawar is dying," referring to the supergiant Saudi oil field, the biggest in the world and the mainstay of the country's production for decades now. Others, notably bloggers on theoildrum.com, have used the Hubbert linearization method to suggest that Saudi production was near a decline, and, several times in the past, a drop in supply from that country was claimed to be signaling a peak.[3]

More prosaically, the issue of whether or not the Saudis would raise production sufficiently to meet expected demand has risen from time to time in the past three decades, partly in response to the announced long-term production ceiling of 8.5 million barrels per day, first bruited in 1978. Some, such as Schlesinger, have actually argued that they would be better off keeping the oil in the ground.[4]

YOU CAN'T ALWAYS GET WHAT YOU WANT

"Ghawar is dying." Could those three simple words signal the beginning of the end for the industrialized human civilization on Planet Earth?[5]

All day, yesterday, we were talking about this revelation as being bad news. That this proves that Saudi oil is peaking.[6]

Forecasts that the Saudis would produce 20 or 25 million barrels per day at some point in the future have frequently been greeted with jeers, even outside of the peak oil community (and predating them). Many organizations like the Department of Energy project global demand; subtract their forecast of non-OPEC supply; and, then, calculating expected supply from other OPEC members, *assume* that the difference will be made up by the Saudis. Since their oil reserves are enormous, it is presumed they would be physically capable of doing so; whether or not they would is not considered.

Predicting Saudi oil production policy in 2025 or 2030 is nearly impossible, however, at least accurately, so it is hardly foolish to question whether they would do so or not. And relying on them to do so to protect the global economy understandably makes policymakers a tad nervous.

But if you try sometimes, you just might find—you get what you need.

The funny thing, though, is that these warnings have persisted for roughly three decades. As we saw in Chapter 2, most forecasters became pessimistic about supply after the Iranian oil crisis in 1979, and foresaw diminishing supply from not only non-OPEC but also non-Gulf OPEC production, such as Indonesia, Nigeria, and Venezuela. As a result, most expectations foresaw ever-rising Saudi production to "fill the gap" between demand and projected supply, even as the Saudis themselves described a policy of not producing more than 8.5 million barrels per day over the long term.

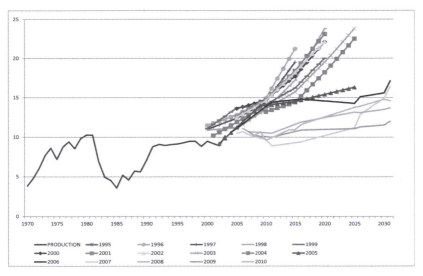

Figure 8.1   Forecasts/Assumptions of Saudi Production Capacity by DOE (million barrels per day) (Energy Information Administration, International Energy Outlook, various years)

Figure 8.1 shows a number of forecasts made by the U.S. DOE over the years, with a general consensus around those numbers. The IEA, in Table 8.1, indicates more recent expectations of Saudi supply. In the past few years, forecasters have felt the pressure from a number of analysts who find the assumption that Saudis will produce whatever necessary to be unrealistic, and it seems as if they have tinkered with demand and supply variables to try to keep the need for Saudi oil growing too rapidly.

Table 8.1   IEA Forecasts of Saudi Production (World Energy Outlook, various years)

	2006	2008	2010	2012
2005	9.1			
2010	9.7		9.6	11.1
2015	11.3	14.4	11.2	10.9
2020			11.5	10.6
2025			12.2	10.8
2030	14.6	15.6	13.2	11.4
2035			14.6	12.3

SHORT-TERM CONSTRAINTS AND LONG-TERM ABILITIES

Still, by seizing on sometimes offhand remarks about capacity constraints by Saudi officials, peak oil advocates have tended to expose themselves to serious mistakes. In Texas, capacity was a question of the resource and the economics, not government restrictions on drilling. Indeed, by setting production quotas as a fraction of capacity, Texas actually encouraged the creation of surplus capacity. When it ran out, this reflected the rising cost of Texas oil and showed that the peak of Texan production was, indeed, near (especially at then-prevailing prices).

The same is most decidedly not true in Saudi Arabia, where the government decides what production capacity it wants, based on estimates of likely demand plus a policy of keeping a "strategic" surplus capacity of about 2 million barrels per day, to deal with unexpected supply disruptions. The government describes this as insurance for the global economy, but of course it also doesn't want consumers to become so concerned about supply that they would switch to other fuels.

This came into play when, for example, Venezuelan production dropped sharply in late 2002 and then Iraqi production was shut in during the U.S. liberation/occupation[7] (Figure 8.2). Neither event had been specifically anticipated by long-term planners at Saudi Aramco (or the government), and so plans to have enough capacity on hand were not in place. And because the engineers at Saudi Aramco tend to be very methodical, they were unwilling to rush investment into place for what was expected to be a short-term problem.

Thus, the comments that there was not additional capacity available in 2005 were undoubtedly quite accurate, and yet said little or nothing of the country's long-term capabilities. After Texas ran out of surplus capacity in 1970, its production declined, whereas Saudi production and capacity have both increased significantly since they supposedly hit the ceiling, being about 1million barrels per day higher by 2014 than when Deffeyes remarked that they were out of capacity.

And the many alarums over declines in Saudi production have been more a demonstration of shortsightedness and ignorance. Saudi production has always been volatile in reaction to market developments, responding to changes in both prices and demand. In recent years, the instances when production declined have been due to apparent weakness in prices or perceptions that supplies were adequate or even abundant rather than technical problems or resource constraints.

Figure 8.2   Saudi Production (thousand barrels per day) (Energy Information Administration, US Department of Energy)

TECHNICAL PROBLEMS

I had never heard the term 'fuzzy logic' before. Hearing the Aramco manager's comment was one of the little events that tipped my thinking about the Saudi Arabian Oil Miracle towards skepticism.

Simmons 2004, 330

That's how the well-trained mind is known to me. . .
What you don't reckon, you think can't be true.[8]

Mephistofeles, Faust

Matthew Simmons was invited to visit Saudi Arabia as part of a good-will tour, unknown to the executives at Saudi Aramco, who were well aware of his recent conversion to peak oil advocacy. While there, he was treated to a standard presentation about the national oil company's technical abilities, with an executive bragging about the use of fuzzy logic to model the Ghawar reservoir.[9]

This, he said, made him realize that Saudi oil fields were in trouble, since he had never heard of fuzzy logic. And so, he undertook to learn more, reading every paper published by the Society of Petroleum Engineers about the situation in the Saudi oil fields. (It's not clear why he didn't look up the term *fuzzy logic* on Wikipedia to find it was just a computer programming method.) The things he learned, he claimed, made him realize that Saudi production was in threat of imminent decline, with catastrophic results for the global economy. He began to present his "findings" to various audiences.

This led to an invitation to debate the situation with two prominent Saudi Aramco executives from the Center for Strategic and International Studies, a prominent Washington thinktank, in February 2004. I arrived at the seminar to find many of the attendees looking over a front-page article in that day's *New York Times,* which summarized many of Simmons's arguments and a few others, relying heavily on information from long-retired Aramco executives.

Simmons put forth his technical arguments (the slides used to be on his company's Web site; however, since his passing, they have removed most of them), and the Saudis responded by asserting that they had no major problems and the resources to easily meet a much higher production level, up to 15 million barrels per day for 50 years.[10] In theory, this should have ended the controversy.

But it was somewhat of a "he said, they said" situation in that the Saudi assertions were not backed up by detailed technical information. Not solely because of normal corporation secrecy, but because the company could hardly produce the type of geological and seismic information in a 20-minute presentation that would be convincing to anyone—especially anyone who already distrusted them. Simmons, in fact, acknowledged this in subsequently calling for the Saudis to allow outside analysis of their fields and audited production of well-by-well information.

(Later, at the debate at the 2004 Offshore Technology Conference, Marlon Downey had said he fully approved of Simmons's call for full information disclosure, adding with a chuckle that his company never complied with such demands.)

The Evidence

Simmons's evidence relied heavily on two particulars: descriptions of problems in the Saudi oil fields in SPE papers, and U.S. government documents from the 1970s that claimed that Saudi capabilities were overblown and the fields had been damaged by bad reservoir management.

The latter are interesting but primarily described the loss of field pressure due to the rapid expansion of production in the late 1960s and early 1970s, which, it might be argued, was bad from both an engineering and economic standpoint. Simmons goes much further, claiming that once a field dropped below the "bubble point," when natural gas dissolved in the oil evaporated out, the field became "inert" and the oil could not be recovered.[11]

When I was repeating this at a presentation at Michigan Technical University, an audience member shouted, "Repressurize!" Exactly. A drop in field pressure, while it is not desired, is not only not insurmountable; it doesn't present much of a challenge to a well-funded organization like Saudi Aramco. And that is what they did.[12]

Another issue raised was the high water cut in Saudi fields, *high* being a subjective term in practice meant 36% in 1999, increasing from 26% in 1993–before declining to 33% and stabilizing. Unfortunately, he does not provide any context for this, but he simply assumes it was bad, which, combined with other evidence, makes him wonder "if Aramco insiders genuinely appreciate how fragile Ghawar's oil production must now be."[13] The truth is that the global average water cut is 75%, meaning Ghawar's is actually rather low.[14]

Finally, Simmons provided a field-by-field review, describing a variety of technical difficulties he found in SPE papers. Considering the Khurais field, he notes that the quality of the Arab-D reservoir "deteriorates towards the field's periphery" and the aquifer is too weak to sustain production, necessitating reinjection of associated gas.[15] Also, a gas reinjection program in 1983 was "far from trouble-free" and maintaining the valves was a constant struggle. And there was a lack of consistency and uniformity in the rocks, which meant that production increases varied a lot by well.[16] He concludes that the odds of Khurais reaching the planned 800 thousand barrels per day of production are "relatively long" and states two pages earlier that the previous peak of 144 thousand barrels per day in 1981 was "likely Khurais' all-time peak output."

In 2009, Khurais began producing at the level of 1.2 million barrels per day, 10 times higher than Simmons thought possible, providing another case where peak oil advocates foolishly make extreme statements that are proven false, rather than simply raising concerns about potential problems.[17]

THE CRITICS OF SIMMONS

I am aware of four primary rebuttals of the Simmons's claims from non-Saudi sources, including this author. A short article by Jim Jarell, who is trained as a petroleum engineer, noted a number of technical mistakes in his assertions, including his interpretation of the use of water for repressurizing Ghawar as a problem, instead of a solution.[18]

Sadad Al-Husseini, a former Saudi Aramco executive and a convert to peak oil advocacy, took issue with Simmons's characterization of Saudi Aramco as relatively clueless about petroleum engineering.[19] His interpretation seems to come down on the side of technological optimists, treating the advanced methods used by Saudi Aramco as a sign of their technical competence, not rising, let alone insurmountable, challenges.

In more detail, Lou Powers, a well-traveled petroleum engineer who worked for Saudi Aramco during the 1970s, when they were planning a massive expansion of production, provides a technical description of the main Saudi fields in his new book *The World Energy Dilemma*.[20] He notes that there were no technical barriers to going to 20 million barrels per day, but a policy decision was made not to do so at that time. His views on the management of the fields also contradict Simmons's by considering the application of the newest technologies to be evidence of good engineering rather than indicative of looming disaster.

Analysis for Dummies

For my part, I noted a number of flaws with Simmons's logic, such as referring to "secret" government studies that his assistant located in a nearby library. Our intelligence agencies might have their faults, but you can rarely find classified materials in libraries (rarely; there have been cases).

Also, at one point, he shows field reserve estimates from the American Aramco partners in 1976, suggesting that they should still be valid, and implies that current official reserve estimates are seriously exaggerated.[21] But elsewhere, he states that production declines after 50% of reserves have been produced. This would mean Ghawar, for one, would have gone into decline decades ago.

More problematic was Simmons's tendency to simply note problems and difficulties in various oil fields and then leap to the assumption that they could not be overcome. Indeed, it appears in one case where I double-checked the paper he cited, the authors had actually described their method of dealing with the fractured geology, but he neglected to mention that. Novice analysts too often think that stating a fact and a conclusion is the same as research, without demonstrating how one leads to another, and the basis of this book is nothing more than that: oil field problems mean looming catastrophe.

Table 8.2   Estimates of Saudi Petroleum Resources (Lynch, Michael C., "Crop Circles in the Desert: The Strange Controversy Over Saudi Petroleum," Occasional Paper 40, The International Research Center for Energy and Economic Development, Boulder, Colorado, 2006)

	Date	Proved Reserves	Undiscovered Oil	URR
OGJ	2013	266		
ASPO NL	2002	194	14	300
IHS Energy	2004	294		
Saudi Aramco	2004	260		
USGS	2002		87	371
Riva		116	57	269
Halbouty	1970			
Masters	1983		41	
Masters	1981		57	
Masters	1990	255	41	351
Masters	1994	259	61	374
Campbell	2004	144	18	260

ACTUAL DATA AND ANALYSIS

Obviously, analysis of the Saudi oil situation is hindered by not only all of the problems described in Chapter 13, but by the fact that the oil sector is in government's hands and is considered a strategic industry. This limits the data available more than in most Western countries, although almost no one publishes the kind of data for which Simmons has asked.

Table 8.3   Estimates of Ghawar Field Size (bln bbls) (Lynch, Michael C., "Crop Circles in the Desert: The Strange Controversy Over Saudi Petroleum," Occasional Paper 40, The International Research Center for Energy and Economic Development, Boulder, Colorado, 2006)

Source	Date	Size
IHS Energy		138
ASPO		
Simmons		
Saudi Aramco	2004	
Halbouty	1970	75
Laherrere	1995	52.5
Campbell	2004	85
Lyles	2004	114

Even so, there are some data available, all of which have generally been ignored by peak oil advocates, possibly because they don't confirm their arguments. First and perhaps most tellingly is the disagreement by virtually all analysts who are not members of the peak oil community that Saudi reserves are abundant and not a constraint on their ability to produce. Tables 8.2 and 8.3 show various estimates of both national reserves and those for the Ghawar field from the primary sources, although nearly all rely heavily on Saudi data itself (and some are relying on others in making their estimates; e.g., BP and EIA rely on OGJ). The clear disparity at both the field and national levels implies that the peak oil advocates' reserve estimates are not reliable. This is not to say that the Saudi estimates are completely, precisely reliable, but they seem much more valid than those of Simmons or Campbell.

There are other indicators that provide information about the state of Saudi resources, using publicly available sources. Consider that the United States has 1,800 drilling rigs operating in mid-June 2014, while Saudi Arabia has about 100. While the United States has five times the petroleum-prospective area, this means that despite being the world's most mature petroleum basin, it is drilling three times as much as the Saudis.[22] Since the Saudis have ramped up drilling heavily recently, it is easy to understand the Saudi's contention that the country has not been heavily explored. In fact, the total number of wells drilled in Saudi Arabia is trivial compared to the United States: a ratio of 1 to 500, roughly.[23] Overall, although the geology of the two countries is not very similar, the scale of the difference is so large to make the claim that Saudi Arabia is completely explored nonsense.

Even better, the *Oil & Gas Journal* annually reports the number of operating oil-and-gas wells in each country, which allows the productivity per well to be calculated. Figure 8.3 shows estimates over time for the past six decades, and while productivity has declined, it remains far above nearly every other country in the world (especially outside the Middle East). For comparison, the average onshore well in East Texas produces about 10 barrels per day, while shale oil wells that are considered productive produce 1,000 barrels per day, usually for less than a year.

For an economist, the best indicator of scarcity is cost. Historically, every estimate of full-cycle production costs in Saudi Arabia has shown it to be extremely cheap, beginning with Adelman's 1968 estimate of 6 cents a barrel (30 cents in 2010 dollars) and continuing to recent IEA estimates, which show them as being among the world's cheapest, with only Iraqi oil seeming to have lower costs, and that is presumably due to roughly three decades of underproduction as a result of war and sanctions.[24]

Recent development of "difficult" fields supports the idea that Saudi oil is relatively cheap. The Khurais redevelopment project, begun in 2006, was completed at an estimated $8–9 billion.[25] However, the denominator, capacity, was an equally enormous 1.2 million barrels a day, which translates into $7,500 per daily barrel of capacity, or roughly $10 a barrel. Similarly, the Manifa field was developed for $17,500 per daily barrel of capacity, or more than $20 a barrel.[26]

Compare that to fields analyzed in Chapter 16, where costs have been many times that amount for recent projects elsewhere, and it becomes obvious that Saudi costs are fairly low, contradicting the idea that their oil is "difficult" or scarce.

Consumption

The most recent aspect of the controversy over Saudi oil production is the focus on their exports, rather than production, because in recent years, Saudi oil consumption has soared to meet electric power demand. As one environmentalist put it, "What is not well known is the degree to which Saudi Arabia's massive oil exports are threatened by its demographics and a probable decline in its aging supergiant oil fields."[27] (The reference

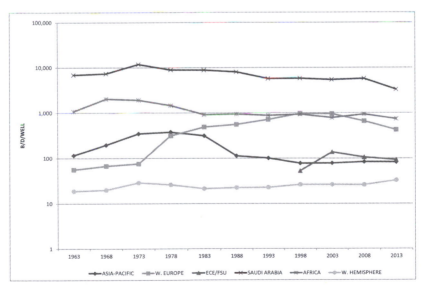

Figure 8.3   Production per Well Over Time (log scale) (Oil & Gas Journal, Worldwide Reserves and Production, various issues)

to probable decline in aging supergiant oil fields should be a clue that the author is not up-to-date on this issue.)

The concern stems from the fact that Saudi oil consumption has risen half a million barrels a day in four years (20% or 4.5% per year, from 2009 to 2013), but more a concern for Saudi budget planners than the world oil market. The problem arises not from inexorable demand increase due to a booming economy (as in China) or population, but from (a) very cheap electricity prices that don't discourage wastage, (b) very cheap natural gas prices that suppresses exploration and development, and (c) a very high demand peak in the summer.

The combination means that the government has found itself using lots of oil for power generation, which accounts for most of the increase. This could continue for some time to come; however, traditionally, substituting oil away from power generation is the easiest way to reduce consumption and requires "only" some policy changes in Saudi Arabia.

The Export Land Model, however, extrapolates recent consumption trends forever, assuming that the Saudi government either can't or won't change anything.[28] Indeed, there was a similar round of concerns in the late 1970s from analysts who assumed that non-Gulf countries could not increase production and would not restrain consumption, and so would be all net importers by roughly 1990. Only Indonesia fell into that trap, in reality.

Cherrypicking also comes into play in this instance, as the Chatham House published a very good research report, which made precisely the points about policy that are mentioned earlier.[29] The Hunt article neglected to consider the possibility that natural gas prices and production might be increased or that electricity prices could be raised, resulting in conservation.[30]

LESSONS FROM A CASE STUDY

This appears to be a simple case of someone believing in peak oil and then choosing a variety of evidence to support his conclusions. What is rather amazing is that so few have apparently looked at Simmons work critically, noted the many errors and inconsistencies, or acknowledged the all-too abundant refutations. A number of industry professionals have told me they admired the book, but when I ask what specific information suggests that Saudi oil production is near a peak, the silence has been meaningful.

Further, the peak oil community's embrace of Sadad al-Husseini's arguments supporting their cause, while completely ignoring his rebuttal of Simmons's thesis, is particularly illuminating, providing additional evidence of their advocacy trumping their analysis. At the cost of repetition, this is more cherrypicking.

Peak oil advocates must have felt a frisson of joy when the Simmons book seemed to overturn a long-held conventional wisdom that Saudi Arabia has enormous reserves and the ability to keep the world supplied, but this also violates Carl Sagan's dictum that extraordinary claims require extraordinary evidence. The actual evidence is largely misinterpreted or wrong.

If anything, this provides an excellent example of how many peak oil advocates go wrong: first, by writing on subjects that they are not really knowledgeable and making many technical mistakes as a result. Also, very much in the way "ancient alien" proponents go from not knowing why ancients drew lines in the desert or how they moved large rocks to the conclusion that aliens must have been involved, Simmons and others consider the use of unfamiliar methods to be evidence of looming catastrophe.

NINE

The Track Record of Peak Oil Theorists: Missed It by That Much

An economist is shown a method that is said to be working very well for a company, and theorists. He responds, "But how does it work in theory?"

Classic joke, favorite of Ronald Reagan

"Part of the problem that Campbell suffers from is that he's been saying this for 20 years. People have gotten to the point where it's like the boy crying wolf"

Mr. Rodgers[1]

Economists are often accused of preferring theory over reality, but the geologists at the core of the peak oil movement can hardly pretend to be innocent of such behavior.[2] The amazing aspect of the debate is that peak oil advocates have accused their opponents of the same. Specifically, Glenn Morton has created a "scorecard" comparing the track record of Cambridge Energy Research Associates, with that of peak oil analysts (or so he claims).[3] Similarly, as described in detail later in this chapter, Chris Nelder has noted my many bad price forecasts.

This is possibly the most misleading thing produced in this entire debate. Nearly anyone in the oil business knows that virtually no oil price forecasters foresaw the current price run-up well in advance, although like attendance at Woodstock, there are many who would claim it after the fact.

I refuse to forecast oil price as price depends mainly on human behavior and feeling, which are too erratic, but you [Lynch] seem to be sure that oil price will go down to 25 $/b next summer.[4]

But more important, predicting price is totally different from predicting supply, as many peak oilers acknowledge. Jean Laherrere refuses to forecast price for that very reason (although a careful reading of the quote above will suggest that I should eschew the practice as well.) Supply responds to geological, economic, geographic, fiscal, and political factors, while price responds to the supply/demand equation, geopolitical (and other) threats to supply, seasonal factors, and market psychology.

Even more important, the criticisms ignore the peak oil enthusiasts' track record, which has been completely abysmal in terms of supply forecasting. Also, while it claims that CERA would not accept "civil dialogue," they ignore the fact that many peak oil theorists have repeatedly refused to debate this author, as discussed in Chapter 10.

FORECASTING: OUT, OUT DAMN HARUSPEX

I judge the [Saudi] government's chance of survival for the next half-dozen years quite good, and for a dozen years, fairly good. *But there could be a successful revolution there this evening.*

Anonymous Saudi watcher 1979[5]

Seventy-eight years after Abdul Aziz ibn Saud triumphantly carved out his kingdom on the Arabian Peninsula following a quarter-century of warfare against rival tribes, Saudi Arabia is living on borrowed time.

Ilan Berman 2010[6]

In the social sciences, predictions are generally taken with a grain of salt as the influence of human psychology adds a huge element of uncertainty. Economic forecasts, such as those surveyed by the *Wall Street Journal,* vary widely—and no one is surprised.

Thus, the fact that the Saudi regime remains in place no less than three decades after its stability was questioned by the Harvard Business School Energy Project (among many others) is not to inveigh against the scholarship of the forecasters: politics is always unpredictable because of the many complex variables involved.

Thus, when Richard Duncan crowed about having predicted an oil crisis triggered by Middle East unrest during the Palestinian intifadah, he was not showered with praise.[7] Predicting violence in the Middle East is like predicting it will rain in Seattle, without saying when.

On the other hand, economic factors are thought to be more amenable to forecasting than political, even though there remains a significant element of uncertainty, and many peak oil advocates insist that geological factors are driving their forecasts, so that they should be reliable in a way that price forecasts are not, which, in fact, they are not.

HUBBERT

It is worth considering the predictions of M. King Hubbert more carefully, since his correct prediction of U.S. lower-48 oil production is often held up as evidence of the validity of the Hubbert "curve" as a modeling technique. It is true that in 1956 he predicted the peak of U.S. lower-48 oil production correctly, saying it would be between 1965 and 1970, and it occurred in 1971, which is quite accurate compared to other forecasts.[8]

While he almost nailed the date, he did miss on the amount, which was 3.4 billion barrels compared to his projection of 2.8–3 billion barrels (not including NGLs for either number). Not too bad, but far from "scientific" accuracy. And by 2000, when he anticipated that production would be 0.7 to 1.6 billion barrels, the actual (excluding Alaska) was 1.8; he was again low, but pretty good *especially* for 45 years out. Unfortunately, after that, U.S. production has recovered, thanks to deepwater and shale oil, reaching 2.5 billion barrels in 2013 compared to his expectation of 0.5 to 1 billion barrels. As described before, this reflects the inability of his method to foresee the discovery of new areas.

For Texas, he was extremely accurate until the recent explosion of activity in the Permian and the various shale basins. His projection closely matches actual production until just a few years ago. It could be argued that this proves the Hubbert curve will closely approximate production in an area with little political interference and which is relatively mature, but the number of such cases is small.

For the world, he also predicted a peak in 2000, which some peak oil advocates might claim to be relatively accurate, since they believe production peaked in 2005. However, few others would agree with that (certainly not this author), and the production level at which he anticipated a peak, 12.5 billion barrels, proved only about half the actual amount.

Admittedly, he was relying on a conservative estimate of recoverable resources and didn't claim high confidence in his prediction.

Later, when natural gas production was declining because of falling consumption due to high prices, Hubbert extrapolated the production trend to erroneously predict that the remaining undiscovered was less than 200 Tcf.[9] It took about 15 years to find that much, and gas reserves continue to grow.

PROOF IS IN THE PUDDING

Of course the best way to test any scientific theory is by how it relates to reality. Campbell and Laherrere have repeatedly made the point that their work is robust and reliable because of its scientific nature and their access to superior data over earlier researchers. As a result, they have argued that they have not been prone to the problem of earlier researchers, as described in 1995 by this author, namely:

- A tendency toward pessimism in oil production predictions.
- Repeated predictions of a near-term peak for nearly every region, which subsequently had to be moved out into the future and higher.
- Repeated increases in estimates of recoverable resources.[10]

When assessing predictions of long-term oil production, it takes many years to build up a reliable track record. There have been several organizations, such as the U.S. Department of Energy and The International Energy Agency, which have such track records as was discussed in Chapter 2. A few individual forecasters also have track records, having been producing forecasts for as much as a quarter-century, including this author and Robert Esser of CERA. But most peak oil theorists are relatively new to the game, and as such it is difficult to evaluate the quality of their forecasts, although Colin Campbell has been producing articles since 1989, and some of the other forecasts can be judged by considering them in detail.

Predictions of the Peak

But that trick never works.

Rocky

This time for sure!

Bullwinkle

One standard behavior for failed prophets is to insist that their work is good, but the date a little off. This can be seen frequently with millenarians, such as the 19th-century Millerite movement.

When the first [predicted apocalypse did not occur] in 1843, they went back to the drawing board, and they realized that they had made an error of one year by neglecting to take into account the transition from BC to AD, and because of that, they had gotten it off by a year. So they simply moved it forward one year to 1844. So that extended the excitement for one more year. But then at that point came the Great Disappointment, and the movement simply fragmented for the moment.[11]

Peak oil advocates have followed precisely the same path, as I predicted in my 1996 analysis of the method, consistently moving their forecasts of a peak further out and at higher levels.[12] In Chapter 1, it was described how Schlesinger had thought 1979 levels were the peak (apparently unbeknownst to his supporters in the peak oil movement), but he is hardly the only one. In recent years, Henry Groppe believed that the short-term decline in production in 2000 represented the peak, and Ken Deffeyes thought this was likely.[13]

Others, like Total CEO Christope de Margerie, thought it impossible for long-term production to grow as the IEA and others predicted, but found themselves repeatedly raising their estimates of the level at which it would peak, as the quotes below show.

Last year [Margerie] declared that the world would never be able to increase its output of oil from the current level of **85m** barrels per day (b/d) to 100m b/d.

The Economist, 2008[14]

Shortly after taking over at Total, he jolted oil executives at a London conference by stating the industry would be unlikely to produce more than 100 million barrels a day, far below the 120 million or so the International Energy Agency estimates the world could produce by 2030, and which will be needed for Asia's galloping growth. De Margerie now says **90 million** barrels a day is "optimistic."

Time, 2010[15]

New discoveries and technological advances have increased the oil industry's ability to increase production in recent years, pushing

Table 9.1   Campbell's Predictions of the Peak in World Oil Production

Date of Prediction	Date of Peak	Production	Type of Oil	Source
1989	1989	65.8	Total	Noroil 1989
1991	1992	63.0	Conventional	Golden Century
1997	1999	65.0	Conventional	Next Oil Crisis
1998	2003	71.0	Total	*Scientific American*
1999	2003	n.a.	Conventional	Tomorrow's Oil
2002	2000	63.9	Conventional	ASPO
2005	2006	66.0	Conventional	ASPO NL 3/05
2005	2007	83.0	Total	
2006	2005	66.0	Conventional	ASPO NL

global maximum oil production to **98 million barrels** per day for longer than initially expected, Total SA's Chairman and Chief Executive Christophe de Margerie said Tuesday.

DJ Newswires, 2012[16]

First and foremost, have the peak oil theorists avoided crying wolf as they have been charged with doing? Actually, as Table 9.1 shows, they have failed in this regard. Colin Campbell, in particular, first predicted the peak in 1989, arguing it was occurring that year. He has since revised its estimates repeatedly, always moving it out slightly, conforming to the earlier criticism.

In fact, he has produced his own table to demonstrate the changes in his estimates after a memo was produced by this author (then at MIT), noting his repeated revisions.[17] He explains this by quoting Keynes who remarked that upon receiving new data, he revised his opinions. However, since *that was the criticism of the method,* he has hardly resolved that issue, but rather *confirmed* its shortcoming. And he misrepresents his forecasts, leaving out his 1989 prediction that the world was peaking that year, and erroneously claiming his 1991 book predicted a peak in 1998 (he showed production hitting a peak in 1992, plateauing to 1998, then declining).

Testing the validity of production forecasts is constrained by the fact that most forecasters have not produced detailed regional or even global production forecasts in a usable manner. Because Campbell often (but not always) projects the behavior of "conventional oil," using his own definition (e.g., excluding polar oil) and for which data are not readily

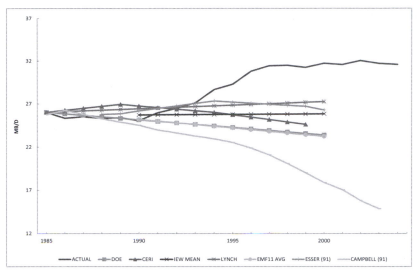

Figure 9.1   Forecasts of Non-OPEC, Non-FSU, Non-U.S. production (US Dept. of Energy; Canadian Energy Research Institute; International Energy Agency; International Energy Workshop; Lynch [1997], Campbell [1991])

available (how much Siberian oil is polar?), it is hard to judge the accuracy of many of his forecasts. And Laherrere produces forecasts in the form of extremely small figures, with 120 years in a one-inch square, most of them in the past few years, making it all but impossible to judge the degree of error only a few years later.

Still, there are some exceptions, and these are illuminating. In 1989, Campbell produced a graph showing his expectations for global oil production, claiming that it had already peaked, and would decline from 65 million barrels per day to about 25 million barrels per day by 2015.[18] Since that time, global production (total, not conventional) has grown strongly, by over 15 million barrels per day, of which perhaps one-third would be considered unconventional, mostly deepwater. (His expectation for a "price leap" in the early 1990s certainly proved incorrect.)

Similarly, Campbell's 1991 book included detailed tables showing production for many countries. In Figure 9.1, the projection for non-OPEC, non-FSU, non-U.S. production is shown, and clearly Campbell's is the worst of any public forecast.[19] (This author, and Robert Esser of CERA, both frequent targets of the peak oil theorists, have the best.)

Interestingly, some peak oil advocates have claimed this to be a victory for Campbell, including Nate Hagens on theoildrum.com, by the expediency of looking at the forecasts for total production, which is actually a consumption forecast.[20] Since world oil consumption dropped sharply with higher prices and the 2008 recession, global production is not very indicative of what could be produced so much as what the world needs.

Although Laherrere's forecasts are often difficult to interpret, consisting primarily of graphs covering a century or two, there are a few instances where projections can be teased out. Laherrere figure shows his 2003 forecast for all liquids with a peak in about 2015, and it is relatively close to actual production, which implies that he expected the 2008 recession, the supply losses from the Arab Spring and so on.

Others have an equally dismal track record. These include the following:

The Uppsala World Energy Group under Aleklett predicting in 2008 that world oil production would fall from 81 million barrels per day in 2006 to between 70 and 84 million barrels per day by 2014, with 75 million barrels per day being the Outlook (and the other ends of the range the "slow" and "fast" cases). Actual was an increase of 5 million barrels per day.[21]

Johnson et al., in a report to the Department of Energy in 2004, put non-OPEC, non-FSU production at a peak, declining from 35 to 28 million barrels per day by 2010; the actual was flat.[22]

Duncan and Youngquist predicted in 1998 that world production would peak in 2006, and that OPEC would be producing more than non-OPEC after 2008.[23] As of 2013, OPEC was producing 36 million barrels per day, and non-OPEC 50 million barrels per day.

Individual Countries

In a classic instance, Campbell projected UK production to decline at a rate of 10% per year, the same as the Forties field (Figure 9.2). This implies that there would be *no additional supplies from reserve growth, new large discoveries, or small and medium fields*. (This corresponds to their theories.) Yet as Figure 9.2 shows, this was not the case—despite a lack of large discoveries. Indeed, as much as 400 thousand barrels per day of 1995 production was from small fields that were discovered before 1980, but not put on production until subsea technology made them recoverable—a clear example of technology enhancing supply, contradicting peak oil theorists' claims.[24]

Similarly, Laherrere produced a later forecast of UK production using the discovery curve Laherrere figure which predicted that production

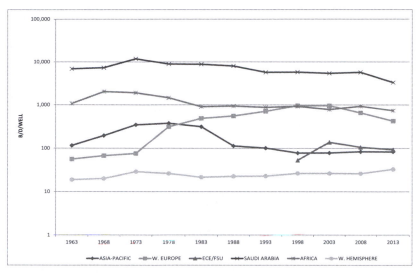

Figure 9.2   Forties Production and the Campbell Forecast for UK Total (Campbell [1991], UK Brown Book, and BP SRWE)

would be under 1 million barrels per day by 2005, whereas the actual is approximately 1.7, or 70% higher.[25] The subsequent sharp decline still left him well under actual; 2014 production was roughly 0.9 million barrels per day, versus his prediction of 0.5 million barrels per day.

The overly pessimistic UK production forecasts are notable because the United Kingdom is the *one of the best understood provinces,* relatively small, well-known geology, with an extremely detailed data available. Nor were any major "unconventional" fields added in recent years, which would skew the results in other countries, such as the United States. Overall, the models used by these two authors have failed badly at the easiest possible test.

Blanchard has bragged about his success forecasting North Sea production, noting that he was accurate for 2010, whereas DOE was almost 100% too high[26] (both from 1999). He does seem to understate the actual slightly, leaving him off by only about 10%.

Production in the Former Soviet Union serves as an interesting litmus test of the bias of peak oil theorists. In 1991, the FSU was known to have many under- and unexplored areas, as well as billion barrel fields that had been discovered but not developed, and was relying on technology far more primitive than that used in the West. Everything should have suggested that the region would see good prospects for recovery, yet not only did Campbell not foresee

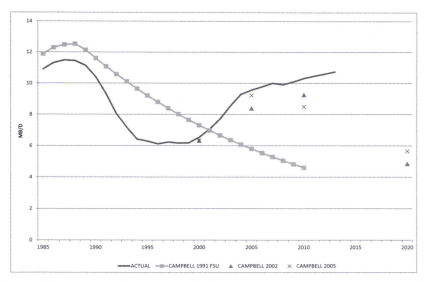

Figure 9.3   Campbell Forecast of FSU Production (1991) (Campbell 1991, BPSR)

bright future for oil production, but he predicted a sharp continuing drop in production based on (supposedly) the geological situation (Figure 9.3).[27]

Laherrere has written specifically on Russia and the FSU, and has repeatedly taken a decidedly pessimistic view, as Table 9.2 shows. And he argues that the level of 9 million barrels per day he noted in his 2005 letter to Youngquist was "an overproducing level as Yukos used a lot of frac (using Schlumberger) and that the decline will be more severe."

Any number of other peak oil advocates have looked at Russian or FSU production and seen nothing but doom and gloom, particularly whenever there was a pause or drop in production. Jeffrey Brown, who has created the "Export Land Model," argued in 2008 that Russia had peaked and supply would decline by "−5.1 ± 2%" per year, reaching between just over 6 and about 9.5 million barrels per day by 2012.[28] Wattenbarger used Hubbert curves to analyze FSU supply, which projected it falling to about

Table 9.2   Laherrere Forecasts of FSU/Russian Production (mb/d)

Date	Russia	FSU	Peak Year	Source
2002	8		2007	Laherrere (2005)
2002	9	12	2010	Campbell (2002)
2005	9	10	2005	Laherrere (2005)

7 million barrels per day by 2000, although he also noted a Russian forecast of recovery to 7.4 to 8 million barrels per day by 2010.[29]

In 2008, when weak prices and high taxes led to a brief drop in production, theoildrum.com carried a piece by Sam Foucher, which saw production by 2013 as being between 8.25 and 8.75 million barrels per day.[30]

In fact, the latest data show Russian production in 2014was 10.8 million barrels per day, although possibly flat but with potential for more increases. FSU production is already 13.7 million barrels per day, with continuing increases, particularly from Kazakhstan. The pessimism of peak oil theorists is clearly extreme and evidence that they are driven by a *Malthusian bias,* predicting a peak and downturn almost *without regard for the evidence.* (Note also the analysis of the Samotlor oil field described in Chapter 4.)

Mexico

So, Cantarell Field is a "poster child" for Peak Oil concerns.[31]

And now, as production falls rapidly in Mexico's largest field, Cantarell, there is little other new production to replace that lost production.
Kurt Cobb, 2013[32]

Mexico and the Cantarell field are indeed poster children for Peak Oil, but not in the way they intended. As production from the Cantarell field declined over the past decade, Mexican production also declined, however representing primarily the result of government inaction. It also shows that the way the industry is able to offset declining production in one area with new production in another, which bloggers like Cobb consider difficult.

This explains why Mexico is one of the few countries where Campbell (1991) was too optimistic, predicting a surge in the 1990s and a peak about 0.5 million barrels per day higher than the actual. More recently, the decline experienced in Mexican production does not differ greatly from peak oil advocates' forecasts, at least the more recent ones. Sam Foucher used two methods that predicted 2015 supply at between 1.5 and 3.25 million barrels per day, where the actual is 2.8 million barrels per day.[33]

Life Cycle Forecasts

A number of analysts have looked at individual countries and try to project future production according to where they reside on the "life-cycle" or pattern of production, which is described as starting with rising production, then reaching a peak and plateau, and then moving to a declining phase. Two examples, done by petroleum industry analysts Rodgers and Smith, exist publicly only in the form of PowerPoint presentations rather than textual work, which means that it is difficult to assess their accuracy, especially since they are only a few years old.[34]

Another case is the work done by Zittel and Schindler, who work for an energy and environmental company whose coverage is primarily on alternative energy technologies.[35] While they did not produce a detailed forecast (as Campbell did in his 1991 book), they did have a summary that contained some fairly precise projections.

First, they disagreed with the EIA projection that Caspian projection would grow from 1.3 to 4 million barrels per day by 2010. They believed it would only reach 2.3–2.8 million barrels per day. The number presently is 3.0 million barrels per day, above the high end of their prediction, and the startup of the much-delayed supergiant Kashagan field should add roughly 1 million barrels per day to production.

Second, they said that production off Angola (which they described as deepwater but clearly meant total) was predicted to rise from 0.8 to 2.5 by 2015, and they thought it would instead be 1.7–2.2 million barrels per day by then. Production in 2014 is at about 1.7 million barrels per day. Recent discoveries should increase production well beyond this level, but for now, their prediction can be considered reasonable.

Third, they predicted that peak production will eventually lie somewhere in the region of 80 million barrels per day, after then BP chairman John Browne said he thought it would be 90 million barrels per day. Production in 2014 was about 88.7 million barrels per day, with between 3 and 6 million barrels per day shut in by political troubles in various countries.

The Uppsala Global Energy Systems Group has done similar work, using models of giant fields that they believe will allow them to project production trends. Their primary error is to assume that resource estimates are static (although they allow for reserve growth), but essentially see fields as following an unalterable decline profile, once a certain amount of oil has been produced.

Given the relative youth of the organization, it has few detailed forecasts, but some stand out. In 2006, Aleklett included predictions for Chinese and Russian production, both of which were forecast as being on the verge of a peak and then sharp decline, as Table 9.3 shows.[36]

EREOI

One of the more puzzling subsets of peak oil theories is the argument that "net energy" is the most important indicator of utility of energy production, meaning that a process that uses a cheap form of energy, such as waste wood, to create an expensive form of energy, such as gasoline, is not necessarily worth pursuing. Essentially, this appears to be a proxy for profitability, rather like the labor theory of value. It is made more complex (and substantially less transparent) by the fact that the "energy invested" includes the energy embodied in the equipment used, the transport of men and materiel, and so forth. This is a complicated procedure, and nearly all research of this nature seems to rely on the estimates made using aggregate data for the oil industry.

But there is a simple test for the accuracy of the work, The original work, by Hall and Cleveland, predicted that the energy needed to extract oil in the U.S. would become negative between 1995 and 2000.[37] In that case, surely drilling would have ceased some time before then.

Natural Gas

For peak oil advocates, natural gas has been kryptonite, and for a very logical reason: production is not closely responsive to prices because markets are isolated. When U.S. natural gas demand dropped in the 1980s, it was due to weak demand in the United States and an inability to export, due to the high cost of transportation. But Hubbert (1980) and Smith and Lidsky (1992) extrapolated the production curve and foresaw little or no production by 2000.[38]

Simmons also looked at international gas production in 2004 and argued that since the four biggest producers had peaked, and "65% of world production is now in decline," therefore global production had peaked.[39] Of course, what he was seeing was weak demand in Europe reducing sales of Russian gas, and some weakness in North American

Table 9.3   Aleklett Forecasts 2006 (Aleklett 2006; BP)

	Peak Year	Level (bln barrels)	Decline From	Annual Rate	Actual 2014
China	2008	1.2	2010	−0.04	1.55
Russia	2011	3.4	2011	−0.053	3.96

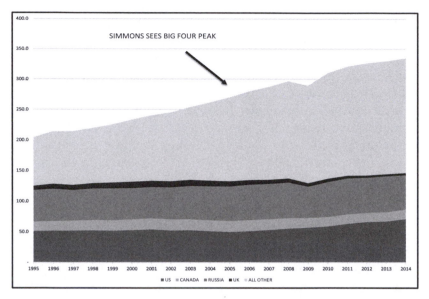

Figure 9.4   Natural Gas Production: Big Four and Others (BP Statistical Review)

production. In line with his arguments about the bulk of oil supply coming from a small number of fields, he presumed that, as the large producers go, so go the total. Figure 9.4 shows that this was very wrong.

More spectacularly, Simmons made a prediction for U.S. natural gas production based on an analysis of well data from Texas. Arguing that most gas came from a few large producers, and that these were declining rapidly, he predicted "a decline in gas supply as little as 1% to 3% now seems almost impossible, once the full impact of a drilling collapse is finally felt. I think the US will be fortunate if the decline is only 10%. It could be far higher."[40]

And in fact, production declined by 3.33% in 2002, but since has grown by nearly 30% since then. Again, this is based on an assumption that somehow all old wells decline, and the replacement wells resemble the smaller producers, which is a misreading of the supply curve (see Chapter 21).

Laherrere was actually more optimistic about global gas than Simmons, which isn't hard given known discoveries, and didn't predict a global peak until about 2025, but he did anticipate a sharp decline in U.S. gas production after 1995, based on historical "discoveries."[41]

RESOURCES

The inability of peak oil theorists to estimate resources also stands out. Having, in their early work, produced estimates that were near the then-consensus of 2 trillion barrels, it appeared that their methods were valid, since they generated results so similar to many other groups. But since then, Campbell and Laherrere have continued to confirm that number, adding some non-conventional oil, like NGLs, even as the consensus has moved on to a level of 3 trillion barrels or higher (Chapter 15), reflecting better understanding of the resource base, better recovery factors, and more access to new areas due to better technology—precisely as resource economists predict.

Table 9.4   Assessments of Reserves and Ultimately Recoverable Resources (Campbell [1997] and [2005])

	URR 1997	Discovered 2005	URR 2005
Saudi Arabia	300.0	258.0	275
FSU	275.0	**281**	307
US-48	210.0	197.0	200
Iraq	115.0	92.9	100
Iran	120.0	**128.0**	140
Venezuela	90.0	81.6	88
Kuwait	85.0	84.3	90
Abu Dhabi	80.0	60.5	65
Mexico	50.0	**52.6**	56
China	55.0	**57.2**	60
Nigeria	40.0	**50.5**	55
Libya	45.0	**50.4**	55
Norway	27.0	**30.0**	32
UK	30.0	29.2	30
Indonesia	25.0	**30.9**	32
Algeria	23.0	**25.8**	28
Canada	28.0	25.1	26
N. Zone	12.0	**12.6**	14
Oman	12.0	**13.4**	14
Egypt	11.0	**13.3**	14

Note: Numbers in boldface are where more had been discovered by 2005 than estimated in 1997 would ever be discovered.

The details of their results are extremely illuminating. On an individual country basis, the amount of *discovered oil* in Campbell's last estimate now *exceeds their 1997 estimates of URR for 32 out of 58 countries!* (Table 9.4 shows the 25 largest countries.) It has taken only eight years for the majority of countries to accomplish what Campbell asserted was impossible. If these estimates do not prove valid for so short a time, how can they be expected to hold true for the long term, as claimed? (Note also that by 2005, total discovered oil was only 40 billion barrels less than the expected ultimate in 1997, although this might be partly due to definitional differences.)

One of the strongest claims made by Campbell in particular is that his research provides a robust estimate of ultimately recoverable resources (which he also refers to is estimated ultimate resource). Yet he repeatedly revised his estimates upward, for a total of 325 billion barrels over 16 years. The rate of increase in his estimated ultimate recoverable resource has been similar to the amount of oil produced during this period, implying, using his tendency for mindless extrapolation, that we will never have a problem.

But a closer view of his work is illustrative: looking at the amount of undiscovered oil, we see that it is dropped by about 50 billion barrels over the last decade, despite large-scale discoveries, estimated at about 120 billion barrels by IHS Energy. Again, though, the reliance on "conventional" oil in Campbell's data obscures the reality.

This is also typical of other estimates, such as the USGS's, where discovered resources grow over time, but the undiscovered level tends to remain the same, as discussed in Chapter 15.

HE CAN DISH IT OUT, BUT CAN HE TAKE IT?

Given that much of what is written about this debate, and my work especially, appears on the Internet, it is not surprising to find misrepresentations of my views and record. First, several have commented on interactions with me:

> Tellingly, Michael Lynch refuses to offer his own prediction of when global oil production will peak, even when pressed to do so. One may assume that he believes that production will continue to increase forever.[42]

Or one could listen to what he said, which is that it is impossible to predict a specific date when production will peak.

When I debated Michael C. Lynch and ExxonMobil representatives in 2006 on a PBS program, I confronted Lynch with the Texas case history; his response was basically to pretend that the Texas case history didn't exist.[43]

No, my response to this, and similar points about other areas peaking, is that global production does not seem to have been effected at all by the peak in Texas, or the United States, West Virginia, the North Sea, and so on.

Most peak oil advocates would quickly point out that my track record is hardly flawless. Chris Nelder responded to a *New York Times* op-ed I published in 2009 by pointing out that:

- In 1999, I said an increase in IHS Energy's URR estimate had put a "nail in the coffin of peak oil" and that there was 2500 billion barrels of remaining conventional oil that could be produced, which he called "absurd" and based on my belief the industry would "somehow achieve a recovery rate of 35%, without venturing to guess how it might be done."
- In April 2004, I predicted prices would fall to $25 in the summer.
- In September 2004, I predicted prices would fall to $30 in the next year.
- In April 2007, my forecast was that prices would drop to the mid-to-low $40s by mid-2008.[44]

While he doesn't give citations for these quotes, they are probably accurate—certainly, I repeatedly predicted prices would return to more normal levels throughout the decade. And the 2007 prediction that prices would drop by mid-2008 was not bad, although more from luck and the financial crisis that occurred.

And his first point, about my 1999 paper, is completely wrong. The point was that discoveries in 1999 and 2000 represented 20% of the amount that Campbell and Laherrere said *would ever be discovered,* a near statistical impossibility. And the IHS estimate was not of URR but discovered to date, which they put at more than Campbell and Laherrere thought would ever be discovered.

Typically, Nelder was completely dismissive of me and my work—of which he admitted he was largely ignorant.

Similarly, Nate Hagens on theoildrum.com states that "[I]t is [Lynch's] nemesis Campbell that appears to have hit the bulls eye" while describing

Campbell's prediction that would oil production would hit a plateau of roughly 70 million barrels per day in 2000 and begin declining in 2008.[45] (Current production is about 89 million barrels per day.)

CONCLUSIONS

Although Hubbert's 1956 prediction of U.S. oil production (ignoring Alaska) proved quite good, as did his forecast of Texan production, nearly every other forecast made by peak oil theorists appears to have been much too pessimistic. All forecasters have been too pessimistic, but peak oil theorists have tended to have the greatest errors. Even in countries with reliable data, such as the United Kingdom, or clearly abundant resources, such as the FSU, they have been rarely provided assessments that are remotely accurate.

What is especially noteworthy is their pessimism about virtually very province, despite the enormous difference in maturity of oil exploitation in various countries. If you always get the same answer, regardless of the data, then *the method is flawed* and will always produce a near-term peak and decline, and *the researchers are biased,* in that they are pessimistic regardless of existing evidence.

The most symptomatic statement is probably from Campbell's 1989 article, in which he denied the likelihood of major new finds, stating, "Some commentators hope that new technology will lead to important deepwater finds. ... [G]enerally the geology of most deepwater tracts is not very promising." That he could make such a sweeping, definitive statement about areas that had not been explored suggests the problem lies with him, not the world's resources.

Ultimately, then, we come to the conclusion that the peak oil advocates have failed to predict supply at almost any level. The exceptions are Mexico, after Cantarell production collapsed, and the North Sea, after the peak and post-1998 oil price crash production decline. But their forecasts in these areas were in line with those of most others, like the IEA.

Certain things stand out: Nearly every country outside the Middle East is seen peaking within a couple of years of the forecast. Azerbaijan and Kazakhstan are prime exceptions. Most misinterpret low production in Brazil, China, and India as persisting, because they assume that geology is the explanation for poor performance, rather than due to restrictions on drilling. (Duncan alone was somewhat optimistic about China.)

Similarly, none think that the OPEC nations—Iran, Libya, or Venezuela —have a chance to reach new peaks even though, again, all have depressed production due to political decisions and/or weak markets, depending on the time period. Venezuela shows how this can be misleading, with nearly all forecasters (peak oil and otherwise) thinking it could not increase production after 1979, only to see supply boom in the 1990s after fiscal and legal reform.

Their pessimism about world oil production stems from an underlying belief that a decline in production in a country or region cannot be reversed, and the extrapolation of trends under the assumption that nothing can change them. Yet repeatedly, these expectations have failed, as fields, countries, and regions have often increased production after a peak. By always erring on the side of pessimism, the peak oil advocates guarantee that their forecasts will invariably fail.

APPENDIX: MY TRACK RECORD

In chronological order: a forecast of price and supply at EMF6 in 1980 was the best produced for that study.

The three-volume study International Natural Gas Trade included chapters on supply by me and M. A. Adelman, which argued that supply was plentiful and relatively cheap.[46] For North America, when others predicted gas prices would rise at 5% per year above inflation, we argued that they were more likely to be flat. In fact, they fell.

In 1986, after the oil price collapse, an MIT paper argued that changes to the market structure meant prices would remain low and volatile, disagreeing with the consensus.[47]

In 1989, the Economist Intelligence Unit published my *Oil Prices to 2000,* which argued non-OPEC supply would be higher than most expected and that prices would decline relative to inflation.[48] The *Petroleum Economist* described that as "heretical," but it proved prescient, almost completely precise on prices. My North Sea supply forecast was too low, though.

My 1996 report for the Gas Research Institute argued that resources were adequate to keep oil prices at $22 a barrel (2010$) for a lengthy period, excluding supply disruptions.[49] Some have argued that this was flawed ignoring the qualification, and others have focused on my prediction of global oil production, not realizing that this is really a consumption forecast. My forecast erred in not predicting lower consumption levels;

my non-OPEC production forecast was too low, but better than others, and far superior to Colin Campbell's.

In 1997, a paper published by the Emirates Center for Strategic Studies and Research estimated that Russian oil resources were cheap and abundant and that production and exports would soar when the fiscal regime was improved and export capacity increased.[50] This proved correct, but did not differ markedly from the conventional wisdom, although it contradicted all peak oil analysts' expectations.

In 2001, I published an article in the *Oil & Gas Journal*, "A New Era of Oil Price Volatility," arguing that lower surplus capacity in OPEC and a tendency toward lower inventories in the private sector should mean much more volatile prices.[51] This proved correct, although my suggestion that prices would be higher—averaging $30 instead of $20—was a tad too low.

In 2007, a working paper titled "Crop Circles in the Desert," refuted the arguments in *Twilight in the Desert,* which suggested Saudi oil production was near a peak and decline.[52] My conclusion proved correct.

In June 2012, at OPEC's International Energy Symposium, I predicted the long-term oil price was likely to be $50–60/barrel, which prompted humor and derision.[53]

In March 2013, another article in the *Oil & Gas Journal* warned that LNG prices were elevated by two factors that were likely to be transient and defeated by, if nothing else, the rise of shale gas in the United States.[54] High oil prices elevated LNG prices due to contractual links, and the Fukushima nuclear disaster in Japan had created a sudden, but temporary, surge in prices. In particular, high-cost producers, like the United States, would be threatened by $60 oil prices. This is now proving to be the case.

In 2014, the long-term oil price forecast I submitted to the Department of Energy, published in its Annual Energy Outlook, was so far below the other forecasts that a colleague joked that people thought I must be drunk.[55] It is obviously too early to suggest that my prediction of prices for the next 20 years is reasonably accurate, but the fact that prices are now expected by many to remain near $50/barrel for five years or more implies that my forecast might prove superior to many others.

The overall lesson is that while short-term price projections are always problematic, and changes in policy, such as resource nationalism, uncertain, in-depth analysis can provide insight and reduce the degree of uncertainty inherent in commodity businesses.

TEN

She Blinded Me with Science

Economists treat the oil and gas resource as near infinite, seeing supply driven by market forces, whereas *Natural Scientists* evaluate the occurrence in Nature, seeing supply controlled by immutable physical forces. C. J. Campbell[1]

Back off, man, I'm a scientist!

Bill Murray in Ghostbusters

One persistent claim among peak oil advocates is that they are scientists (at least some of them) and their work is science. In this chapter, the value of these claims is tested and found wanting.

The great majority of writing on the subject of peak oil has been done by people who are not academics, and are often little more than interested observers. Some are clearly nothing more than hobbyists. This is not in itself condemnatory: consider the example of Michael Ventris, an English architect who had a fascination with Linear B, the Minoan script, and made a major contribution to deciphering it.[2] Theories and research should be judged on their merits, rather than the background of the people making them.

And yet, any number of peak oil advocates have portrayed this debate as one between geologists and economists, which is not really true. In fact, many of those making this argument are not even scientists, like the late Matthew Simmons, a Houston banker. Nonetheless, this has led to a number of observers (e.g., the press or senior corporate executives) to think that the work must be reliable because it is being done by scientists. As one foreign policy commentator said about peak oil advocates, "I originally assumed these fellows knew what they were talking about."[3]

There are several objections to this, including: (a) most of the peak oil writers and "analysts" are not scientists at all; (b) those who are scientists are not trained in forecasting oil supply; and (c) the scientists are frequently violating the scientific method in a variety of ways, such that their work should not be characterized as science. Not everything that comes out of a scientist is science any more than everything that comes from a cow is beef.

EXPERTINESS

Q: Are you saying the S.E.C. under Schapiro is about to catch fraud on Wall Street? A: She has the wrong staff. A: They're a bunch of idiots there. Q: What do you mean? A: The five commissioners of the S.E.C. are securities lawyers. Securities lawyers never understand finance."[4]

Media personality Stephen Colbert coined the term *factiness* to refer to things that appear to be facts but aren't, strictly speaking. (In other words, at all.) Similarly, peak oil writings include a lot of things that appear to be scientific but generally violate the rules or norms of science.

A major problem in assessing writings about peak oil (or really any issue) involves the difficulty of assessing the qualifications of the speaker/writer/blogger. In the days of network television (i.e., when I was a boy), those who provided expert opinions were generally few in number and had respected credentials: a retired general, an academic from a major university, a senior government official. The combination of the explosion of news channels on cable, increasing the demand for expert commentary, the ease of access to the Internet, and reducing the bar for qualification has made it much more difficult to judge any supposed expert.

Perhaps my favorite case occurred during the frenzy over cold fusion, when one of the major networks had a split screen with a physicist from a top university arguing that the research results had not been reproduced, while a professor from a large state university countered that he had reproduced the results, to which the first noted that they had not been reproduced at a school *without a competitive football team*. (The anger on the second expert's face spoke volumes.)

In the peak oil debate, the clearest example of expertiness without expertise is the case of Secretary James Schlesinger, who recently announced that the peak had apparently arrived. He was not only the first secretary of energy (under President Carter), a former head of the Central Intelligence Agency, former secretary of defense and, wait for it, *a Ph.D. economist.* Surely, few would be more likely to understand oil markets? Well, how about a petroleum or resource economist. In fact, Dr. Schlesinger focused his attention on the Soviet threat for much of his career, has few, if any, publications involving petroleum economics or petroleum markets, and has at best marginally more knowledge than an educated layman.

Sometimes the expert's background can inform one as to why they have been led astray. Physicists such as John Holdren and Albert Bartlett decry the dangers of exponential growth, unaware that social systems have self-correcting mechanisms, unlike, say, a rocket spinning out of control or the path of a planet. And geologists like Jean Laherrere and Ken Deffeyeshave repeatedly displayed an ignorance of statistical modeling methods (to say nothing of resource economics).

More recently, Kjell Aleklett has become the president of the Association for the Study of Peak Oil and has created an energy study group at the University of Uppsala, Sweden. This position gives him a certain gravitas, especially in the eyes of those who are unfamiliar with his background, as a physicist who did not work on petroleum or resources until he developed an interest in the issue of peak oil. (His online resume does not list any energy publications prior to 2003.[5]) Roger Bentley, prominent in British peak oil circles, was similarly from outside the field, with a background in cybernetics.

But peak oil writers rarely mention this and instead gloss over the lack of expertise or ignore it completely. That Bentley is from computer sciences is typically not mentioned, and while some describe Aleklett as a physicist, it is primarily in passing.[6]

And apparently it doesn't matter what kind of scientist you are. A recent *Nature* article was written by a physicist and an oceanographer, with experience neither in petroleum resources nor in forecasting. And the type of scientist doesn't seem to matter: one peak oil commenter, described as a "scientist" in one financial press article, is apparently a neurotoxicologist by training.[7]

BOUNDED RATIONALITY

Other examples from the 1970s provide insight into the problem. The CEOs of the major oil companies were all convinced that oil and gas were "mature" industries, that prices could only continue soaring (if only they had listened to Julian Simon!), and that their cash flow should be reinvested not in oil and gas, but unconventional sources such as kerogen and photovoltaics, or diversified into department stores and office equipment manufacturers. (See Chapter 3.)

How can the CEO of an oil company not understand his or her own business? The explanation can be found in the work of Herbert Simon, who argued that while markets and information might be perfect and actors rational, they lacked the time to locate and analyze all information. A CEO hardly has time to analyze and judge writings on petroleum economics, and relies instead on staff and often colleagues or people whom they consider to be experts (i.e., those containing a degree of expertiness). Many staffers can tell stories of senior executives ignoring them not just in favor of advice from a prominent consultant but something told them on the golf course or even seen in a magazine.

And so often, the primary source of an institution's apparent stance on an issue is a report by one or a few staffers. Outside observers are careless in attributing a single article or research paper as representing the view of the entire institution, and, in corporations, a belief or disbelief in a position by senior management is often only their own view. In fact, many oil companies that don't believe in peak oil contain employees who disagree.

HOW TO JUDGE?

So, how can the level of expertise of an analyst be judged? The best way is to put two speakers with opposing viewpoints on a podium and ask them to debate the questions. This is imperfect: the audience is rarely able to double-check facts—or alleged facts. But lacking that, there are some other hints that can be informative.

Debate

Dr. Campbell so dislikes Mr. Lynch that he has declined invitations to appear with him in debates. "It's like asking a doctor to talk about medicine with a faith healer," Dr. Campbell says. Jeffrey Ball, *Wall Street Journal,* September 21, 2004

What's the point of a debate if you won't even back up your blog posts with a bet??? It suggests your words don't mean much. Joe Romm[8]

I will go to Uruk now, to the palace of Gilgamesh the mighty king.
I will challenge him. I will shout to his face: "*I* am the mightiest! *I* am
the man who can make the world tremble! *I* am supreme!" Enkidu, 87[9]

Science is not just a noun but a *process,* that is, a dynamic by which theories are proposed, tested, reexamined, and modified as needed. And it is
nearly always needed, as rarely do theories spring from the heads of scientists perfectly formed, like some full-grown Athena from the head of Zeus.
Indeed, many scientific debates look more like Enkidu versus Gilgamesh
than civilized discussions of substantive differences.[10]

Medicine is the best example of this, partly because organic chemistry
is far more complex than, say, astronomy. It is rare for anyone at a physics
convention to challenge references to the speed of light or the gravitational constant (special cases do exist), but nearly every advance in medicine leads various other research groups attempt to duplicate them, and,
until they have been, the results are typically not accepted as proven.
And medical researchers, among other scientists, never challenge the right
of others to analyze their work. Physicists, too, often find that their work is
not confirmed or even overturned, and while differences of opinion can be
very heated, the right to dispute research findings is rarely at issue.

However, Laherrere, in response to several of my criticisms, posted
online the remarks, "I challenge you to write for a full year without
mentioning Campbell, Simmons and myself, as your papers would be
empty! ... Stop attacking me as I requested a long time ago, stop talking
about me and I will stop to be obliged to say that you confuse things."[11] Similarly, Campbell responded to a letter in the journal *National Interest* criticizing his arguments by saying that I should stop attacking good work.[12]

Yet peak oilers rarely debate the pertinent issues with nonbelievers.
The classic case, perhaps, is the debate that occurred at the OTC
conference in front of perhaps 2000 industry executives, where Simmons,
one of two peak oil advocates, gave a presentation that was tongue-in-
cheek and entertaining (slides of trees in fall labeled "peaking" and in
winter "peaked"), but contained no substantive remarks.[13] As he finished,
he informed the audience that the issue needed serious debate.

Why he chose not to do so himself was never clear, but subsequently,
I was again invited to a debate with him, this time in front of the Society
of Petroleum Engineers, and the sponsors commented that they were having a moderator try to pin him down a little better. Instead, he made the
following comments to the effect:

- Peak oil is inevitable.
- Oil supply is "long in the tooth."
- Much oil comes from fields long in production.
- Many countries are in decline.
- Modern technology drains oil faster.
- Global decline rates are soaring.
- The likelihood of vast new reserves of light crude is diminishing.

There was in fact no analysis shown to back these assertions up, nor did he discuss the technical issues that he had, earlier that year, made the center of his talk in Washington on the problems of Saudi production (Chapter 8). In my response, I noted some of his technical arguments, such as that new technologies were not increasing but rather accelerating production, and asked the audience rhetorically if they agreed with that. Subsequently, he refused to appear with me or debate with anyone, to my knowledge.

Similarly, Campbell has repeatedly not only ignored my substantive criticisms of his work but treated them as personal attacks. The above-mentioned quote from 2004 was in response to a list of substantive questions about his work I provided to the reporter, for example, about the validity of the Hubbert curve, and he apparently had no response to them.

While the holding of numerous conferences by ASPO, both an annual international conference and those hosted by specific national chapters, would imply that the issue has received detailed debate, the reality is quite different. The ASPO conferences are almost completely addressed by believers in peak oil, and outside of ASPO conferences, the topic rarely comes up. Peak oil organizers made much of Daniel Yergin's refusal to appear at an ASPO conference in Boston, presumably to defend his company's research and disbelief in the peak oil hypothesis, ignoring the fact that he was not in charge of its supply analysis. Interestingly, the organizers were familiar with my work, and I'm not far from Boston, but received no similar invitation. (I have been invited to attend ASPO from time to time, but never to speak.)

And much of the work ignores the arguments of their critics. Robelius, in a doctoral thesis supervised by Aleklett, mentions that the Hubbert model is heavily debated, citing this author, but does not mention the criticisms or explain why he ignores them. Hirsch et al. solicited my "forecast" for the date of peak oil production, and even cited my work criticizing peak oil, but without mentioning, let alone addressing, the criticisms.[14]

Conspiracy

"[T]hey [the IEA] think that you should remove your analysis from the Web." I was stunned and I asked, "Who are 'they'?" There was a short silence and then he answered, "Personally, I do not have any problems with the analysis but they think you should remove it."[15]
More than three quarters (77 percent) of those surveyed believe there are signs that aliens have visited Earth, and over half (55 percent) think *Men in Black*-style agents threaten those who report seeing them.[16]

Those who deal with bad science have remarked on the manner in which bad research is often defended by its proponents by claims that disagreements are due to a conspiracy among the "powers that be" or "mainstream science." Since this typically stands in for actual scientific debate, it is a good indicator of the weakness of the arguments.

This is certainly the case with the peak oil controversy. Refusal to accept specific bits of data or research and peak oil advocates's theories is attributed to a desire to hide the truth from the public—and not just by fringe elements of the debate, but core proponents like Campbell, Laherrere, Kunstler, and Simmons.

Some of the claims are not unreasonable, disputing the validity of data, but many others go beyond what most would consider appropriate for a policy debate. Heinberg's argument that "Evidently growing public concern about the inevitable decline in world oil production has *rankled some powerful people,* who've been knotting their ropes in search of a bit of favorable data (declining oil prices, rising production) to use as the pretext for a public lynching."[17] Kunstler sees media unwillingness to spread bad news because "The major media, hard pressed by declining revenues and the extremes of competition on cable TV and the Internet, are in thrall to corporate advertisers who expect cheerleading for the status quo in return."[18]

Or Fridley's explanation why former Secretary of Energy Chu didn't lead a public discussion of peak oil: "[Chu] was my boss . . . He knows all about peak oil, but he can't talk about it. If the government announced that peak oil was threatening our economy, Wall Street would crash. He just can't say anything about it."[19]

In his book *The Believing Brain,* Michael Shermer has a chapter describing the belief in conspiracies, and he explicitly mentions peak oil.[20] He argues that of those who embrace such theories "their pattern-detection filters are wide-open. . . . Add to those propensities the *confirmation bias* and *hindsight bias* . . . and we have the foundation for conspiratorial cognition" (p. 209).

Bad data

Two particular issues have caught the ire of peak oil advocates, reserves and resources data and estimates. The USGS has come in for particular criticism, as it represents geologists (supposedly the source of peak oil "science") and has done what many consider the best resource estimates. While many like me, criticize specific elements of their work, peak oil advocates argue that it is dishonest. For example, "He [Deffeyes] specifically criticized the USGS's world energy assessment, claiming that the USGS 'cooked the books'."[21]

Andrew MacKillop makes a similar, but more general, claim about resources: "the real 'bottom line' is simple: the amount of recoverable oil now left in the ground is significantly less than what the oil industry would have us believe. The name of the game is confidence, as in any confidence trick, but the global economy is the patsy."[22]

And since the failure of global reserves to decline despite supposed low discovery rates has refuted a major tenet of peak oil theory (that reserves are not evaluated upward over time), many have simply insisted that the data are false. Prominent UK scientist David King "added that oil companies were consistently overstating the scale of their reserves."[23]

Laherrere extended this to other agencies more generally: "Once in the 80s, BP in the annual statistical review changed the official UAE reserves estimates by their own data (being the operator of almost every fields), but the next day they were obliged to destroy all their reports and to replace their values by the official data. Since, BP avoids not to repeat this mistake! It is the same problem with most official agencies, official data from other nations are difficult to deny."[24]

Motivations

Related to the reliance on conspiracy theories to avoid responding to substantive arguments, many accuse those they disagree with of only representing their self-interest. This becomes problematic when the interests are not homogenous or not logically represented by opposition to peak oil.

Most consistently, peak oil advocates accuse those who disagree with them of being in the pocket of the oil industry, including me, but also the prominent consulting firm Cambridge Energy Research Associates and its CEO, Daniel Yergin, the Pulitzer Prize–winning author. As Kunstler put it, they are "professional propagandists . . . CERA is the main public relations shop for the oil industry."[25]

Of course, this completely contradicts the many assertions by Campbell especially that the major oil companies "acknowledge" peak oil, or the fact that some, such as Total, supported the theory. More troubling, why would the industry hide its challenges and problems? Presumably so that people would continue to use the vanishing resource. Yet wouldn't they actually use the looming crisis to seek government assistance? Or, as in the 1970s, look to move away from a dying industry? And how to explain that the Saudis, whose oil production was supposedly about to drop sharply, were consistently trying to quell the very facts that would send prices higher?

Others see an effort to blunt the adoption of renewable energies.

As mentioned by the Swiss MP and member of the Swiss Parliamentary Energy Commission, Dr. Rudolf Rechsteiner, the IEA has effectively been *delaying the change to a renewable world*. Remarkably, the so-called global oil watchdog acted more and more like a zealous oil industry lobby.[26]

Conspiracy of Silence

Finally, the bizarre claim that the news is being suppressed recurs among the peak oil believers, with Kunstler saying "The major media, hard pressed by declining revenues and the extremes of competition on cable TV and the Internet, are in thrall to corporate advertisers who expect cheerleading for the status quo in return."[27] Heinberg is more vague but insists "Evidently growing public concern about the inevitable decline in world oil production has rankled some powerful people."[28]

Others are even more fantastic: "Coll's heavily researched book backs up Dredd Blog's long-time characterization of the exercise of power in Western Civilization with the Dredd Blog caricature of 'MOMCOM' (Military Oil Media Complex). Why we gave up the old notion of 'MIC' (Military Industrial Complex), as explained in the Dredd Blog post 'MOMCOM: Mean Welfare Queen,' is because it is so yesterday, it is so 1950's reality."[29]

All of which ignores the huge attention paid to peak oil, despite the fact that it has been widely refuted by the vast majority of experts and organizations. *New York Times,* the *Wall Street Journal, Newsweek, National Public Radio,* and many others have covered the story and still continue to refer to peak oil in stories on the industry. Indeed, the media's affection for crises and bad news is much more commonly remarked on than their wish to lull the public into complacency about the world, making this claim nonsensical.

Ultimately, then, we are told that there is a conspiracy to cover up peak oil that includes the OECD, the IEA, the governments of the United States and Canada, the USGS, CERA, most of the oil industry, and many independent analysts (including this author). The media, which has given significant coverage to the issue, is nonetheless said to be participating in this.

MUCH IS NOT SCIENCE

Although some peak oil advocates also embrace pseudoscience like cold fusion, the primary, serious publications on peak oil do not fit into that category. However, there are many characteristics of what Emily Willingham[30] calls fake or pseudoscience, including lack of expertise and reliance on claims of conspiracy. Robert Park, in describing *Voodoo Science,* notes that there are cases where apparently sincere researchers, when challenged, retreat into ad hominem attacks.[31] Certainly, those are apparent in the peak oil debate in abundance.

So, despite the claims that peak oil originates in "natural science" made by Campbell and some others, and the belief among many that the work is rooted in geology, the great bulk of it resembles demography more, the study of patterns, and curve-fitting better suited to economists and other social scientists. In the next chapter, it will be shown that not only is the peak oil world devoid of hard science, but the soft or social science work being done is actually quite deficient.

ELEVEN

This Is Science? Poor Research Quality

Given that most people do not study the peak oil analysis in detail, they can hardly appreciate just how poor is its quality. A little of the work has in fact been published in refereed academic journals and at least one book by a university press, but much of it is aimed at a popular audience and large volumes of work have been self-published on the Internet.

Peak oil research is not just questionable, but riddled with problems and shortcomings. Several specifics, such as their ignorance of other research and imprecision, are discussed in this chapter, while some of the more technical methods and arguments were covered earlier.

GET THEE TO A LIBRARY

In fact, none of the peak oil theorists appears to have any background in resource economics, which is the primary field that seeks to explain the behavior of resource supply. Campbell (1997) goes so far as to make comments about economists not understanding depletion when it's clear that he is not familiar with the work on the subject, beginning with Jevons, Ricardo, and continuing on to such modern-day experts as Adelman, Gordon, Radetzki, and Tilton. These authors have discussed the impact of depletion and explained its effects mathematically far more elegantly than the simplistic discussions of the peak oil theorists such as Campbell.

But there is also work by geologists who attempt to model basins and field size distributions, which is all but completely ignored by the peak oil theorists, even though they are attempting to do the same thing. Only minimal reference is made to work by such as Arps and Roberts, let alone the modern work of Ahlbrandt and Klett, Drew, and others.[1] Possibly, they are seeking to avoid dealing with questions that they find awkward, that is, things that refute their primary methods or assumptions.

Nor has any of the work on oil supply forecasting been discussed, except to denigrate that done by economists. Yet the field is not only rich, but there is a body of work analyzing the forecasting of petroleum supply, including by this author.[2] Were they aware of this literature, they would realize that pessimistic forecasts of oil supply have been the norm for the past three and a half decades, and the cases they hail as supporting their expectations are in fact only more instances of a long line of Malthusian outlooks.

Indeed, as we saw in Chapter 1, James Schlesinger's declaration that the peak oil debate was over did not bring forth the recollection that he had made a similar warning three decades ago, and most peak oil advocates who list predictions of the peak, like Hirsch (2007), seem unaware that some, like Campbell, have made numerous predictions of earlier peaks that failed to come true.[3]

GET THEE TO THE INTERNET

Some specific instances of ignorance demonstrate clearly the lack of research. The most astonishing is Simmons noting that a Saudi Aramco official described using "fuzzy logic" in modeling the Ghawar reservoir, which he states made him suspicious that they were having technical problems, since he had never heard of fuzzy logic. (Maass repeats this in his *New York Times* article and book.[4]) Fuzzy logic is nothing more than a programming method developed in the 1960s, a fact that could have been easily checked on the Internet.

Similarly, Roberts describes his jaw dropping when he heard that the work cut at Ghawar was approximately 35%, when a Google search would have quickly shown him that this is in fact less than half the global norm, let alone for a 50-year old field.[5]

And much of the basic data go unnoticed (or ignored). Oil production data are readily available online, yet it took years for Campbell and Laherrere to admit that most nations' historical oil production does not follow a bell curve, and Goodstein is unaware that mineral production often fluctuates, stating it never recovers once it declines.[6] URR estimates by any recent number of experts are typically ignored by peak oil theorists, who focus only on earlier, more pessimistic estimates. And of course, none of the authors ever refutes the basic tenets of mineral economics theory, since none of them seems to be familiar with it.

In one amazing situation, Campbell actually proved to be unaware of articles published by this author, even though they twice appeared jointly with his work. When given him a copy of one such article by a reporter, he described it as "Stand by for more vitriol" a full year after it was published in the *Oil & Gas Journal* in tandem with his own article.[7] (Readers should not be surprised to learn that he has never addressed the substantive criticisms of peak oil theories, including his own work.)

CAN HE CHERRYPICK THE DATA?

One sign of poor research is a tendency toward what is called *cherry-picking* or the selective choice of information. Politicians regularly practice this, as do some lobbying groups, but it is considered unscientific, rather like a medical study claiming complete effectiveness by the artifice of not reporting negative results. When this happens, in fact, the practitioners come under severe and well-deserved criticism.

In fact, much of what is ignored appears to be willful, that is, peak oil theorists not only are refusing to respond to criticisms of their work (except with ad hominem attacks), but they simply do not mention data of which they must be aware. There is a huge literature describing methods of raising recovery in existing fields, but the peak oil theorists continue to insist that this is impossible, citing the occasional case of a field whose production has not been successfully increased—at least not yet. Laherrere (2006) is a good example; see Chapter 7.[8]

Similarly, sources such as the ASPO newsletter as well as the articles by peak oil theorists often cite data and references that are pessimistic about oil supply, while ignoring the optimistic ones. One case is the Cantarell field in Mexico, where statements about its impending decline are often cited as an example of problems in the oilfield, but no mention is made of the stories stating that the country expects to offset its decline, *if the budget is sufficient.*[9] A very important qualifier.

Audiences can also be cherrypicked, as when Matt Simmons addresses a group of foreign policy experts in Washington he includes numerous technical references like "dewpoint" that they would not understand.[10] Later that year, speaking to the Society of Petroleum Engineers, he did not mention any of his technical arguments about how low pressure causes fields to become inert, or superstraw technologies that he alleged was damaging fields.[11] Instead he made vague references to the challenges facing the industry.

In medical research, if a new drug helps a portion of the patients it is tested upon, it is considered effective, and the peak oil theorists such as Laherrere make a similar assumption, where they observe a pattern of behavior in some instances and rely on that as proving a theory. However, geophysics is not medicine—if a theory is often violated, it should be considered invalid or at least incomplete. Laws of viscosity or the strength of a diamond bit, for example, do not vary from nation to nation.

But Laherrere, in making arguments about the validity of his field size estimations, assumes that a trend observed in some fields proves that the trend is always correct and unchanging. He has shown numerous graphs with field production, all following a pattern where production settles into a specific trend and moves toward a point on the x-axis, where production would cease. He argues, effectively, that this is caused by geology and cannot be altered by, for example, new investment or technologies, and therefore yields precise field size estimates. Nowhere does he seem to mention that this pattern is not universal, and in fact, he earlier published graphs of fields whose production violates his rule, like Wilmington and East Texas, then later omits them.[12] Aleklett, at least, notes that some fields examined do not behave as predicted, though he is coy about how many or why.[13]

The Hirsch report is also illustrative in this regard, in that it ignores all the false reports of peak oil, including Campbell's work, and only publishes forecasts for a peak subsequent to the report.[14] The fact that the predictions have repeatedly proven false does inform us to some degree as to the credibility of the analyst, particularly where Hubbert's original success is often described as validating his theory, but no mention is made of the many erroneous predictions made by Hubbert, Campbell, and others. (And in the original report, my own comment that no peak was visible was included, but omitted in later publications, though possibly simply for space considerations.[15])

REFERENCES (OR NOT)

Although the book of collected writings by Campbell published in 2003 has an extensive bibliography, this is somewhat unusual.[16] Much of the work by peak oil advocates has no references or citations, including for some of the more important claims. For example, arguments that the size of fields such as Oseberg or Prudhoe Bay have not grown from their original estimates have been made without any reference giving to the sources of those original estimates. Deffeyes also publishes a number of graphs showing discoveries without providing a citation for the data.[17]

Similarly, Jean Laherrere often includes references to works or data that are not otherwise referenced. For example, he provides a graphic from a study by British Petroleum that he believes shows that United Kingdom does not experienced field growth, but there is no clear citation to the actual study.[18] This is presumably carelessness, something even I have fallen prey to it, but that occurs regularly in his work.[19]

And other work, existing only in the form of PowerPoint presentations such as those by Michael Rodgers and Michael Smith, naturally contain no citations.[20] Since they rely on estimates that involve disputed data, especially the ultimate recovery of oil from given countries, this is a glaring omission (see Chapter 9).

In the case of Laherrere, this possibly reflects the fact that much of his work has been copied and regurgitated many times, and he may have simply misplaced original references. But it is nearly impossible to tell precisely where he is getting data used in many of his papers, beyond the fact that he originally relied on what are now IHS Energy data for discovery sizes and presumably field production and new field wildcats. Some of the data cited more recently are coming from Wood-Mackenzie, which makes it even harder to know when data given are his own estimates using production curves, IHS Energy's data, or Wood-Mackenzie's.

In an e-mail to me, he excused this on the grounds that IHS Energy did not report their data sources, yet the company is not claiming to do refereed, academic-quality scientific research but generating a commercial database. Similarly, since Laherrere's entire research is premised on the superior quality of his data, it would seem that acknowledging the specific sources is vital.

IMPRECISION: VERBAL AND QUANTITATIVE

Aside from the lack of sources, there is a tendency to print figures that are all but unintelligible, either because of the numerous data points included or, more often, the small size and extreme coverage, often more than 100 years of data in a graph not much more than 2 inches square. Laherrere is the primary offender, although Deffeyes is guilty as well, while most other peak oil writers are notable by the absence of any research.

There is also a tendency to confuse comments by individuals with official positions. For example, when an industry newsletter states that some officials in Kuwait admitted that their reserves were exaggerated, the *ASPO Newsletter* referred to it as "Kuwait Admits to Exaggerating its Reserves" (ASPO 2/06). Any scholar would know better than to rely on a single individual's comment, particularly one appearing in the general press, as anything more than suggestive, and certainly would not imply that it represents an official position.

(The general press often commits this error as well, referring to research by an individual as being by the institution where that person resides.)

Interpol has been instructed to *pay attention* to piracy of this video content.

Peak oil theorists frequently argue that there is broad agreement with their work within the petroleum industry, but an examination of the actual comments by industry officials shows they are often misrepresented. When the Shell CEO says "easy oil has probably passed its peak" this is reported as "Shell follows Chevrons leads in admitting to Peak Oil in as many words". (ASPO 2/06) Chevron has a public relations campaign that talks about the difficulty of finding more oil (www.willyoujoinus.com), that Campbell converts to "Chevron Admits to Peak Oil and Depletion" (ASPO NL August 2005), and he says "Exxon Accepts Peak Oil" (ASPO NL July 2005), simply because an executive used a graph of discovery trends. Contrast that with the *Wall Street Journal* article, "Producers Move to Debunk Gloomy 'Peak Oil' Forecasts," about a conference featuring the major oil companies that was specifically intended to dispute the claims about peak oil.[21]

This has led some, such as Hirsch, to mistakenly think that the industry generally believed in the peak oil arguments when, in fact, they have largely dismissed them as foolish. Some peak oil theorists pressured the Department of Energy to look into the matter, and it asked the National Petroleum Council to consider it (along with other challenges facing the oil industry), and it listened to both sides of the peak oil debate (including me) before dismissing it in a couple of pages.[22] (The report's primary conclusion: it's hard to produce oil and getting harder. Send help. Shocking, I know.)

Another instance occurred during a debate over the question of peak oil. As presented on "The Oil Drum" by Jeremy Leggett, the poll of the audience gave him a clear victory: "The result seems to suggest that the rank-and-file practitioners hold a very different view of peak oil from the BP/Shell/Exxon etc. top tables." And what did the poll involve? Whether or not peak oil was "a concern."[23] Surely, if astrophysicists were polled as to whether an asteroid collision with the Earth were "a concern" there would be broad agreement; but asked if they expected one within the next decade, the answer would be very different.

Even a casual reading of the peak oil articles and speeches should raise red flags given the heavy reliance on vague and imprecise terms. References to the end of "easy" oil, or Saudi oil fields being "tired," or Burgan in Kuwait being "exhausted" or the need of the industry to run faster to stay in place, are effectively meaningless.[24] What precisely is easy oil? Do "tired" fields require 5% more electricity each year to maintain production, or 75% more water flooding? A billion dollar gas injection program? Perhaps a double espresso? Will production drop by 5% per year, 8% per year, 1% per year or simply end when a field is "exhausted"? The terms used provide no insight.

And even in the published (often Internet) graphical work, such as that for the Netherlands below, the scale of 80 years and more tends to reduce even fairly major changes and events to such a small scale (the original graph was about three inches square) that it is impossible to know what precisely is occurring. In the case of many of the figures published for national production forecasts (Campbell 2002), it is also difficult to tell what are historical data and where the forecast begins.

Similarly, a number of graphs are produced—again primarily by Laherrere and Deffeyes—where data points are apparently used to demonstrate a pattern (i.e., fit a curve), but it is never clear when they are being eyeballed and when analyzed using statistical methods, such as regression analysis.

In looking at recovery factors, Laherrere simply shows a cloud of data points and a curve in their midst, with an enormous dispersion. And in fact, by erasing the curve created by Laherrere and drawing a simple line through the middle, we get something that appears to have a roughly similar level of correlation. (Without the actual data, it is impossible to do serious statistical work.) Aleklett published a similar graph produced by Meling, showing recovery rate by size, actual (29%) and predicted (38%); the source is a PowerPoint presentation with no date or sources.[25]

And also note the lack of a time component to the graph. If field size estimates were stable, it wouldn't matter, but this is one of the major controversies in the field, and, as Chapter 7 shows, the vast majority of analysts disagree with the contention that field size estimates are robust and do not evolve. This means that size could be a determining, but omitted, variable explaining recovery factor.

And while Campbell produces a summary table, showing national resources and peak dates, he does not explain the degree to which they have been produced by, or in cooperation with, Jean Lahererre or, again, by what methods. Especially since they have subsequently denounced IHS Energy data as unreliable, the extent to which they continue to rely on its data would be useful to know.

WORLD FAMOUS IN POLAND

Although many argue that the media and the oil industry have been trying to cover up peak oil (Chapter 11), the reality is that it has received attention—and been found wanting. Indeed, some peak oil advocates wax eloquently about the phone calls they have received or the talks they have given, as if this was proof of their stature and the reliability of their arguments. Aleklett says, "I received a phone call that was, to me, astonishing. A lady introduced herself and told me that she worked for . . . Sweden's Military Intelligence and Security Service."[26]

And peak oil theorists often note that they have spoken at prominent conferences, to organizations like OPEC and the IEA, and been contacted by the CIA, among others, implying that their work had credence. And it does mean that their work had enough credence to at least be examined by those groups, but the majority of them—at least those with experience in energy—subsequently ignored the work.

It's impossible to say what, precisely, led to any given person or organization to decide to issue an invitation, but it is easy to see that most would feel an obligation to consider the theories and arguments. That said, it is worth noting that OPEC, the IEA, and others have not made it a practice of listening to peak oil advocates on a regular basis, but typically held one hearing to establish whether the theories are worth considering.

Amazingly, the oil industry is regularly said to believe in peak oil. Heinberg quotes a 1999 Goldman Sachs report that says, "The great merger mania is nothing more than the scaling down of a dying industry in recognition that 90% of global conventional oil has already been found."[27]

HOW HARD IS IT?

Rubin dedicates an entire chapter to the changing oil supply picture, with his main argument being that oil companies "have their hands between the cushions" looking for new oil, since all the easily recoverable oil is either gone or continues to be depleted.[28]

The most common evasion among peak oil advocates is the claim that "cheap" and "easy" oil is gone, that the remaining oil is "difficult" or "expensive." This will be shown to be largely fallacious in Chapter 18, but it is worth mentioning here because it evades the real question: when did the cheap oil go? What is the difference between cheap and expensive oil? Why was it so sudden? (if considering high prices). Doesn't that imply that other causes explain the price increases?

Serious scholars would try to quantify these trends, not simply divide them into two categories, like black and white swans. Adelman calculates changes in the cost of U.S. oil production over the course of decades, and elsewhere, shows estimates of global production costs for four different years, covering three decades.[29] Recently, IHS-CERA has begun putting out a very explicit estimate of cost inflation in the oil industry.

Simmons deserves special attention here for his word usage, although a more thorough consideration of the Saudi situation is presented in Chapter 8.

When my daughter was little, I taught her to say that "My Daddy is old and tired," in the appropriate circumstances. Oddly enough, these words are used to describe some oil fields, such as Ghawar. However, the precise meaning of these words is never made clear: Presumably old in terms of producing years, since all oil fields are "old" in geological terms.

But this is all but irrelevant. Prudhoe Bay and Forties are "young" compared to Ghawar, for example, both having been discovered two decades after the supergiant Saudi field, yet their decline began within a few years after production commenced, and Ghawar shows no sign of decline as of yet despite the "Ghawar is dying" comment of various peak oil advocates, including a bumper sticker to that effect.[30]

What would be of interest is how much of the oil-in-place remains, what level of capacity is installed, and what the operators plan to produce. Prudhoe Bay still has about half of its oil-in-place remaining, but without much greater investment, production will continue to decline. Ghawar, on the other hand, is constrained more by the decision of the operator to produce gradually over a long period; production could theoretically be expanded but at the expense of future production levels.

In his 2005 book, Simmons described many of the oil fields in Saudi Arabia, apparently drawing on SPE papers written by Aramco engineers, and saw no chance of production increasing. The problems that would prevent this are not clear, nor are the motivations. In a lengthy review of the world's largest field, Ghawar, he said, for example, that "the discovery of this complex faulting ... must have been mind-bending." (p. 174). Intense study of the field led to a picture that was "not pretty" (p. 175). And when describing Saudi Aramco efforts to produce 300 thousand barrels per day for 30 years from one zone, he describes this as "naïve" (based on his experience as an investment banker) (p. 177). Ghawar's productivity is described as "fragile" (p. 178).

He speculates that the failure to develop Khursaniyah, Abu Hadriyah, Fadhili earlier is surprising because of the need to reduce the "strenuous burden" on the bigger fields (p. 222). It's not clear what either "strenuous" or "burdens" means, and how they would differ from normal operations. No reason is given to suggest that there are even minor increased efforts involved in maintaining production in those fields.

HOW INFINITE IS IT?

The word infinite comes into play occasionally, sometimes from optimists but most often from peak oil advocates. It has a very definite mathematical meaning, but in this case, the imprecision comes from the modifier that is sometimes added—or omitted. For instance, Campbell (1997, p. 68) references a geologist's work that relies on what he calls a "near 'infinite' resources, quoted at 4 trillion barrels." The use of the word "near" is appropriate in a sense: he avoids saying infinite, which no precise number could be. At the same time, as Chapter 15 shows, 4 trillion is not even an unreasonable amount, and certainly nowhere "near" being infinite.

There is a counter argument: in Adelman and Lynch (1997) we said that petroleum resources were "effectively infinite." The qualifier might be considered a weasel word in the sense that it was hardly precise, yet it also has a specific meaning: the resource is so large that it is not a constraint in any relevant time frame. "Relevant" then becomes the debating point.

QUALIFIERS

All of these stories were full of the words "could" and "might. "

Michael C. Ruppert[31]

This brings us to the importance of words that qualify a statement. It can be very frustrating when an analyst says, for example, oil prices "could" drop or that new supplies "might" make a difference.

This underlines a major difference between peak oil advocates and most other analysts: a degree of certainty where no certainty is possible. Given that no mineral or energy resource, no nonrenewable resource, has ever become scarce, how can so many be so certain that the peak is imminent (or past) for petroleum?

Looking back, a certain amount of precision is to be expected: how much did discovery size decline, to what extent has drilling become more efficient, and so forth? But when discussing a new discovery (the subject of Ruppert's complaint) it only makes sense to qualify the likely outcome. And the politics of oil, or the trend in prices, is uncertain even at the best of times, so analysts can be excused a degree of caution.

I know a Guy

Interestingly, while Colin Campbell and Jean Laherrere attack government data on oil reserves as unreliable and "political," or deride optimistic announcements about discoveries, prospects or production plans, they have no such compunction when citing government officials' comments about problems in the oil fields. In any number of instances, Laherrere pulls out references to field decline rates or national decline rates and takes them as carved in stone, completely reliable and unalterable.

Because so much peak oil "content" appears on the Internet, it becomes very common for casual references to make their way into the debate. Thus, when the question arose concerning how much exploration had been carried out in Saudi Arabia, several weighed in with comments like, "In the 80s I actually talked to some of the guys doing the seismic exploration in the Rub Al-Khali. It was, in their words, the last unexplored area of Saudi Arabia. But no more, it has now been thoroughly explored."[32]

This is interesting, since there are actual data, in terms of both the number of wells being drilled and a map showing their locations.[33] And virtually no one who isn't a peak oil advocate has such a pessimistic view of Saudi resources.

It is also amazing that peak oil advocates will often dismiss an institution, then laud one of its members for their affiliation. Campbell (1997) has a chapter called "Economists always get it wrong," but later praises economists who agree with him. Several peak oil advocates have disagreed sharply with the 2000 USGS assessment of world petroleum resources, but then praised Les Magnoon, who talks of the "Big Rollover."[34]

One amazing instance is the reliance on the work of Sadad Al-Husseini, a prominent former official of Saudi Aramco. He first addressed this issue by completely contradicting the arguments Matt Simmons made about Saudi Arabia's oil situation.[35] Subsequently, however, he produced his own writings on global oil resources and has often been cited by peak oil advocates as holding extremely alarmist beliefs about world oil, who claim he is reluctant to say these publicly.[36] It is not clear how seriously that was meant, but if in jest, it still does not explain why he is taken seriously on the issue of non-Saudi oil, where he has no special expertise, but his thoughts on Saudi oil are ignored.

Jargon

Pardon me, venerable venator; as classification is the very soul of the Natural Sciences, the animal or vegetable, must, of necessity, be characterized by the peculiarities of its species, which is always indicated by the name.

James Fenimore Cooper, *The Plain*

Since many reporters are not familiar with the technical jargon of the oil industry, they commonly confuse terms such as reserves and resources, exploration and development, and depletion and decline. But also they may make the mistake of ignoring the ceteris paribus nature of a comment: a field *will decline without further investment* (or tax breaks) becomes transformed into a field *must* decline.

It is also the case that any number of complaints about the difficulty of operations can be found among current and former employees of any industry, but these are usually treated as golden by peak oil theorists (and sometimes reporters).[37]

Creaming Curves "Valid, Robust"

Creaming curves have been long estimated and used by geologists, but primarily focusing on areas like Alberta and Texas, where there is a very good, lengthy record of discoveries.

But the shortcomings of the method have long been recognized by geologists. Drew (1997) noted that the method had performed well in predicting the number of discoveries in the shallow Gulf of Mexico, but not the size, because data on size were unreliable, as discussed in Chapter 7.[38] More recently, Ahlbrandt and Klett examined the creaming curves for the Neuquen Province in Argentina and noted that, despite the maturity of the play, the curve shifted upward approximately 40% in only five years, suggesting that the creaming curve method is not very robust.[39] No estimate produced by it can be treated as final.

And the method is typically applied only to a specific geological play, recognizing that each play has a unique curve with different sizes of fields. Combining the basins in Saudi Arabia's Eastern Province with those in Oman or Syria, as Laherrere does when he reports a "Middle Eastern" creaming curve, simply doesn't work because the curve is heavily influenced by various political decisions, not just geology.

More important, Laherrere doesn't seem to understand that the fact that the creaming curves give different estimates at different points in time suggests that there is some missing element to them. Statisticians would consider this as evidence that the method is not "robust" meaning it cannot be relied upon to deliver reliable results. As a geologist, Laherrere apparently has no experience with statistical methods or tests of validity.

Discoveries

The constant concern about falling discoveries is almost entirely based on a misunderstanding of what the term means. While its usage is common and the meaning seems perfectly obvious, the nature of underground resources translates into uncertainty about the size of any given discovery, and the industry copes by being conservative in its estimates, but also treating "discovery" as a technical term referring to the amount thought to be in the field, and recoverable with existing technology. To treat the announced size of a discovery as representing the ultimate recovery that will be achieved is a very basic mistake.

Yet this mistake is actually unusually common, even in the industry, with people like T. Boone Pickens not seeming to understand it.[40] Others, like Colin Campbell, have tried to work around it by arguing that recovery factors don't increase, which is not only false, but that he also implicitly acknowledges this elsewhere in his work (Chapter 7). Not surprising, those who replicate the curve of historical discoveries do so naively thinking it is the final answer, given its repetition by so many.[41]

Being a novice, Michael C. Ruppert falls prey to this type of mistake, creating his own definition of "discovery estimates" and "total reserve estimates."[42] He apparently doesn't know that both represent estimates of the *recoverable* reserves, and there is no such thing as the latter, since they are continually revised.

RESOURCES: CONFUSING RECOVERABLE WITH TOTAL

Again, the industry (and especially petroleum geologists) have a specific technical definition in mind when they use the term resources, particularly when talking about potential global resources (see Chapter 17). As in the case with the term *discoveries*, they do not usually attempt to estimate the entire amount, but rather the *recoverable* amount. This is explicit in the term *URR*, or *ultimately recoverable resource,* but many peak oil advocates use an alternate term *EUR*, or *estimated ultimate resource,* which mistakenly omits the qualifier "recoverable."

The primary effect is to disguise the way estimates of URR have increased over time, from less than a trillion barrels to nearly four trillion now, which highlights a crucial mistake made by peak oil advocates: arguing resource estimates, by field, country or globally, don't change.

Conventional Oil

When I began working on petroleum economics in the late 1970s, the term *conventional* was often used interchangeably with "recoverable." If oil was not recoverable, it was unconventional. This included "deepwater" fields and enhanced oil recovery, meaning the more exotic techniques like CO_2 floods, which were then quite expensive.

Now, the term *conventional* is typically used to refer to oil that has a specific chemistry and was produced with conventional/historical methods from a noncontinuous deposit. An oil field that is too small to produce at $25 a barrel, for example, would not be considered unconventional, merely unrecoverable. (Note that Jean Laherrere makes a similar point in Campbell 1997, saying that he means oil that can be produced at $25 a barrel).[43]

Campbell, however, takes a more restrictive view of conventional, by excluding "Arctic" oil (by which he presumably means deposits above the Arctic circle), for example, even where the fields are what all others would categorize as "conventional" fields, like Prudhoe Bay.

His categories of unconventional oil include the following:[44]

- Oil from shale
- Oil from tar sands
- Heavy oil
- Oil from enhanced recovery
- Oil from infill drilling
- Oil in very hostile environments
- Oil in very small accumulations

Aside from the vagueness of some terms (I, personally, would consider East Texas a "very hostile environment" but not everyone would agree), he includes resources that no one else ever categorizes as unconventional, specifically oil in small accumulations and infill drilling. The latter is an interesting case, since throughout the history of the industry, infill drilling has been a common practice and the source of significant amounts of production. It is also one source of reserve growth, the nemesis of Campbell and Laherrere, which might be why he attempts to exclude it from his analysis.

Real Oil: Crude plus Condensate

As discussed in Chapter 12, a number of peak oil advocates have seized on crude plus condensate as the important supply, not petroleum or total liquids. Amazingly, many of them seem to think that this is normal, when in fact, the distinction is largely unknown within the industry.

Shale or Not?

Quite a number of peak oil advocates have argued that recent optimistic forecasts of shale oil production are flawed because kerogen is expensive and has a gluttonous need for water and energy. This is rather bizarre, as one of the primary topics of interest in the oil industry at present is the booming production of conventional oil from shale deposits. This is not kerogen and is in fact quite cheap (see Chapter 19) and requires substantially less energy than kerogen production needs. While the duplicate use of the word *shale* to refer to kerogen and "tight oil" is confusing, no one with any serious knowledge of the industry would make this mistake.

Hirsch et al. (2010) is one of the most surprising to make this mistake, given his background in the industry.[45] And Aleklett, whose book appeared in 2012, surely should have noticed the boom in the United States, which began roughly in 2008. On theoildrum.com in March 2013, criticism of peak oil caused several bloggers to raise the issue of shale oil, which six of the respondents clearly mistook for kerogen, and only one actually mentioned the current development of shale oil.[46]

CARELESSNESS AND MISREPRESENTATIONS

Some of the errors found in the peak oil publications are rather startling. For example, David Goodstein states bluntly that once mineral production has started to decline it cannot recover.[47] Yet world oil production dropped sharply in the early 1980s, only to recover. The same has been true of U.S. natural gas production, oil production in Russia, the United Kingdom, Venezuela and many other countries and minerals. A peak and permanent decline does occur sometimes, but it is hardly the rule— *nor is it necessarily forced by geology.*

And the very fact that the great majority of countries do not show a bell curve production, but so many of the peak oil theorists still maintain that its production *must* follow such a trend, is rather astonishing.

Similarly, much of the writings have referred to the fact that (some) peak oil theorists are geologists, while those who disagree with them are (flat earth) economists. This is clearly untrue, as any number of geologists, prominent and otherwise, have made it clear that they disagree with the peak oil theorists' methods and conclusions, including William Fisher, Michael Halbouty, John Edwards, Peter McCabe, Marlon Downey, and Thomas Ahlbrandt. In fact, the peak oil theorists are much generally less professionally prominent than their geologist opponents.

There are also far too many cases of statements that misrepresent the positions of others. In the case of reserve growth, peak oil theorists have repeatedly attacked the USGS for estimating reserve growth and pointing to the fact that earlier studies, headed by Charles Masters, did not do so.

Furthermore, the present USGS study contradicts Masters in the previous studies who rightly rejected significant field growth outside the USA.

Jean Laherrere[48]

Yet Masters and his group openly acknowledged reserve growth and even attributed the mid-1990s increase in Middle East reserves to this phenomenon.

In fact, however, we believe the fully grown reserves concept has been captured for only about three quarters of the reported crude oil reserves (US, USSR, and Gulf OPEC). For the remainder of the non-OPEC countries, when considering possible future production capabilities (see below), statistical growth additions, as calculated fractions of the US oil field growth experience, are introduced to compensate for the field growth omissions in the tables.

Masters et al. 1990[49]

Not only that, but where the USGS described the amount of oil that was technically available to the year 2050, peak oil theorists from Campbell to Skrebowski have described this as predicting a pattern of discovery. A graphical representation of the supposed prediction by USGS has been widely duplicated by writers who apparently do not realize the source is not the USGS or that it misrepresents their research.

Rookie Mistakes

Needless to say, there are numerous areas of disagreement about future oil supply, even among those who are not peak oil advocates, and that is hardly evidence that any particular piece of research or publication is invalid. But peak oil writings stand out for the frequency of basic mistakes that are made.

This is evidenced by many mistakes that would not be made by someone familiar with the subject. Generally, most either don't know the history of oil supply forecasting, don't understand the oil industry, are unfamiliar with economics, and sometimes even are largely ignorant of peak oil research.

Internet comments are naturally the biggest source of such errors, presumably because writers don't feel obligated to exercise the kind of caution that the author of an academic paper or even a magazine article does.

Errors of Fact

Peak oil advocates make an enormous number of errors because many of them are unfamiliar with the history and economics of the oil industry, petroleum geology and reservoir engineering, or supply forecasting, but also because most of them are relying on other writers and not cross-checking the facts. David Goodstein, for instance, mentions that once mineral production declines, it never recovers, which is both absurd and easily refuted.

Simmons (2005) expresses puzzlement that production at Qatif dropped sharply in 1982, when he says "Saudi Arabia was straining to keep its total oil production at all time highs" (p. 216). The reality: the market was collapsing and Saudi Arabia cut production by 7.5 million barrels per day to keep the price from dropping. Similarly, he expresses concern (p. 153) at the drop in Ghawar's production from 1997 to 1999 along with a rising water cut, apparently forgetting the 1998 oil price collapse.

And many have repeated the claim that OPEC used reserves to set quotas, resulting in the apparent inflation of reserves in the late 1980s. In fact, OPEC was considering such a move, but never actually implemented it, largely because of realization that using data for reserves (as well as production capacity, population, etc.) would leave the "scientific" quotas open to manipulation. This, in particular, is a picayune complaint, but it does serve to highlight the way so many writing on the subject are not knowledgeable and blithely repeat what other peak oil advocates have written.

Didn't Know

My first estimate in 1989 was 1.65 Gb, but I did not then appreciate the industry practice of understating the initial reserves of a discovery.[50]

C. J. Campbell (1997), 93

"[Campbell and Laherrere] have shown that such reserve growth is largely illusory and is derived partly from unverified and inflated reserve reports of OPEC countries."

Heinberg[51]

As discussed at length, the question of reserve growth is an important element of the mistakes of the peak oil advocates who believe that the resource constraints are determining, but this is hardly a new issue. Adelman discussed the matter in many places, and it has appeared in a variety of other analytical sources (Drew, Ahlbrandt, and Klett). But for someone in the business of petroleum geology for four decades, it is somewhat incredible that Campbell would have been unaware that initial reserve estimates were regularly revised and typically upgraded.

Ruppert (2009) is not surprising in his repetition of "Contrary to popular belief, not all oil is recoverable" (p. 23). Simmons treats his ignorance of "fuzzy logic" as evidence of technical problems in the Saudi oil fields, and others, such as Maass, repeat it, apparently unaware that it was nothing more than a programming language (see Chapter 8). Roberts says, "The hairs on the back of my neck stood up. Ghawar's water injections were hardly news, but a 30% water cut, if true, was startling. Most new oil fields produce almost pure oil or oil mixed with natural gas—with little water."[52]

Simmons also projects his own ignorance on others, saying that "I began to finally grasp that all these supply experts were forgetting about depletion!"[53] As Chapter 14 will discuss, depletion was well known to the experts; he misunderstood an optimistic forecast for one that was excluding depletion.

Ignorance

In the days of the Internet, many researchers have found themselves with a historical blind spot: if it's not downloadable, it might as well not exist. Reports written before 1990 are almost never on the Internet, although many newspaper and academic articles are now, and the U.S. Department of Energy has scanned quite a number of older documents, but few peak oil advocates seem to have any significant library of pre-Internet texts.

This comes through in a number of publications, such as a recent paper by researchers at the IMF. It shows a number of forecasts—all postdating 2000.[54] Similarly, many peak oil theorists refer to the IEA has having an optimistic bias about oil supply, which is far from correct. In terms of long-term forecasts, the IEA has tended to be too pessimistic; only after the 1998 oil price collapse and the change in non-OPEC production trends, did their short-term expectations prove too optimistic.

Similarly, Aleklett makes a number of claims that reflect his ignorance of the literature, including the claim that he and Campbell published the first peer-reviewed article on peak oil in 2003, ignoring Edwards (1996) and Bentley et al. (2000). (Roland Watson, apparently an investment analyst, not only mistakenly claims that Campbell's first book was published in 1997, he gets Deffeyes's name wrong.)[55]

Simmons displays his serious ignorance about geology when he states that "it would be natural to assume that most, if not all, of the great field's reservoir properties were now thoroughly understood. This is clearly not the case" (p. 159). Actually, the reservoir is what we would call "underground" and consists of matter that would be considered "opaque" or "not transparent". Looking at seismic is very different from looking directly at rock, and the study of a field that is 380 by 20 kilometers in size has hardly been done to the point where every fracture, every zone, has been identified.[56] Older fields like Kern River are moderately well understood, but constant study is being undertaken to try to improve the knowledge of the geology.

A case that confirms several of the weaknesses of many of the peak oil advocates publications revolves around the 1997 debate held at the IEA. Many believe that Campbell and Laherrere won the debate, and the IEA "rejected Adelman and Lynch's arguments."[57] This is based entirely on Richard Duncan's claim that "this represents a significant reversal of the IEA position: 'This is a real stand-down for them because until recently they were in the Julian Simon no-limits camp.' "[58]

Two problems: none of the peak oil advocates who accept this conclusion have ever checked it. Instead, all are repeating Duncan's claim and, when Steve Andrews asked online for support for the claim, the only response was Laherrere quoting Duncan.[59] On top of that, the IEA was not "in the Julian Simon no-limits camp," but had instead been too pessimistic about non-OPEC supply (as were most forecasters), as described in Chapter 2. The primary change was moving some of the projected supply increase into an explicit "unconventional" category. And by the publication of the 2000 *World Energy Outlook,* the IEA was again forecasting that conventional oil production would grow from 75 to 110 million barrels per day in 2020.

> If you haven't seen it, it's new to you.
>
> NBC promo for reruns

> [M]ost energy experts have little idea that the overproduction of Saudi Arabian oilfields had ever been discussed.
>
> Simmons[60]

Simmons must be considered the champion, though. Aside from making the claim that "we didn't know about depletion," he talks about the "secret" U.S. studies about Saudi oil in the 1970s, studies published by the U.S. government, one of which was reprinted *in Petroleum Intelligence Weekly*. His comment, quoted above, apparently reflects the fact that *he was unaware*—I read them as a young researcher.

Further, he not only describes a *New York Times* story on the subject, but actually mentions later in the book that an assistant found one study in the University of Houston library. (Presumably, he originally described the studies as secret and later found them, but didn't edit out the earlier remark before publication.)

And Aleklett, in his book and some lectures, informs audiences that oil doesn't flow from reservoirs like water from a bottle, but rather seeps out as water from a sponge. Assuredly, many lay audiences probably are unaware of this (one local DJ has asked me directly about it), but it is hardly novel to those of us in the field.

Ruppert is fantastic in claiming, in his 2009 book *Collapse*, that Simmons's book on Saudi oil was not only "the premier desk reference book" on Saudi reserves but states bluntly that it "hasn't been challenged." Given that a number of writers, including myself, have pointed to the many mistakes and shortcomings in his book, it would seem that Ruppert has not followed the field of peak oil work (despite writing about it) or is so tightly wrapped in the peak oil circle that he is unaware of any discord.[61]

Yes, We Have No Oil Data

Perhaps because of the emphasis by Campbell and Laherrere to their access to the superior database of Petroconsultants (now part of IHS), and Simmons' complaints about lack of field and well data for Saudi Arabia, there has been very little attempt by peak oil advocates to find other sources of data. As a result, many of the secondary writers seem to assume that there is no such data.

Granted, the data that are available are often incomplete and poor, as Chapter 18 discusses, but there are certainly ample data that can be illuminating—and most of which contradicts the claims of peak oil advocates.

One of the more amusing omissions is in Matthew Simmons's book on Saudi oil production. After some criticism that the draft didn't contain much research, the published version of the book included an appendix that listed many of the Saudi oil fields, production data, discovery date, and so forth. None of this data, however, supported his thesis that they were experiencing problems maintaining production.

POWER OF THE PRESS: BIG DISCOVERIES

Mexican President Vicente Fox has announced the discovery of a new deep-water oil field, which is believed to contain 10bn barrels of crude.[62]

Another mistake common with lay researchers is to think that a given quote represents proof, or at least strong evidence. No experienced oil industry analyst would take an offhand remark by a government minister as more than suggestive. If the president of, say, Mexico announces a major oil discovery, experienced observers take it with a large dose of salt, waiting until the explorationist makes a clear statement. In the case cited above, not only were the results preliminary, the well had actually found a gas field.

Similarly, when Chevron announced in 2006 that it had made a large oil discovery in the deepwater Gulf of Mexico, early reports were that it might contain as much as 15 billion barrels of petroleum. The CBS headline used the word *reserves* inappropriately, but the story did not make that mistake, leaving readers confused no doubt.[63] A careful reading of the CBS story would show that the area where the discovery was made was thought to hold up to 15 billion barrels, not the field that was found, and follow-up reports confirmed this, meaning the size of the discovery was much smaller and the 15 billion barrel estimate was conjecture, similar to USGS estimates of undiscovered oil.

The opposite end of the spectrum can be seen in Ruppert's dismissal of announcements about giant discoveries offshore Brazil, noting press claims of a 33 billion barrel field that were criticized as being premature. In fact, the field, now called Lapa, probably only contains a billion barrels or so, but the Santos basin appears likely to contain more than 33 billion barrels of oil.[64]

POLITICAL SCIENCE (AS OPPOSED TO POLITICAL SCIENCE)

Many people are well versed in politics and as such consider themselves as knowledgeable as the typical political scientists. Because politics is so uncertain—given the human element and huge number of variables—predictions are nearly always of limited use, and numerous interpretations of any given event can be made. The temptation to read political meaning into various actions has, over the centuries, been more tempting than the Sirens of Ulysses.

One of the worst examples in the petroleum business has been the case of Peter Schweizer's book *Victory,* which hypothesizes that President Reagan asked the Saudis to bring the price of oil down to put pressure on the Soviet Union.[65] They raised production in late 1985, and the price promptly fell by about 40%, causing significant economic distress in a number of places including Russia.

This has become a favorite among some oil industry circles, because they like to believe that the price of oil wouldn't have collapsed without some political intervention, and especially among some U.S. independent companies, there seems to be a desire for victimhood. (Others have blamed the futures traders on Nymex for deliberately damaging the industry.)

Campbell gives credibility to this thesis, with a section in his 1997 book analyzing the intersection of political and oil market developments, including suggesting that the UN embargo against Saddam Hussein's Iraq was partly to placate the Saudis, by keeping competing oil off the market.[66]

The 2014 oil price drop led to a number of similar conspiracy theories, some more credible than others, such as the Saudis wanting to deter investment in competing oil supplies like U.S. shale oil. However, others have pointed to: the United States wanting to punish Russia for its incursion into Ukraine, the United States, and Saudis wanting to weaken Iran, or the United States wanting to hurt Venezuela's government.

A Rig is a Rig is a Rig—but Some Are Oil and Some Are Gas

Experienced oil industry analysts understand that oil and gas are completely different markets because the cost of transporting and distributing natural gas is extremely high compared to that for oil. As a result, it is not uncommon for nations to have large amounts of oil, but little gas—and vice versa. Kuwait, for example, has long been short of natural gas and has had to import it when possible, but rely on oil instead at times. Saudi Arabia is in a similar unenviable situation.

Thus, we have novices like Ruppert point to a huge (120%!) increase in rigs operating in Saudi Arabia in 2006, without realizing that most of them were intended to find gas so that the country could save the oil it was using to generate electricity.[67] (If you've ever been in Dhahran in the summer, you will not begrudge them one watt-hour for air conditioning.)

UNSCIENTIFIC RESEARCH

Certain aspects of the work by peak oil theorists make it possible to argue that they violate the norms of science to the point that their work is more demagoguery than research (including Colin Campbell's frequent use of religious terminology—priests, heretics, etc.). One is the tendency toward contradictory arguments, depending on the point to be made. Another is the manner in which many conclusions are unsupported by evidence, merely asserted following statements of—often unrelated—facts.

Logical inconsistency

To obtain a valid discovery trend, we need both valid numbers and valid dates. This information cannot be obtained from public data but is held in industry databases, such as that maintained by IHS (formerly Petroconsultants) in Geneva.

Campbell 2004

The problem is that now scout companies do not want to upset the national oil companies, (NOCs) which are their new clients when many international oil companies IOCs have disappeared. So now, scout companies accept NOC political values and lose reliability.

Laherrere (2006), 12 (NB: IHS Energy is
the scout company referred to.)

The poor quality of the research by peak oil theorists is exemplified by a tendency to make statements lacking in connection to earlier work. A wonderful example occurred recently, when Kjell Aleklett criticized the report by CERA for the fact that it did not publish its data. This criticism was reprinted in the *ASPO Newsletter*, edited by Colin Campbell, *who himself has not only repeatedly bragged about his reliance on a proprietary database but the very same database.*

Another issue concerns the controversy over reserve growth. Campbell and Laherrere have insisted that *reserves should be backdated*, booking all revisions to the year in which the field was initially discovered. By doing this, discoveries appear to have been much higher in the past than in recent years. By showing reserve additions, that is discoveries plus net revisions, the rate of success appears much more level and it becomes obvious that there is no recent drop-off.

But the very act of backdating reserves refers to *a correction for later revision,* that is, *field growth*, and Campbell and Laherrere insist that there will be no such future revisions. In effect, they are claiming that prior to the point where they performed their research, reserve growth occurred; once they did their research, reserve growth ceased.

So, on the one hand, Campbell and Laherrere attack economists for not backdating field revisions to the year of discovery, but then claim that the snapshot data they used, which included such revisions, would no longer be revised in the future and insist that the IHS database enabled them to make robust resource estimates, but now claim that IHS is untrustworthy—after the companies' experts contradicted both their claims about its database and their conclusions.

And the recent turn against IHS is informative, if not amusing. Having bragged about their access to the Petroconsultants database, which they claimed was highly reliable, once IHS made it clear that they disagreed with both the assumptions and conclusions of the peak oil theorists, they began to accuse it of having corrupted their database to suit their clients' needs. For example, they claim that the high Saudi oil reserve estimate was made to satisfy Saudi Aramco, which had become a client of theirs (Laherrere, 2006). Not surprisingly, no evidence is provided for this or indeed any of the other ad hominem attacks on their critics.

Simmons, for his part, makes a number of fairly odd remarks, including references to secret government reports *that he found in university libraries*. He also remarks on the fact that claims of problems in Saudi oil fields were ignored in the past, while citing newspaper stories in major publications. Most odd is his heavy reliance on the claim that a drop in field pressure below the bubble point will leave the field inert; yet he cites studies claiming that precisely that happened in the early 1970s—*without major oilfield damage occurring*.

But How Does It Work in Theory?

Another contradiction is the habit of claiming that the Hubbert method must be valid because Hubbert correctly predicted the peak in U.S. oil production. And certainly, a good forecast is evidence that must be considered, but peak oil theorists tend to ignore the fact that their own track record has been abysmal, whereas others, such as Robert Esser and this author, have much better track records (see Chapter 9).

Yet there is little mention of their track record. In his 1997 book, Campbell refers to his 1989 article, without noting that he called for peak oil that year, let alone explaining the egregious error. Indeed, peak oil advocates often refer to other publications without noting the errors, or mentioning bad forecasts, avoiding the subject altogether. For example, Deffeyes has praised Simmons work, but doesn't mention the technical mistakes. Hirsch (2005) cites my papers critiquing the work of Campbell and Laherrere without mentioning their arguments.[68]

You Can't Get There from Here: Asserted Conclusions

As Robert Skinner has noted, "Observation is not analysis," but all too often, peak oil publications are filled with observations masquerading as analysis.

These include:

- Most of the world's oil comes from a handful of producing fields.
- The Middle East holds most of the world's oil reserves.
- The industry has to run faster to stay in place.

One thing that seems to have gone unnoticed among the debate is the degree to which many conclusions are not supported by the facts presented. In a number of cases, peak oil theorists present specific facts without explaining what precisely their relevance is to the debate or the industry itself, often proceeding directly to the conclusions. In a quantitative sense, the worst example is Laherrere's tendency to show discovery of giant fields and a production curve without explaining how he connected the two other than putting them on the same figure.

Qualitatively, Simmons is the worst offender, noting conditions in the Saudi oil fields and concluding that a peak is near, without explaining the connection or the causality. Are these problems different from those experienced in other fields, other countries? Are they insurmountable? How are they different from problems and challenges that have arisen historically, in Saudi Arabia and other areas? If the geological conditions in Saudi Arabia are so challenging, how has their oil sector performed so well over all these years?

Or that reliance on a few fields for production means a near-term decline can be expected. Why should that be? Is the implication that previously, the world relied on many smaller fields for its supply? If not, why didn't such reliance result in a decline earlier? Others point to the decline in the Cantarell field as evidence of an imminent global production peak, as if dozens of other major fields hadn't peaked and declined without such a result. Concentration of production in a few fields is just an example of what is known as the power law, which is that a few cases generate much of the activity; for instance, a small number of criminals commit the great majority of the crimes.

A number of observations have been made and repeated often, with the assumption that the connection between the fact stated and the conclusion that peak oil is near is obvious. This is rather like a Sid Harris cartoon where step two is a box saying "Then a miracle occurs," and in fact, as a student at MIT, a friend claimed he used the same method when developing mathematical proofs: work forward as far as you could, and assert that your last step led "QED" to the conclusion. (QED is scholarly talk for "it's obvious"; literally, it means "thus it is demonstrated.")

Half Empty, Half Full or Both?

Chevron has injected steam into the reservoirs, coaxing the sedimentary rock into giving up million of barrels of heavy oil that was too thick and sticky to retrieve using the technology of decades past.[69]

It is amazing the way optimists and pessimists can look at the same thing and draw opposite conclusions, especially when it comes to technology. Any number of peak oil advocates have looked at the technological marvels that the oil industry employs and interpreted them as evidence of the extraordinary effort required to retrieve the scarce resource remaining. Others see it as evidence that the industry is overcoming obstacles and developing ever-more advanced methods to get oil cheaper, cleaner, and easier.

In a related approach, explanatory variables are often assumed, rather than demonstrated. The lack of discoveries is taken to mean that there is no oil in the ground, rather than political or economic reasons for not drilling. A drop in production is assumed by virtually all peak oil theorists to be caused by geology, not higher taxes, political interference, accidents, or even seasonal factors. Alaskan production rises in the winter and falls in the summer, for example, due to the impact of temperature on the natural gas processing equipment, and Canadian drilling slows in the spring as the ground unfreezes and becomes muddy. Maintenance for North Sea fields is performed in the summer, causing a regular drop and recovery. But Deffeyes, Groppe, Simmons, and others all point to month-to-month changes as reflecting the underlying long-term trend, which is why they misinterpreted the temporary peak in November 2000 as possibly being the permanent peak, and subsequently May 2005, now surpassed by February 2008, and so forth.

SAUDI NUMEROLOGY

There, in the middle of this mall is the Washington Monument, 555 feet high. But if we put a one in front of that 555 feet, we get 1555,

the year that our first fathers landed on the shores of Jamestown,
Virginia as slaves.

In the background is the Jefferson and Lincoln Memorial, each
one of these monuments is 19 feet high.

Abraham Lincoln, the sixteenth president. Thomas Jefferson, the
third president, and 16 and three make 19 again. What is so deep
about this number 19? Why are we standing on the Capitol steps to-
day? That number 19 — when you have a nine you have a womb that
is pregnant. And when you have a one standing by the nine, it means
that there's something secret that has to be unfolded.

Louis Farrakhan, **National Representative of the Honorable**
Elijah Muhammad and The Nation of Islam, speech
at the *Million Man March*, October 17, 1995.[70]

A more detailed look at Campbell's assertions about Saudi oil is telling about the manner in which he approaches research. In February 2004, at the Center for Strategic and International Studies, Baqi and Saleri (2004), commented that Saudi Arabia could produce at a plateau of 10–15 million barrels per day for 50 years, which would consume 68% of their proved and probable reserves.

Campbell's response: "This sounds utterly implausible. The statement of using 68% of *Proved & Probable Reserves* sounds as if it really means 68% of *Proved & Probable* oil-in-place. Also, claiming static production until 2054, which is an odd date to select, sounds suspiciously like a *Reserve to Production Ratio of 50*. It simply divides remaining reserves by current production, ignoring natural depletion,"[71] emphasis in original.

From there, he calculates that 50 years of production at current levels of 3.1 billion barrels per year would be 155 billion barrels. Adding this to past production of 97 billion barrels yields 252 billion barrels, which he notes is suspiciously close to official proved reserves of 260. From this, he assumes that the official reserves figure includes past production, even though the Saudis have published precise figures showing past production, current reserves, and oil-in-place. (He conveniently ignores this.) Instead, by *assuming* that 68% is the recovery factor, he calculates that oil-in-place is 370 billion barrels. Then, stating that a more reasonable recovery factor is 50%, he argues that there is really only 88 billion barrels of remaining conventional oil. This would, by the way, mean that *the Saudis should be past peak* under the traditional Hubbert-style model, and thus experiencing declining production. Instead, they have repeatedly shown the ability to raise production quickly when circumstances warrant, and the level remains quite high.

This is an astonishing performance, worthy of the most naïve high school student. In the first place, the Saudis clearly state that their discovered oil-in-place is 700 billion barrels. He discards that figure without explanation. Secondly, the 50-year figure apparently refers to a planning horizon, known as a "round number," and is not "suspicious" in the least. The 68% figure is quite clear cut, meant to represent exactly what the Saudis said it was, the proportion of proved and probable reserves produced under the scenario described. There is no justification to modify it from the stated proportion of reserves recovered by 2054 to a recovery factor (proportion of oil in place that can be extracted). Simply put, his entire calculation is fanciful and selectively ignores the published data and resembles the numerology of Louis Farrakhan more than any scientific research.

TWELVE

Motivations

[Lynch's 2009 *New York Times* op-ed] was a politically astute effort, and I'm sure that whoever pays him for his work appreciated it as they fight for their survival against a turning tide of public opinion.[1]

Without a doubt, some peak oil advocates are simply concerned about the potential problems that peak oil could create. However, there are all too many who seem to have, at best, multiple motives in embracing the argument—and its various proposed remedies. Self-interest is the most obvious reason, but there are also moral or religious arguments, many of which support the theory that beliefs drive conclusions.[2]

Some pessimists, actually, would seem to be acting out of honest conviction, since peak oil would not benefit them. Many in the airline industry appear to believe in the peak oil theories, despite the fact that high oil prices would be to their detriment, and it is not clear why groups like the medical profession would care. Instances like Goodstein's embrace of peak oil in support of promotion of nuclear power is an oddity, where it is not very clear why peak oil would be a good explanation for such a policy. Apparently, he's unaware that oil is not a major source of fuel for power generation.

Of course, the U.S. natural gas industry spent years arguing that prices were too low, which had the contrary effect of discouraging large consumers, such as power generators, from investing in equipment that would have used gas, but again, there are mixed motives for this, with managers wanting to believe that higher prices would deliver good financial results even while the industry wants higher sales.

Similarly, some have been puzzled by the idea that the Saudis would be acting against their own interests if they were actually facing the kinds of problems described by Simmons, and embraced by most peak oil advocates. By denying that they face resource constraints, they are calming markets and "talking down" the price of oil, which would be nonsensical if they expected an imminent peak in their production followed by much lower production, which, presumably, they would want to be offset by higher prices.

Noting that environmentalists or oil industry executives support peak oil *does not*, as argued in Chapter 11, *imply that the theories are wrong*, but they do help to explain why some people would embrace theories that are wrong. And examining similar controversies and other sources of belief can provide some insight into how people can go astray.

RAW SELF-INTEREST

Although most of the petroleum industry does not support peak oil, there are those who have, including Total, the French oil company, and possibly Schlumberger, as well as various executives scattered throughout the industry. This might be hard to see as self-interest, since it implies that governments should accelerate efforts to wean their economies away from oil, which is hardly in the industry's interest.

Alternatively, the self-interest becomes clearer when the broader argument made by many in the industry, that the "easy" or "cheap" oil is gone. This is essentially a cry for assistance, be it through greater access to areas for petroleum development or tax breaks for investment, R&D, and so forth. At the minimum, think of it as a *cri de ceour* for recognition and appreciation of the difficult job facing the industry rather than acceptance of the peak oil theories. Chevron's "Will You Join Us" is one such example, where some peak oil advocates have interpreted their campaign as "acceptance" (see Chapter 11 on imprecise words) of peak oil, which it clearly is not.[3]

Oddly enough, this is not the first time Chevron used developments in the oil market to further its agenda, at least in my opinion. In 1986, after the oil price collapse, Chevron was in the forefront of those pushing for an oil import tariff in the United States, ostensibly to protect U.S. oil production from low prices (translation: foreign competition) and thus ensure

U.S. energy security (cynical translation: their own profits). In discussions with a reporter from the *San Francisco Chronicle*, I noted that the United States had been vulnerable before prices fell, but Chevron had seem disinterested then, suggesting to me that they cared more about low oil prices than energy security.

[This actually led to an amusing interlude, since the news story appeared just before the CEO of Chevron arrived at MIT to receive an award for his fundraising efforts, and the story noted that I worked at his alma mater. One of my bosses was at the luncheon in the CEO's honor, and said that someone in the MIT president's office had greeted him with, "Who the hell is Mike Lynch?" I was rather surprised (in the days before the Internet, you didn't know what was in a newspaper unless you bought it, or had a wooly mammoth deliver it to your home), but shrugged and said that this represented my honest opinion, and to his credit, my boss was fully supportive.]

Competing Energies

That the threat of peak oil has been trotted out by those seeking assistance for alternative energies is hardly surprising, nor illogical. To a degree, it is simple salesmanship: buy my fuel/technology because you won't have access to oil soon, or it will be prohibitively expensive. Stories about those embracing somewhat exotic technologies for their homes, such as fuel cells, or geothermal, often talk about their wish to become independent from the vagaries of the oil market and rapacious large energy firms.

Goodstein is an obvious example, since the main point of his book *Out of Gas* was that we needed to encourage nuclear power because peak oil is threatening.[4] He also pointed out the concentration of petroleum reserves in politically unstable nations, pollution problems, and so forth, most of which are reasonable concerns.

One of the early (and valuable) peak oil Websites, www.oilcrisis.com, was created by Eric Swenson, who is in the renewable energy business. Joseph Romm, who blasted me in the *Huffington Post* (see Chapter 10) promotes renewable energy technologies and energy efficiency. The various analysts affiliated with the Post Carbon Institute are concerned with global climate change, and as such, have embraced peak oil as part of their cry for an evolution to a new energy system. For his part, Amory Lovins speaks in favor of energy efficiency and biofuels as valuable for their own sakes, rather than necessitated by peak oil.

In other cases, the motivations might be mixed. The airline industry would certainly not benefit should peak oil occur, but it is possible that some within the industry are promoting biofuels, and so rely on peak oil arguments to bolster the cause. In the U.S. Defense Department this appears to be the case explicitly. In the Defense Department, the Office of Naval Research has a budget aimed at developing new fuels, and as such, should be expected to support peak oil.[5]

But this is hardly true of all. The Uppsala University Global Energy Systems Group, which took over leadership of the Association for the Study of Peak Oil, would seem to have no particular vested interest one way or the other, given that they are, professionally, physicists. Ken Deffeyes of Princeton exhibits every sign of being simply a curious scholar. And other writers and researchers show every sign of simply exploring what appears to be an interesting and important issue.

HATRED OF THE OIL INDUSTRY

In many parts of the world, the oil industry seems to be not just unloved but hated. In the United States, this has been attributed to the depredations of John D. Rockefeller, for example, although similar distaste of Bill Gates has not made the computer industry as a whole a target of opprobrium. A dislike of large enterprises is also common among Americans, though it doesn't extend to Apple. Leftists often claim that the petroleum industry has acted as colonialists in many countries, or that its environmental damage is relatively unique. The perception that it actively opposes most environmental legislation is obviously a factor as well.

There is a possible economic impetus, namely the manner in which gasoline prices in much of the world fluctuate wildly. In the 1970s, there was a widespread belief that oil companies had created the oil crises, or at least the lines that arose at gasoline stations in 1974 and 1979. Recent events show this animosity continues. In Massachusetts, my adopted state, the 2012 Senate race saw the challenger, Elizabeth Warren, making repeated references to tax breaks for oil companies, even tying them to the problem of student debt. This might seem to be odd, but I suppose that a well-paid college professor like her might wish to divert blame for high college costs to another target.

And after Hurricane Sandy, when urban areas around New York were severely disrupted, politicians called for investigations into cases of gasoline price gouging, but no one suggests investigating farmers when food prices rise during a drought.

PILING ON: OIL AND THE JEWISH QUESTION

In nineteenth century Europe, three university students were asked to write an essay centering on elephants. The Italian wrote, "The Love Life of the Elephant," the French student penned an essay entitled, "How to Cook and Elephant," the German sought concentrated on "The Military Uses of the Elephant," and the Jewish student wrote, "The Elephant and the Jewish Question."

Just as many in the 1970s wrote about oil in relation to their area of expertise (agriculture, exchange rates, North-South relations, etc.), so a number of authors who are not experienced in energy or oil matters have started with the presumption of peak oil and then turned to their own interests.

Admittedly, I haven't read Michael J. Lynch's book[6] on prison reform or even the chapter on peak oil and its impact on the issue (base prisons locally, perhaps? Combine with the Locavore movement, maybe), but there are quite a number of advocates who have raised the peak oil issue to promote agendas that are only indirectly related.

A classic historical case is the working paper, "The Future of the Automobile in an Oil Short-World" by Christopher Flavin, Lester Brown, and Colin Norman, all advocates for, shall we say, the "soft energy path."[7] (Not that there's anything wrong with it, as Seinfeld would say.)They note high oil prices, provide many quotes from oil industry executives about the end of the oil age, and then discuss a variety of solutions, or at least mitigating policies. These include bike paths and mass transit, which I would certainly not oppose. To their credit, they acknowledge the electric car was not yet viable, a case where rationality appears to overcome the expected personal biases.

And to be frank, many of their proposals have merit above and beyond the likelihood of peak oil, or even permanently high oil prices. I wouldn't support a bike path through the middle of a cemetery, for instance, but if you've ever been in Tokyo on a Sunday afternoon and walked around the Ginza shopping district when cars are banned, it's a very pleasant environment.

Khmer Blanc: Urbanization Advocates

A classic case of using an issue to promote one's own agenda comes from James Howard Kunstler, who has written four books on peak oil (including two novels) and speaks frequently about the coming changes it will engender.[8] He comes at this issue as an urban planner, whose professional focus has been to denounce suburbanization and the evils thereof: He hates suburbs, roads, low population densities, strip malls, and fast food restaurants.

(Mea culpa: I don't like corn bread, but do not feel obligated to write books encouraging the banning of corn cultivation. Yes, I'm from the South, but it's just too dry for my taste. Except in a casserole with chicken, but that won't be part of my next book either.)

He has seized on the possibility of peak oil to bolster his arguments that people should live in city centers, relying on bicycles and mass transit, which, by an incredible coincidence, coincide with his own preferences. He sees urban centers as incubators of culture, diversity, energy efficiency, and so forth, which of course ignores that the actual history of urban centers includes horrible trash problems, disease, crime, poverty, and mob violence, demonstrating once again how bias can lead to cherry-picking of data and history. One can almost imagine him in charge, reversing the policy of the Khmer Rouge and forcing people at bayonet points to return to the cities.

Locavores

Oddly enough, peak oil has been embraced by some with the opposite agenda of "eating local," which is said to have many benefits, including the provision of fresher, healthier food, but also reducing the energy component of food, particularly for that which has come from great distances. To a small degree, this movement fall prey to the "single-variable" mistake, that is, thinking that distance is the only important effect. Local food left in the fields too long, or in a distribution center, might not be fresher, and although one would presume Locavore farmers are also organic, that is not necessarily the case.

That said, my family participates in the community farming movement and there are many other benefits in purchasing locally sourced food than just reducing diesel fuel consumption. The flavor of the vegetables we get definitely seems better than the supermarket ones (although I might get in trouble for not giving locally grown carrots to our guinea pig.)

It doesn't require the most discerning reader to realize that this conflicts with the urbanization movement promoted by Kunstler, among others. The amount of "local" produce that a city like New York can consume is naturally going to be limited, unless there is a large scale movement to urban gardening constructs. Perhaps the Locavores will clash with Kunstler's followers, in a kind of reverse Hunger Games.

The Anticonsumer

Live Simply, so that others may simply live.

Bumper Sticker

A subset of peak oil advocates are motivated by a desire to promote a simpler lifestyle, something usually associated with religious sects. Since religion by definition involves spirituality, this is hardly surprising: few sects promote material acquisition (the Cargo Cult of Papua New Guinea is a prominent exception) and within most religions, there are significant subsets who decry materialism.

Some cases are extreme. Simeon the Stylite lived on a pillar for 37 years during the 5th century in what is now Turkey. It was said to be originally about 4 meters high, although the locals later built taller ones for him. Apparently, he had become disheartened at being constantly approached by supplicants and pilgrims, who distracted him from his meditations ("I can see my house from here!") and sought to distance himself from them. I sympathize when receiving junk phone calls, but using caller ID seems a much better solution than standing on a pillar.

And in medieval times, the so-called anchorites practiced an extreme form of seclusion, having themselves walled up in small cells in monasteries, with only a small opening for food to pass through, allowing them to focus on spiritual questions. Sadly, most modern ascetics have eschewed this practice, and prefer to interact with, and proselytize, the rest of us, as for example through the above-quoted bumper sticker.

Indeed, many religious sects emphasize a simple life style, from the English Puritans to the Islamic Salafists, arguing in favor of devoting time and effort to improvement of the self, not acquisition of material goods. Of course, not a few religious leaders have been found preaching more than practicing. Mullah Omar, leader of the Afghan Taliban, was said to have a preference for Range Rovers and satellite TV, for instance, something not very compatible with the ascetic lifestyle he preached.

This is not to say that an element of hypocrisy can't be found among the non-religious proponents of the simple life. Many commune dwellers seeking a simpler lifestyle nonetheless relied on others (parents mostly), just as the anchorites survived on donated food. And if you go to a store that specializes in "natural" or "organic" food, you will also almost always also find spring water that has been transported from a distant land such as Italy—don't even ask about the carbon footprint.

Symptomatic of the intersection between anti-consumer sentiment and environmentalism, Stan Cox published a book attacking the reliance on air-conditioning in modern society, which was embraced by many—and castigated by many more.[9] Without a doubt, air conditioning is often wasteful and many of us are profligate energy users, but most of us can think of situations where no air conditioning would be brutal.

Civilization and Its Opponents

[The Luddites] are attempting to bear witness to the secret little truth that lies at the heart of the modern experience: whatever its presumed benefits, of speed or ease or power or wealth, industrial technology comes at a price, and in the contemporary world, that price is ever rising and threatening.[10]

The Luddites had a somewhat different take on the issue of modernization, opposing mechanization of their work more than societal progress generally. The Industrial Revolution changed the nature of work, and especially reduced the role of craftsmen, who rebelled at being moved into mindless machine-tending, but also did not perceive that the change would ultimately mean more employment, not less. Napoleon actually remarked on this, when confronted by textile workers protesting mechanical looms.[11]

More recently, it was proposed that post–World War II Germany be returned to a pastoral state by Secretary of the Treasury Walter Morgenthau, who thought a lack of industry would make it incapable of aggression in the future. Oddly enough, some Nazi thinkers, as part of the racial stereotype of the Aryan peoples, also believed that agriculture should be given priority over industry, as it enabled bringing the people closer to the land.

At the other end of the spectrum, Mao Tse-Tung, leader of communist China, felt that economic progress would weaken the revolutionary spirit of the populace, which was one of the reasons he launched the Cultural Revolution and opposed others, like Deng Xiaoping, who argued for economic reforms. Deng was accused by Mao (and Chinese leftists) of being a capitalist roader, but in the end he won out because the Chinese people wanted economic progress and material goods, not chaos and revolution.

To some degree, this type of attitude overlaps with groups like Locavores who promote not just fresher local food, but are often unhappy with the industrialization of agriculture, including the huge transport infrastructure necessary to move perishables long distances, as well as the energy-intensiveness of modern farming methods. However, the Locavore movement is much broader in its orientation and motivations, and also more rational than the Luddites.

Opposition to hydraulic fracturing of shales to produce oil and gas often falls into this category. Although there are questions about the environmental impact of the practice (see Chapter 19), much of the opposition appears to be emotionally driven, with reports of pollution and health effects clearly exaggerated, if not falsified. Moreover, many of the opponents are simply complaining about the "industrial" activity in their area, with large movements of trucks and running of generators, sometimes in an otherwise quiet environment.

Apocalypse Now

> Beloved men, realize what is true: this world is in hast and the end approaches; and therefore in the world things go from bad to worse.
> *The Sermon of the Wolf to the English*, 1014[12]

Although the implications of peak oil for the oil industry appear obvious, many peak oil advocates go far beyond that to predict vast problems for global civilization, and even the "potential extinction of mankind." (Campbell/Hanson) Most mainstream peak oil advocates do not seem to have adopted such views, but it is not hard to find Internetizens who issue dire warnings, and even some survivalists point to peak oil as portending the end of civilization. Don't believe me? The Simpons's episode "Homer Goes to Prep" includes him imagining the various threats that concern doomsday preppers, one of which is, you guessed it, "peak oil."

As Amos Funkenstein remarked, "Every culture seems to harbor thanatic views of an ultimate catastrophe and hopes of rebirth of a new world."[13] These appear to have originated in the ancient near-East, and the Zoroastrian view of the ongoing struggle between good and evil, which would have a final climatic battle. This concept is particularly prevalent in the Western religions, albeit in different forms, with the Jews apparently picking it up during their time in Babylon, the Christians from Jewish sects, and the Moslems from the predecessor faiths of the Christians and Jews.

(Other religions, particularly Eastern ones, see a cycle rather than a beginning and end, and often prefer a balance between opposing forces instead of conflict between good and evil. Most recently, the brouhaha over the "end" of the Mayan calendar actually refers to the Mayans view of cycles and eras, and thus the end of the current one, according to their measurement of time. Taking down my Betty White 2012 Pinup calendar was a sad event for me, but it never occurred to me that the world might end when the calendar was used up.)

The Jews have had a number of prophets who foresaw, if not the end of the world, cataclysm, particularly during times of troubles, such as the Roman occupation of Jerusalem or the Spanish expulsion of the Jews after 1492. Daniel was a famous precursor, quoting Gabriel as saying, "I will show you," he said, "what is to happen later in the period of wrath; for at the appointed time, there will be an end."[14]

More commonly, the appearance of the Messiah was thought to have occurred on a number of occasions, such as around the time of Jesus (when he was one of many the Romans crucified for such pretensions), as well as others such as Shabbetai Zevi in the 17th century.[15] That notable attracted quite a following, but when offered the chance to prove his divinity by standing up to an onslaught of arrows, took the alternative choice of accepting a job in the Ottoman bureaucracy.

Although some scholars see the book of Revelations as representing a coded message about the then Roman empire, most Christians see it as prophesying the Judgment Day, the final battle between good and evil, and the raising of the good to heaven. The many symbols and cryptic remarks used by John (the author, not the apostle) have left much room for interpretation, and the basic idea of a "time of troubles" presaging Judgment Day has caused many to interpret their own times as the end days. After all, there are always troubles.

FALLING ACORNS

This is no social crisis, just another tricky day for you.

Pete Townsend

Aside from some of the specifically religious premonitions of the end of the world, many of them came in response to dire events. The ancient Celts thought the disastrous battle of Magh Tuireadh presaged the end of the world, as did Pythia's prophecy of doom in the Voluspa. The English Civil War convinced many that the end of the world was at hand as did, more appropriately, the Black Death, which killed as many as one-third the population of Europe. Even the most sanitized description of that era sounds more like the end of the world than has been seen before or since, with the possible exceptions of the Mongols sweep south and west or the Holocaust.

There is clearly a subset within the peak oil community who embrace the concept of a collapse of civilization, to a degree resembling the "survivalist" community, which argues for a Mad Max style of apocalypse. This is hardly new, having been pre-dated, in modern times, by the "Swiss scientist Alphonse de Candolle ... [who] ... envisaged a strangely modern hypothesis: that of decline through the inexorable depletion of sources of energy"[16]—this, in 1873!

The confusion of episodic events with the apocalypse is very similar between millenarians and peak oil advocates, but this is often not so much a case of Black Swan events troubling people, as people unfamiliar with swans and thinking they must be predictors of doom. Most of the problems peak oil advocates perceive as proof of their theories would not even warrant a note in the oil press.

Probably the best example of the tendency to see falling acorns as the end of the world is the iconic Website www.dieoff.org, which was an early Internet promoter of not just the peak oil concept but a more general societal collapse. Things such as war and famine are treated not as common events but somehow unusual and thus evidence of an impending collapse. Examples include:

- Matt Simmons predicting mass starvation in the United States because of low gasoline inventories (which was actually just him misreading the data).[17]

- Campbell, Simmons, and others seeing the reserve downgrading by Shell in 2004 as symptomatic of "geological constraints," or "global oil reserves may be dangerously exaggerated" rather than simply an individual case of misclassification of reserves.[18]

- The Russian decline in production in late 2008 as representing the ultimate peak, instead of high taxes and low prices discouraging production.[19]

- The abrupt decline of the Cantarell field is often described as a sign of the impending oil production peak even though many fields globally have peaked historically.[20]

There are always problems in the oil fields, just as any field of human endeavor.

- The Jubilee field offshore Ghana could not achieve its planned peak production, because clay fines clogged up the development wells. But after using a larger-size screen that would allow the particles to pass through, the company now expects to reach its higher production rates.

- The Yibal field in Oman was expected to see greatly improved recovery, but water encroachment prevented this, and some, including Simmons, believed this meant that horizontal drilling would not be useful on a global level.[21]

Be Good: Eschew (fill in the blank)

Frequently, apocalyptic views have been promoted for other reasons, including many of the Jewish prophets, seem to have embraced the apocalypse as a warning to people to end their wicked ways. In ancient times, this meant not cleaving to fashionable religions, such as Baal-worship. Now, it means not driving an SUV.

The two thought trends meet in the slogan, What Would Jesus Drive? The implication is that he would be too "ethical" to drive a large vehicle. Actually, perhaps he'd use a minivan to hold the apostles. Kidding aside, he did espouse a simple lifestyle, which does not seem to have caught on with most of his followers (visit any Gothic cathedral) any more than "turn the other cheek" ended violence. If anything, where the cobbler's children have no shoes, the ascetic's children often ride in an Escalade.

SOMETHING THERE IS THAT DOESN'T LOVE AN OIL BARREL

The evidence suggests to me that peak oil initiated with research by well-meaning individuals, but then became caught up in a wide variety of issues and movements who found it suited their messages and goals. Few of them understood the issue, or apparently have paid much serious attention to the questions involved. Instead, they have repeated any number of clichés and adages as truths, and launched ad hominem attacks on those who disagreed with them.

This is a classic case of people believing what they wanted to believe without regard for the underlying realities. Some arguably cross over into dishonesty, but ignorance and naivete are more often the order of the day.

THIRTEEN

Real-World Problems

The fact that the geological challenge facing the oil industry is being misinterpreted and exaggerated does not mean that there are not obstacles that need to be overcome. These can be heard at any industry gathering and have been described in books such as 1994's *Crisis in the Oil Patch,* by Hodel and Dietz, John Hofmeister's 2010 *Why We Hate the Oil Companies,* and the 2007 report by the National Petroleum Council entitled, *Hard Truths: Facing the Hard Truths about Energy.*[1] The last was actually commissioned due to political pressure from peak oil advocates, but ended up devoting only a few pages to that topic, while going on to real-world issues.

These include an increasingly unstable geopolitical environment (written before the Arab Spring) and the attendant political risk, coping with increased greenhouse gas emissions, infrastructure needs to meet the expansion of supply, and the retirement wave of skilled personnel which the industry expects.

THERE'S NO RISK LIKE POLITICAL RISK

The upheaval of the Iranian Revolution in 1978/79 and the attendant oil crisis brought the idea of political risk to the attention of the general public, but it was hardly news to the oil industry. Although the industry

bemoans the near-constant political attacks and describes the recent problems as severe, the reality is that access to oil prospective areas is much
greater now than at any time in the past four decades.

Real and Perceived Risk

Revolutions, whether considered good (Libya 2011) or bad (Iran 1979)
by Americans, are not actually the biggest political threat to the oil industry. Mexico is the prime example of the misunderstanding: when the
government nationalized American oil interests in 1937, production had
already declined by 75%—*because of onerous taxes*. Unilateral contract
revisions, even cancellations, occur far more frequently than revolutions
and are probably more costly overall. Demands that the industry undertake additional spending for infrastructure or social support are also quite
common and, if made after the contract has been signed, reduce a project's profitability.

The other major misconception is that the less developed countries are
the unreliable regions. Although few OECD nations would be considered
socialist in the mode of Cuba or even Venezuela, some have a long history
of adopting policies that have damaged the industry. Italy and France set
up national oil companies specifically to compete with the United
States–United Kingdom majors, and Norway often requires that oil field
developments be delayed to minimize inflationary effects on the
economy. Statoil, which is still two-thirds owned by the Norwegian
government, acts as operator for most fields in the Norwegian North Sea.

The prize-winner, though, must be Pierre Trudeau's Canada, a country
that was hardly revolutionary or Marxist, yet was antagonistic to the
(mainly American) foreign oil companies in the late 1970s. As prime minister, he set up a national oil company, using gasoline taxes to fund it, and
encouraged the Canadian oil industry to buy out its American counterparts
(and the banks to loan the money to do it). His administration also
restricted natural gas exports, as we saw in Chapter 3, to save it for "future
generations."

In fact, if you compare this to the case of Angola in the late 1970s, you
should be flummoxed to realize that the Angolan government, made up of
former Marxist guerillas with only a few years in power, was far
more welcoming to the international oil industry in the 1980s, including
American companies. The primary difference between them and Canada
is that Canada could afford the luxury of anti-business policies, the economic losses being subsumed in a much larger economy.

But....

Again, the lack of historical context exaggerates the problem of political risk. For the oil industry, political risk actually precedes the development of petroleum: in the 19th century, whaling operations were often targeted, primarily with the thought of disrupting a valuable economic activity. And during the U.S. Civil War (or War Between the States, as my elderly schoolteachers in Richmond referred to it), Southern cavalry under General Jones attacked Burning Springs, West Virginia, destroying oil field equipment and thousands of barrels of oil in 1863.[2]

The 19th century domination of the oil industry by production in America meant that political risk was largely confined to domestic politics, but as the industry moved in 1871 into Indonesia (then the Dutch East Indies), in 1910 to Mexico, in 1914 to Venezuela, and especially Russia in 1871, political risk became ever-present for the industry. As an example, in 1905, young Iosif Djugashvili was the chief labor organizer in the Baku oil fields for the Bolsheviks. He was later better known by the nom du guerre Joseph Stalin and he was probably less than commodious to his industry counterparts.

Why the Complaints?

There is no doubt but that the industry saw backward steps recently, including Argentina's renationalization of YPF in 2012 and Bolivia's 2006 renationalization of YPFB. Before that, Russian efforts to restrict foreign upstream investment and Venezuela's changes in contracts to increase its ownership in projects and raise taxes caused much hair-tearing and beating of breasts in the industry.

One thing that received too much negative attention was the false dawn of a Saudi upstream opening in the late 1990s. Then Crown Prince Abdallah announced that the country would allow some participation by foreign companies, raising industry hopes of access to their enormous oil resources. Instead, those hopes were deflated when it proved that the government wanted assistance with developing its natural gas resources, a much less attractive prospect than the kingdom's enormous oil resources. At around the same time, American companies that had been considering investing in Iran, which was attempting a limited opening to upstream investment, ran into a roadblock as the Congress passed the 1996 Iran–Libya Sanctions Act.

However, the fact remains that the real political trend has been toward lower overall risk and greater access to upstream opportunities. By far the biggest shift came during the 1970s, with the nationalizations of holdings in many of the OPEC nations, including Iraq, Iran, Kuwait, Saudi

Arabia, and Venezuela. Every reduction in "access" since then has been dwarfed by these changes, which reduced the multinational oil companies from controlling the bulk of the international oil market to merely being large corporations.

True, the 1990s saw a strong move in the other direction, with the opening up of the countries of the former Soviet Union, the *apertura* in Venezuela, and the limited opening of India and Brazil for foreign exploration. This convinced many that oil had become just another commodity and would be far less subject to political interference in the future. Alas, the past decade proved that to be overly optimistic.

Overall, although there are still limitations to where the industry can drill and often interference by politicians and governments in operational decisions. Even in the United States, with its free market economy and relatively conservative politics, in places like Florida and California, Republican governors Jeb Bush and Arnold Schwarzenegger opposed offshore drilling, bowing to political pressure.

The Obama administration, for all its supposed anticorporate bias, has actually done more for the petroleum industry than its predecessor, which was led by the "oilmen" George W. Bush and Richard Cheney. Most notably, opening up areas of the Gulf and Atlantic Coast to drilling, and permitting new exploration offshore Alaska, despite strong environmental opposition, rates up there among counterintuitive policies like Nixon opening diplomatic relations with China.

All Political Risk Is Local

This highlights a major trend that is certainly annoying, though hardly devastating. In many countries, the local citizenry is much more empowered politically than a few decades ago. To some degree, it can be said that this reflects more individualized politics and greater participation, as in many wealthy nations. But even in Latin America, India, and China, opposition to developments can derail petroleum operations. Environmental fears, demand for jobs, opposition to land seizures, and even religious opposition can make it hard for a development to proceed.

> The residents argued that the [solar] 3.3-megawatt installation proposed by Boston-based nonprofit Citizens Energy Corp. on 50 acres of open farmland would be ugly, devalue their properties, and cause health problems.[3]

But this is hardly a problem unique to the petroleum industry. Even in Massachusetts, with its staunchly liberal politics, there is widespread opposition to biomass plants, wind power projects and even photovoltaic plants. In China, petrochemical plants are only one of many projects that face opposition, both out of environmental concerns but also land acquisition practices.

Nimby (not-in-my-back-yard) is faced by nearly all sectors of society, but usually the wealthier the society, the more trivial the concerns. Where Chinese protest against a chemical plant that might release toxic emissions, Americans will object to an apartment house that will make local parking more scarce.

Why Is This Industry Different from All Other Industries?

"We're reaching new thresholds of outrageous profits, and outrageous gas prices," said Rep. Markey, Chairman of the House Select Committee on Energy Independence and Global Warming. "Exxon already is well outpacing last year's record earnings, and increasing oil prices will continue to keep Big Oil on a record profit pace. These profits are a powerful reminder that American families are paying for these profits while simultaneously subsidizing billions of dollars in tax breaks for Big Oil."[4]

Even though the company appears to pay about 10 percent of its pretax income in taxes—when the federal corporate tax rate is 35 percent—Cook said, "We pay all the taxes we owe—every single dollar." Senator John McCain, the committee's ranking Republican, who had earlier labeled Apple "a tax avoider," was soon swooning over Apple's "incredible legacy."[5]

The increase in NIMBY and even BANANA (build absolutely nothing anywhere near anything) movements should not be too surprising, as it began in the United States several decades ago, arguably with opposition to the nuclear industry and major highway projects.

Part of the problem is the dislike of the petroleum industry: "Big Oil" in the United States, and "Big US Oil" in many other countries.[6] This has been attributed variously to residual anger at John D. Rockefeller's monopolistic behavior, the perceived abusiveness of large oil companies, including those in the United States, specific instances like the Exxon Valdez oil tanker accident and the BP Macondo disaster, but also a general

American dislike of the large and powerful and preference for the under-
dog (e.g., Apple vs. IBM).

But there has also been a growing citizen involvement in policy decisions and a reluctance to trust the government to make important choices,
such as the placement of infrastructure, which has since crossed over to
the private sector investments. Whereas much of the history of the United
States saw a desire for "progress," the public now takes a more nuanced
view of developments, opposing projects that appear to be poorly
designed such as highways through residential neighborhoods.

Of course, there are also those who are opposed to nearly any change, to
the point where a local restaurant closing can be opposed on sentimental
grounds, or the building of housing is seen as disrupting lives (i.e., parking, which seems to arouse more activism than threats to health). People
like this can be very hard to deal with, but are fortunately a small minority.

And as described in Chapter 12, there are those simply opposed to
modern, industrial society and so will oppose nearly any large project or
those related to industrialization, such as energy.

Others have concerns that it is possible to address. It is quite common for
developers to meet with local residents and change their plans in response
to complaints. A friend of mine once described how a planned apartment
building was modified, dropping a small park that residents worried would
be a magnet for teens after dark and reducing the number of floors.
One wonders if the builder had anticipated requests for such a change and
added floors to the design that could then be cut to satisfy complaints.

This problem has been attributed by many to the failure to build new
refineries in the United States for over three decades, and there is no doubt
that placing a refinery is extremely difficult politically in most parts of the
country. But the reality is that the United States has little need for new
refineries; capacity has increased for years due to what is known as
"capacity creep." Refiners add components, expand others, and debottle-
neck choke points to increase the size of existing capacity, but relatively
gradually, which rarely arouses notice or opposition.

The flip side can be seen on the upstream, where the industry has been
banned from many areas, such as offshore California following the 1969
Santa Barbara oil spill. This was followed later by bans on drilling off
the east coast of the United States and parts of Florida, bans that were supported even by conservative Republicans like Florida Governor Jeb Bush.
Indeed, it was Democratic President Barrack Obama who proposed
opening some of these areas up to exploration in 2010–just prior to the
Macondo disaster.

The end result is that, just as with resource nationalism above, NIMBY and pubic resistance generally takes some opportunities off the table for the industry, creates delays in other cases, and adds costs in many, nevertheless, the industry has managed to cope, primarily because there are enough areas where drilling and refining can be carried out. The growth of shale production makes this even truer, at least for the upstream.

There is no guarantee that opposition to industry projects won't become more and more widespread, but it appears as if the majority of the public is capable of understanding what constitutes acceptable risk and what represents appropriate tradeoffs. The shale industry is a perfect test case, as there has been major opposition to the process of hydraulic fracturing, most of it uninformed, but many communities have refused to ban the practice. Enough so that, while an individual driller might have losses from being excluded from a given lease, overall there are far more opportunities than the industry can exploit.

THE GREAT CREW CHANGE: A WAVE OF RETIREMENTS

In what has been called the "Great Crew Change," the older generation of geoscientists and petroleum engineers who were hired before the sweeping lay-offs of the 1980s are now approaching retirement age and will soon leave the world of work. But it is still unclear who will replace them. The pool of potential talent is too small, and companies are scrambling to cope with the crunch.[7]

Numerous commentators in the industry have warned about the looming wave of retirements of petroleum engineers who entered the industry in the 1970s and are now approaching retirement. U.S. universities graduated only about 600 petroleum engineers (350 undergraduates and 250 graduates) as late as 2005, when the industry has 30,000 employed.[8] In other words, as long as only 2% of the workforce retires each year, the industry would be in good shape. Seems unlikely.

This is obviously not a new concern, nor is it limited to the oil and gas industry. As far back as *1962,* when two engineering professors wrote about the industry fears concerning low rates of graduation and especially hiring of petroleum engineers:

[There] has been much prediction and self-examination by the petroleum engineering profession-and by the educators in particular-since Jan., 1958, when a number of petroleum companies temporarily

stopped hiring petroleum engineers (and other employees). ...
The shortage of petroleum engineers that developed in 1954 is well
known.[9]

They actually calculated that the number of graduates should be sufficient, assuming a 40-year career life and thought that "The story usually
heard reminds one of the nursery story of Chicken Little who convinced
her associates the sky was failing."[10]

Solutions

In some ways, this is similar to the fears about an inability to meet
equipment needs, which date back to the 1990s, when low drilling rig
rental rates led a (relatively) young Matthew Simmons to decry the ability
of the industry to meet future needs. In fact, there are a number of fixes
beyond taking a college freshman and educating him or her as a petroleum
engineer or geologist with an advanced degree over the course of six to
eight years.

The "Great Shift Change"—resulting in a coming knowledge and
experience gap for the industry—spurs "efforts to codify many rou-
tine analysis and decision-support processes and, where possible, to
automate them" because new recruitment is unlikely to fill the talent
gap completely.[11]

Many industry workers who are not technically petroleum engineers are
performing petroleum engineering. A background in mechanical engineering, for example, can be applied to many aspects of oil production,
although it's hardly as easy as making a saucier into a pastry chef.
(I think.)

Better technology also makes a difference, including software that
makes interpretation of seismic information much easier, which is to
say, requiring less labor. Use of the Internet also reduces the travel needed
to examine field operations, and greater automation of equipment results
in fewer workers at many skill levels.

That said, many of these innovations don't so much reduce the need for
skilled labor as they enable the industry to perform better. Three-
dimensional and four-dimensional seismic are good examples where
much better information enables the industry to find and produce oil more
efficiently, but may not mean fewer geologists and engineers are needed.

And in fact, the past decade has seen what labor shortages mean. Although relatively few deepwater developments announced lengthy delays because of a lack of skilled personnel, it is likely that lead times have become longer and supply has suffered accordingly. However, this is something the industry must, and has, dealt with from time to time. A problem and a challenge, but not a barrier.

REGULATION

The oil and gas industry complains regularly about its regulatory burden, and with a certain amount of justification. Robert Bradley's magisterial 1995 two volume *Oil, Gas & Government* describes in exhausting detail many of the efforts to regulate the industry, and the problem has only worsened in recent years.[12]

A good part of this represents evolving science: better medicine (including better data) has allowed us to understand the impact of exposure to various elements and chemicals, and recognize the environmental damage that can occur from, for example, oil spills. For the first half century of the industry, little thought was given to preventing oil from flowing out onto the ground and into streams—except for the loss of revenue. Anyone who did such now would meet with outrage.

The industry certainly faces numerous regulations at all steps of its operations, and many of them have been poorly designed or imposed haphazardly. This is true of all industries, and represents the fact that few regulatory regimes are designed from scratch, but are rather done piecemeal as problems are recognized—or agreed upon.

And indeed, the industry has spent $899 per person from 1990 to 2013 on environmental issues, including R&D but primarily making cleaner products and keeping refining, exploration and production environmentally sound—or at least in compliance with regulations. Those two sectors accounted for 76% of industry environmental spending in 2013, over $12 billion in total.[13]

Our old friend context provides some insight, however, since total oil industry investment is on the order of $350 billion, meaning that only 3.5% of the industry's investment is the result of regulation.[14] To look at a specific example, the 2010 Macondo disaster led to increased requirements for offshore oil development. Given the existing expense for deepwater fields, where the greatest volumes are, the additional expense of adding a backup blow-out preventer is not enormous ($45 million compared to $3–5 billion for a large field development).[15]

And small drillers have been threatened with having to post bonds for new wells to cover abandonment costs, reflecting a large number of historical wells that might be leaking oil into the environment, which is of a concern for them, but would have little impact on overall production levels.

Most of the regulatory focus, at least in terms of costs, seems to be on oil products and emissions from consumption. The most recent is the EPA's Tier 3 gasoline specifications that require a reduction from 30 ppm sulfur to 10 ppm. California's demand that the carbon content in gasoline be reduced will probably be much more expensive.

And accidents involving oil-filled trains, such as the horrifying 2013 disaster in Quebec, have meant increased scrutiny on transportation and requirements for safer rail cars, but haven't meant any slowdown in the delivery of crude oil, at least so far.[16] In all likelihood, new railcars that are less prone to puncture, combined with increased processing of crude to remove volatile elements, will satisfy the new regulations with only a moderate impact on the industry. As an example, double-hulled tankers were required by the United States after 1989's Exxon Valdez accident offshore Alaska, to much complaint from industry, but little operational effect.

SHORT-SIGHTED CORPORATIONS

A study of political economy often includes lessons about the Anglo-American way of finance versus the Continental approach. The former, dominant in the United States and Great Britain, involves the raising of funds through the sale of stock, while, in the latter cases, companies rely more heavily on bank loans for their capital. The result is supposedly that firms in Continental Europe are much more able to focus on the long term, because they have a cooperative association with their financiers, and can work together over an extended period, with a better understanding of operations and goals and less tendency to panic at every problem. Anglo-American firms, on the contrary, are presumed to sacrifice long-term goals for short-term results, so that R&D spending will tend to be cut in order to boost short-term profits, for example.

There is a significant element of truth to this stereotype and most people in the private sector surely know someone who has seen a manager cut a budget to make current financial results look better. When managers are frequently rotated, or companies give bonuses based on quarterly or annual results, this can be particularly seductive.

But at the corporate level, companies that pursue such a strategy ulti-mately find themselves in the "ashbin of history," as they lack the compo-nents necessary to maintain their operations in the long term. Drug companies without new drugs in the pipeline are the most obvious exam-ple; even though development of most medicines takes many years from conception to marketplace, the industry manages to do so while coping with the vagaries of the stock market and the demands of investors.

Similarly, the oil industry is full of companies that have no trouble en-gaging in the development of long-term projects, such as deepwater oil fields or LNG export plants. Some investments, like greenfield refineries, face uncertain markets on completion and yet the industry has never had trouble making the investments. This is not to say that some of these will prove less profitable than expected once completed, but again, this has not prevented companies from making such investments as are needed.

Where's the Money?

The oil industry is facing major challenges with its future investments necessary to meet the world's growing energy needs. The Organization for Economic Cooperation and Development's (OECD) International Energy Administration (IEA) forecasts that total energy investment will need to be as much as $20 trillion (in 2005 dollars) from 2005–2030 and that oil and gas investments will need to increase by more than $8.2 trillion.[17]

Rising costs (see Chapter 16) have exacerbated concerns about the industry's ability to obtain financing to develop the needed capacity. In part, this reflects the enormous numbers involved: $4.3 trillion over twenty five years, according to an IEA estimate, a number that the National Petroleum Council considered a real challenge.[18] Hard to question.

Except, of course, in context. The annual amount works out to only $170 billion. The industry regularly spends $100 a year in the United States alone, even though it represents only a tenth of global oil produc-tion. Overall, the industry has been investing over $300 billion a year for years, spending $678 billion on exploration and development in 2013, according to a survey by Barclays.[19]

National oil companies face a different kind of challenge, inasmuch as they rarely rely on cash flow for their investment budgets. Typically, their cash flow (after operating expenses) goes to the national treasury, and their capital budget, after approval by the government, comes from the same source. This creates a number of inefficiencies, as when drilling costs changes and the budget is frozen. Labardini argued that the negative effect also included higher borrowing costs, because of the financial uncertainty.[20]

Overall, the idea that the industry could not access needed capital is shortsighted, ignoring the fact that attractive investments, especially in a mature industry, will always find willing lenders, even where some companies have insufficient revenues to fund their operations. Indeed, the shale boom has shown that it is all too easy to find money that probably should not have been invested.

INSUFFICIENT REFINERY CAPACITY

The debate is raging in full swing: the dearth of new refineries in the US. . . . "Why not just build new refineries and scale down the price of oil," our readers continue to ask us. Yes, it's a fact- no new refinery has been built in the US in the past three decades.[21]

Another complaint that surfaces from time to time is the truism that the United States has built no new refineries in decade, another example of a fact that is misleading. This is often cited as an example of an unfriendly business environment for the oil industry, but most industry executives readily concede that they have no interest in building new refineries.

True, the number of refineries has dropped from over 300 in the early 1980s to just less than half that number now, even though oil demand is if anything slightly higher.[22] But total capacity is virtually the same! Again, a beautiful example of how the raw numbers are confusing and even contradictory, and only a detailed knowledge of the situation explains them.

There are two separate and independent reasons for, first, the decline in the number of refineries and second, the maintenance of overall capacity levels. Under the system of import quotas and then price controls, domestic refining was rewarded by the so-called entitlements system, which meant that small, otherwise inefficient refineries could be profitable because of government support. This led to the creation of many "teapot" refineries, which closed rapidly with the deregulation of the crude market when Reagan became president in 1981.

At the same time, the larger, more efficient refineries grew as companies added units, debottlenecked existing facilities, and upgraded equipment. This has the benefit of both being cheaper than building completely new refineries and avoiding the problem of seeking permission for the location of a refinery, which is thought all but insurmountable in the United States.

Although the United States does not refine enough crude to meet demand, this reflects not any lack of capability on the part of the industry, but rather optimization of operations, as the nation imports only about 1 million barrels per day of petroleum products, mostly in small amounts of the many different types. Only about half are finished products, the rest used for blending or upgrading, meaning that the U.S. refining sector falls short of demand by about 3%. There is not a crying need for new refineries; instead, this has become a political football for the unenlightened.

AGING INFRASTRUCTURE

The United States has long had a "crisis" in its infrastructure, particularly roads and bridges, especially if one listens to the construction industry or the American Society of Civil Engineers.[23] (Again, the obvious self-interest does not negate the validity of the argument.) Partly, this is a funding problem, but primarily occurs because politicians tend to be shortsighted compared to industry, despite the common complaint that the stock market makes corporations overly focused on quarterly results, as discussed earlier.

However, Simmons made the suggestion that the oil industry faced a looming crisis in the offshore due to corrosion.[24] It's not clear why he chose to highlight this specific problem, but the industry itself has not expressed concerns on the subject. If anything, it sometimes seems as if the newer equipment is more prone to problems such as bad welding and incorrect installation of valves.

The primary exception would be the nation's natural gas pipelines, many of which are in urban areas and difficult to maintain. Also, the use of increased pressure appears to have caused scattered leaks and explosions, one of the worst of which occurred in New Mexico in 2000, killing 10 campers.[25] Many more accidents, however, involve bad valves and leaks in the distribution sector, including from appliances.

CONCLUSION: SITUATION NORMAL, ALL FOULED UP

The industry faces numerous challenges, but too many have bought into the idea that they are somehow novel or insurmountable. Every problem has appeared at various times in the industry, and some, such as the need for massive investments, are only apparently serious if the observer is ignorant of the context.

Naturally, there is no place where fiscal policies and regulations don't require reform and optimization. If nothing else, constantly evolving conditions require revisiting existing laws and rules.

However, I'm reminded of the wisdom of an old friend, Ted Eck of Amoco, who remarked that since politicians couldn't afford to be seen favoring the oil industry, meaning that any move they made to help it required an offsetting penalty of some sort, he would prefer that the government did nothing.

FOURTEEN

The Red Queen: Running Faster to Stay in Place

[A] whole new Saudi Arabia [will have to be found and developed] every couple of years to satisfy current demand forecasts.
Robert Hirsch, quoting Sadad Al-Husseini, 2005[1]

[J]ust to stay even we need the production of a new Texas every year, an Alaskan North Slope every nine months, or a new Saudi Arabia every three years. Obviously, this cannot continue.
Jimmy Carter, 1977

Many experts are concerned about the decline in production from existing oil fields, and have often pointed to it as a major obstacle to maintaining supply, going back at least to the speech by President Carter. But the fall in prices in the 1980s relaxed those concerns, outside the specialized group of long-term forecasters, until it was revived again by peak oil advocates.

The issue is quite clear: unlike a manufacturing operation, or even hard-rock mineral and energy producers, the fluid dynamics of oil flow means that production at any given well and field will decline over time. Thus, for production to even be maintained requires a significant amount of investment and new capacity. Supply curves like those from the IEA's World Energy Outlook, estimates that nearly 100 million barrels per day of gross capacity will need to be added by 2030 to raise the level of net capacity by 45 million barrels per day.

And some are concerned that the rate at which fields decline has increased in recent years. Simmons referred to this as "superstraw" technology, and his firm actually did some of the early research, publishing a report in the late 1990s that showed that natural gas wells in the Gulf of Mexico were declining much faster in their first years than in the past. The increase was gradual but the rate has become quite high. The implication was that insufficient drilling would see a faster decline in production than in decades past.

THE NEW DEBATE

As discussed previously, this issue became particularly salient when raised by Matthew Simmons, who had remarked that "we didn't know about . . . ." By which he meant that he was ignorant of the issue. And when Simmons looked at other areas, he found that high decline rates existed in fields like Yibal in Oman and in Texas natural gas fields. The Yibal case in particular led him to view "Yibal as a relevant case study of the risks involved in using modern oilfield technology to drain easily produced oil more quickly."[2] In his 2005 book, he pointed to a number of other fields, such as Prudhoe Bay in Alaska and the Forties field in the United Kingdom as examples of how high decline rates led to production collapse.[3]

There were just two problems with this. First, the fields he cited in the book were not relying on what he called "superstraw" technologies, but merely showing normal decline. For offshore fields, a decline rate of 15–20% is not at all unusual, partly because it's more difficult and expensive to do regular infill drilling or enhanced oil recovery.

Second, the "collapse" in production observed at certain fields was not terribly relevant to national production levels. Figure 14.1 shows both production from Forties and total U.K. production, the latter not entering long-term decline for two decades after the so-called collapse in Forties oil supply. (The 1988 drop was due to the Piper Alpha field disaster.)

Geologists

Blanket statements like "economists don't understand depletion" should never be taken seriously, although they might contain an element of truth in them, as in this case. The reality is that few people understand depletion, including most economists. On the other hand, depletion has not only been discussed by economists going back to David Ricardo (1772–1823) but my mentor M. A. Adelman wrote the specific equations demonstrating the impact on costs of depletion.[4]

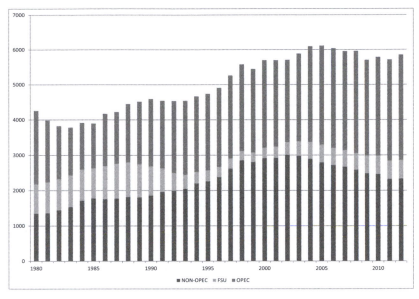

Figure 14.1   Forties Field Decline and UK Total Production (BP for Total, UK Brown Book. Sadly, Annual Data for Field Production in the UK Ceased to be Published a Decade Ago)

Recall, of course, that many peak oil advocates are neither geologists, resource economists nor familiar with forecasting supply, so it is no surprise that those like Simmons or Aleklett were shocked to realize how much effort was maintained "just standing still."

No Time before the Present

Now, here, you see, it takes all the running you can do, to keep in the same place. If you want to get somewhere else, you must run at least twice as fast as that!

Red Queen, in *Through the Looking Glass*

The trouble with conventional oil is that companies are always on a tread mill which is running faster and faster. Eventually it runs too fast and they begin to fall off it. It's the old "Red Queen" effect.

Theoildrum.com, September 10, 2012[5]

And once again, novice analysts are found to be excluding historical context. Graphs like Figure 14.1 don't seem to be doing that, inasmuch as they include some historical data, but the point is that they actually show only future depletion, omitting historical depletion that was offset by new investment. This allows for an exaggeration of the impact of depletion, which has been going on since the first oil well was drilled. The famous line that the industry "must run faster and faster just to stay in place" ignores the fact that it has been doing so for a long time—and actually *gaining ground*, not just staying in place.

WHAT IS DECLINE?

Many analysts, and especially the bulk of peak oil advocates, observe production trends at the field level and call them the decline rate, which is technically incorrect. The "natural" decline rate is the drop in production from depletion, but what is observed is the combination of the natural decline rate and the offsetting factor of new investment, which adds capacity. This can be in the form of new wells drilled or water injection, for example, and it is often lumpy—nonexistent at some times, heavy at others. In theory, this additional investment is strongest when prices are high, but there are many complicating factors—and no serious data collection on such investment. Most companies only announce new field developments or major projects to redevelop an existing field, but everything else is off the radar.

A rough estimate can be made at the aggregate level, by looking at the ratio of production to reserves, which is what economists refer to as the "decline rate" and apply that to production (which usually represents capacity). This shows, that the amount of capacity lost to depletion has not increased seriously in recent years. The message is that, again, depletion appears to be a serious problem only when one ignores the historical context. Alarmists worry that the world needs to 64 million barrels of day needed to be added by 2030, which they claim is 20 million barrels per day more than in the preceding 23 years. But using conventional estimates, the industry added 60 million barrels per day in non-OPEC capacity from 1990 to 2010, and about 110 million barrels per day overall during that period.[6]

Unfortunately, there are no global capacity estimates, but given that production has increased by roughly 20 million barrels per day in the past two decades, the implication is that annual increases in net capacity are around a million barrels a day per year. In other words, the industry has added something like 7 million barrels per day per year for the past decade in gross capacity—beyond what alarmists treat as indicative of a looming catastrophe.

Bottom-Up Analysis

Another approach is to look at well production figures; however, these data are unavailable outside of the Gulf of Mexico. What other analysts have done, including the IEA is look at production trends in large fields.[7] This has led them to believe that the depletion rate has increased significantly, and presumably explains why non-OPEC production is no longer growing as it did before 1998.

But again, these estimates represent the interaction of a number of variables, including more intensive exploitation, that can be driven by prices and which are not apparent by simply observing production trends.

Needless to say, the uncertainty about proved reserves, especially at the national level, feeds into this problem. Aside from the Middle East, both Mexico and Russia have used different definitions for reserves than the SPE "proved reserves" concept (Chapter 15). And while Saudi Arabia, as a whole, has a 1.6% decline rate, because most of the production comes from a few fields, their decline rate in fields actually producing is higher than the national average.

Better Not Faster

This is also part of the never-ending struggle between petroleum economists and petroleum engineers. The former want to maximize net present value, the latter, total recovery. This leads to very different desired production paths, with the economists preferring more production quickly, all else being equal. That is, at least to the point where costs are being driven up too high.[8]

But the difference is normally not huge, especially since petroleum engineers work for companies that want to profit maximize, and they factor that into their plans. Still this helps explain why a giant oil field at a state-owned oil company can show a long production plateau, while a similar field in private hands will usually have an early peak and long decline.

The impact of new technologies and profit maximizing can be seen in the production of small fields in the North Sea. Many of these were discovered but left fallow, as the amounts of oil were too small to justify building a platform. But floating production systems meant that the platform need not be scrapped on field shutdown, but moved to a new location, making it economic to produce all of a deposit's oil in a few years, which equates with a very high decline rate. (These smaller fields have not been included in the studies cited below, however.)

Fears of rising depletion do appear to be ignoring this important effect: to what extent do new, more productive wells offset the more rapid decline of older wells? Obviously, this doesn't negate the worry that a drop in drilling means a more rapid production decline than in decades past, but it goes a long way toward explaining why the increase in decline rates does not seem to have caused a major slowdown in overall production.

FAILURE OF THE MODEL

Although the general question of the impact of depletion is a legitimate one, the degree to which it has been seized upon by peak oil advocates as well as those more conventionally pessimistic about oil supply needs to be addressed. But there is a difference between the two, which again confirms the pessimistic bias of the peak oil advocates.

Many peak oil advocates have referenced decline rates, as mentioned, but few have actually done much explicit analysis with them, although most of their oil production forecasts presumably take decline rates into account. In those instances, however, they usually rely on "decline rates" to refer to production trends, not the "natural decline rate" which excludes the impact of additional development.

Separating the effect of inaccurate estimation or forecast of decline rates is not easy, but in a few cases, authors are emphasizing the use of decline rates in their forecasts. McCarthy (2008) shows global production at a peak in 2010 and, with a 4.5% decline rate assumed and using the Megaprojects database, arrives at a figure of about 75 million barrels per day by 2015, an error of roughly 10 million barrels per day.[9]

Similarly, the Uppsala Global Energy Studies Group with its extensive analysis of decline rates predicted that global crude would decline sharply.[10] Actual production is now slightly above the level in the "Standard High Case" from the Robelius work, but this is partly due to his failure to expect the 2008 recession and subsequent weakness in oil demand.[11] A later forecast by the Group shows crude and condensate production dropping from just over 70 million barrels per day in 2008 to about 62 million barrels per day in 2015, whereas it has actually increased by 5 million barrels per day.[12]

There are also a couple of instances involving natural gas. First, Simmons looked at per-well natural gas production in Texas, and concluded that United States natural gas production was on the verge of collapse, dropping by at least 10% and up to 20% per year.[13] This was far beyond any decline seen historically, and did not in fact prove true–although partly due to the surprising development of shale gas. Dry gas production did drop in three of the next 10 years, but the biggest drop was only 3.5%.

More generally, the application of a "rate of decline" to countries by analysts like Campbell has resulted in bad results overall, because of the presumption that a "postpeak" country will decline continuously. As Chapter 9 described, this has proved repeatedly false and helps to explain why so many such forecasts have proven woefully pessimistic.

ESTIMATED DECLINE RATES IN POSTPEAK COUNTRIES

The idea that decline rates are fixed and that trends don't change is particularly erroneous, and the data support this. Aside from the misinterpretation of political factors affecting supply as geophysical in nature, such as the collapse in Soviet production as described in Chapter 9, the reality is that depletion can and often has been overcome. Cases include Colombia and Oman, both said to be "postpeak" by the peak oil advocates but both of which have seen new record production levels in recent years, the result of increased incentives for upstream investment.

Empirical Results

When the issue of decline rates was at the fore, several studies were done to try to quantify them, most notably the IEA in its 2008 *World Energy Outlook*. Table 14.1 shows the results for fields that are assumed to be postpeak, and broken down by region and size. The most important caveat revolves around the OPEC countries, where many fields have suppressed production because of lack of market demand (or a desire to avoid crashing prices). Otherwise, it is worth noting that there is a huge divergence from area to area, and according to size. The lowest decline rates are mostly in the supergiant fields, while fields in the smallest category—large—tend to have the highest decline rates. This presumably reflects the fact that it is usually easier to develop the largest fields more intensively, or that they are likely to have priority for additional investment.

Table 14.1   Decline Rates by Area and Field Size (IEA WEO 2008)

	Super-Giant (%)	Giant (%)	Large (%)	Total (%)
OPEC	2.3	5.4	9.1	3.1
Middle East	2.2	6.3	4.4	2.6
Other	4.8	5.0	10.2	5.2
Non-OPEC	5.7	6.9	10.5	7.1
OECD North America	6.4	5.4	12.1	6.5
OECD Europe		10.0	13.5	11.5
OECD Asia		11.1	13.2	11.6
East Europe/Eurasia	5.1	5.0	12.1	5.1
Asia	2.1	8.3	6.6	6.1
Middle East	2.2	6.5	7.4	2.7
Africa	1.5	5.2	8.8	5.1
Latin America	8.4	5.2	6.9	6.0
World	3.4	6.5	10.4	5.1

They also found that the decline rate increased over time, as shown in Table 14.2. The increase shows as steady in all regions, except for OPEC in the 2000s, which almost certainly reflects a combination of production problems in some countries and others, like Saudi Arabia, that produced large fields at relatively constant rates (see later in this chapter).

These findings are problematic in several ways. First, OPEC decline rates are heavily influenced by their efforts to stabilize/influence oil prices, as many countries have, at times, shut in or slowed production to avoid driving prices down. Thus, the decline rates for those countries are not very indicative of geophysical or technological elements.

Excluding OPEC countries gives us a better idea of the concerns about rising decline rates, which have roughly tripled in a half century; this is not to say that the increase has been at an alarming rate, but rather steadily rising over a long time.

Table 14.2   Decline Rate over Time (IEA World Energy Outlook 2008)

	Pre-1970s (%)	1970s (%)	1980s (%)	1990s (%)	2000s (%)	Total (%)
OPEC	2.8	3.5	4.6	7.5	5.0	3.1
Non-OPEC	5.9	6.8	8.3	11.6	14.5	7.1
World	3.9	5.9	7.9	10.6	12.6	5.1

In its analysis, CERA's findings were as follows:

• The aggregate global decline rate for fields in production is approximately 4.5 percent per year.
• Individual field decline rates can range to over 20 percent.
• Annual field decline rates are not increasing with time.
• Decline rates are a function of reservoir physics and investment strategies.[14]

This does not differ greatly from the IEA's findings, except for the important conclusion that annual field decline rates are *not increasing with time.*

The Uppsala Global Energy Systems Group has done its own estimates of decline rates using data for giant fields, and it approximately agrees with the IEA's, with any disparities apparently due to differences in data coverage and definition.[15]

WHY DO THE NUMBERS VARY SO MUCH?

Even a casual perusal of Table 14.1 should leave everyone confused as to why the simple act of producing oil can have so many outcomes. Without question, there is no "standard model" for oil production from a field or its decline rate. This is reminiscent of the experience with oil reserve growth that was described in Chapter 7, where the theory did not match actual behavior.

Aside from randomness, there are good physical but also policy reasons why the numbers deviate so much. First and foremost, the type of rock (particularly its porosity and permeability) and the viscosity of the oil will go a long way toward explaining the rate at which the oil can be extracted. Light oil should flow faster and thus such fields should decline faster, while heavy oil fields can be expected to show a long steady decline. Geology with good porosity, such as sandstone, will have higher decline rates while heavy oil should have lower ones.

Of course, categorizing the decline rates by geology and oil chemistry would seem to provide a scientific model for predicting decline rates, but reality is, as always, more complex. There are many geological factors besides the type of rock that determine porosity, including micropores, fractures, and intercrystal pores, so that treating basin types as homogenous is an oversimplification.

Another important element historically is the difference between onshore and offshore fields: onshore fields could be developed slowly, and depletion could be offset with a constant drilling program. With off-shore fields, drilling stepout wells was difficult and expensive, at least historically. Now, the combination of horizontal drilling and subsea installations means that after a platform has been installed, additional drilling is much easier than it used to be. In other words, the decline rates for offshore fields should be converging with the rates for onshore fields.

But the most important element for our study would be time, or more accurately, the technology used to extract the oil, which evolves over time. Needless to say, any given field operator might not be using the latest technology but in aggregate, the correlation should be pretty good: more modern fields use better technology and extract the oil faster. This is seen empirically, as decline rates have grown over time.

RE-ESTIMATION OF DECLINE RATES

The other major problem with the argument that field decline rates are getting extreme concerns that fact that decline rates are not independent of price. In the years before the studies cited earlier were performed, the price had experienced a major collapse, going from $27/barrel in mid-1997 to $14 in late 1998. Major projects, like deepwater field developments, are usually not affected by short-term price movements, but smaller, incremental expenditures like infill drilling and new gas or water injection wells will tend to be curtailed, meaning the estimates of decline rates might be too high.

The test for this would be if production had recovered in some of the fields measured by groups like the IEA and CERA in their decline rates studies. For those without access to commercial oil field databases (including your humble narrator), there are nonetheless some examples that can be considered. The most obvious is the Forties giant field in the UK North Sea, cited by Laherrere as an example of the unalterable nature of production trends (see Chapter 6). In fact, since Apache bought the field in 2003, it has invested in adding reserves and maintaining production, so that the current production rate is roughly the same as it was a decade ago.[16] This means that the decline rate has gone from 8.8% a year (as of 2004) to 6.1% per year as of now.

(The Samotlor field in Russia, cited by Simmons as a prime example of the impact of rapid decline rates has also seen production flatten out. After declining sharply for 15 years, it has shown no decline since 1999. However, it is somewhat of a special case, given the transformation from Soviet oil practices to more modern efforts.)

Modelling future field behaviour is done by extrapolating the histori-cal production data with an exponential decline curve. This does not take dramatic deviations into account and *assumes that declines will continue approximately exponentially.* (emphasis added)[17]

And recently, the Thistle field has been redeveloped, bringing produc-tion back to *twice the level of a decade ago*, which violates the assumption in the work by Hook et al quoted above. There, they found that the Thistle field was "well-behaved" because its production curve followed a fairly stable exponential decline pattern for approximately two decades, so the authors apparently felt confident that this would continue. They acknowl-edge that other factors like new technology can change the pattern, but don't seem to incorporate that in any of their forecasts. In actuality, it means that the 16% per year decline rate for Thistle measured at the time of their study would be about 8.5% now.[18]

INPUTS AND OUTPUTS: FIELD DECLINE IN THE REAL WORLD

All that has been demonstrated is that, in the worst case, more drilling and investment is needed to meet demand, in theory and on average. Witness, for example, cases like Iran and Venezuela, where production has not suffered egregiously from high decline rates: 8% and 20%, respec-tively. Although both countries have had problems maintaining production in recent years, this is primarily because of political and man-agement issues, not due to higher and insurmountable decline rates.

It also reported major project delays and accidents as "evidence that the Saudi Aramco is *having to run harder to stay in place*—to replace the decline in existing production." While fears of premature "peak oil" and Saudi production problems had been expressed before, no US official has come close to saying this in public.[19] WikiLeaks cables: Saudi Arabia cannot pump enough oil to keep a lid on prices. (emphasis added)

The Guardian, February 8, 2011

Peak oil advocates often point to the fact that Saudi Arabia's Ghawar field has a decline rate of 8%, necessitating 400 thousand barrels per day of new capacity each year to offset, implying an imminent crisis. Yet the Saudis have had no trouble doing so, and peak oil advocates, by looking forward without the historical context, somehow thought this would be too challenging for the Saudis to deal with.

This takes us full circle to the decline rate studies of a decade ago. Their conclusion was not that rising decline rates would cause a production collapse, but that higher drilling levels were needed to maintain production. The same is true on the global stage—except that global resources vary so wildly that the location of drilling is often more important than anything else. Shifting a dozen rigs from Argentina to Iraq, for example, would greatly boost capacity additions for the same amount of work.

And, as Chapter 19 will show, repeated predictions of a sharp decline in shale production (especially gas) have been completely confounded by improvements in extraction methods. Such a massive leap in productivity is unlikely to occur in conventional production, but it demonstrates, yet again, that too many analysts tend to start by adopting either an optimistic or pessimistic position and then seek supporting evidence.

Thus it is that, with the initial arguments about geology and peak oil all refuted, many have turned instead to the issue of decline rates. This has served them well for two reasons: the data are all but unavailable, and the analysis is difficult to accomplish. But more important, it is a real concern for oil analysts, even those, who, like myself, reject the "peak oil philosophy." But the data show that there is no reason to consider this more than a challenge that has been dealt with—and continues to be dealt with.

OMITTED VARIABLES AND EXTRAPOLATION

Something encountered repeatedly in the peak oil debate is the disconnect between inputs and outputs, specifically the frequent description of supposedly insurmountable problems that have no apparent effect on production. The many problems described in Chapter 13 are good examples, but decline rates are another. All countries experience decline in their oil fields, but many do not then have continually falling production. Recalling that peak oil advocates often insist that, once in decline, a country cannot subsequently increase production and frequently highlight the number of "postpeak" countries as evidence of imminent decline, their repeated failed predictions of imminent peak are explained. Similarly, with assertions by Aleklett and Laherrere that historical decline patterns predict future production rates.

Thus, Campbell (1997) assures us that Colombia peaked in 1999 with a 3.8% depletion rate. Production that year was 838 thousand barrels per day, and, in 2013, 14 years after his projected peak, production was 1,004

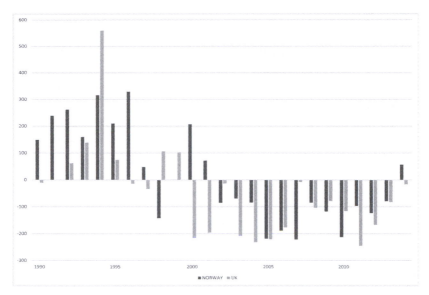

Figure 14.2   Annual Change in Production, Norway and UK (BP Statistical Review)

thousand barrels per day, meaning an increase of 1.3% per year. In other words, the difference between the natural decline or depletion rate and the "rate of decline" or production trend was greater than the decline rate.

Figure 14.2 shows annual changes in production for Norway and the United Kingdom, two areas that have been hailed as representing the best examples of postpeak producers. The only pattern visible is the lack of pattern. Examining individual fields, such as offshore Norway, the same conclusion is drawn. Not only do oil field production trends not follow similar patterns, as analysts like Aleklett et al. argue, but nearly every aspect varies, including time to peak, number of peaks, and rate of increase and decrease at different points in their production histories.

WHY

Again, the problem of omitted variables rears its ugly head. Although field size matters and the location (onshore vs. offshore) does affect decline rates, over the course of a decade and more, a number of other variables are important. The biggest issue would appear to be economics, which is a combination of the oil price and the local fiscal policy, both of which fluctuate. If the oil price rises, more investment for greater recovery can be expected, while when prices fall, this is presumably among the first budget items to be cut or delayed.

Similarly, the government's decision to allow tax breaks for new investment in old fields, something often seen in the North Sea, will lower the rate of production decline, while no such provision would tend to result in higher decline rates. Since these tax provisions tend to be applied at discrete times, they cause the investment pattern to be lumpy, and make it hard to interpret trends without detailed knowledge of fiscal policies in a given country.

And new technology has clearly allowed the redevelopment of many oil fields, expanding the EUR or recoverable reserves and changing the decline rates, despite the insistence of peak oil analysts that such does not occur (see Chapter 7).

CONCLUSION: IMPORTANT BUT NOT SERIOUS

The primary concern that rising decline rates will result in a production plateau appears to be seriously overstated. Although decline rates have increased in some areas, notably non-OPEC production, the issue is hardly new and has not changed dramatically, but rather evolved gradually. And the industry has a long habit of replacing the capacity lost to depletion, while increasing production generally. The biggest problem would appear to be ignorance among peak oil advocates and a lack of detailed analysis, allowing this to be treated as a major problem, not an ongoing challenge.

FIFTEEN

Resources

Aside from the belief that petroleum resources are finite, few people know much about either the amounts thought to exist or questions and controversies involving those estimates. This chapter will discuss the evolution of resource estimates and especially highlight the conservatism in their development.

THE FIRST WAVE: MORE BELIEF THAN KNOWLEDGE

> John Archbold, a top executive of Standard [now Exxon], was told [in 1885] by one of his company's specialists that decline in American production was almost inevitable and that the chances of finding another large field "are at least one hundred to one against it." ... "Why, I'll drink every gallon produced west of the Mississippi!"[1]

In the early days of the oil industry, expectations of depletion were common, reflecting a combination of ignorance and pessimistic bias. Geology was still a young science and petroleum geology reflected this, as did the fact that little work had been done to examine the subsurface in most of the planet. Given that, it is astonishing that so many seemed to have firm opinions about the limitations of the resource, as the case Yergin notes earlier.

Indeed, Clark and Halbouty noted that the American railroads were slow to convert to oil, a cleaner, more manageable fuel than coal, because of fears that discoveries and production might cease.[2] It took the discovery of the gusher at Spindletop in 1901 to convince them that supplies were abundant enough to make the effort.

And even then, there was an episode of "peak oil" enthusiasm, highlighted by a 1919 *Scientific American* article that predicted a peak within a decade.[3] Aside from blind pessimism, these fears were typically driven by a focus on the known resource and the assumption that existing knowledge was, if not perfect, very close to it. Obviously, this seems foolish for those in the early years of the industry, but many erroneously believe that current estimates have now come close to a perfect understanding of the globe's resources. Such is not the case.

ESTIMATION METHODS

Many of the estimates of resources are not published in academic form or are no longer easily available, and the methods used cannot be known. However, of the modern work that can be considered rigorous, specifically the USGS studies, we have a good idea of how the estimates are made. Some were mere back of the envelope calculations, especially in the early years when there was little information available about much of the world.

For the first half century of the industry, it would seem as if the state of the knowledge was so poor that most estimates were rank guesses. The long history of pessimism about the resource base reflects the approach of: I don't know anything, therefore assume nothing exists. This is somewhat reminiscent of the paleontologists who had a grand total of one area chock full of fossils from the Cambrian era, and had almost no other evidence, and so concluded it was a unique "explosion of life" that ended abruptly—until they found more fossils.

More recently, estimation has become more sophisticated, especially with the use of creaming curves in well-established basins, such as the United States and Canada. Naturally, few other areas have such a lengthy, detailed exploratory record, so they are supplemented with seismic studies and educated guesses from professionals working in the various areas. Mathematical techniques like Monte Carlo simulation are often used to reduce uncertainty.

Even so, the method is limited by the many blank spots on the map, including those that haven't been drilled and some where limited seismic is available. The USGS notes that it specifically excluded 10 total petroleum systems from its massive 2000 study because of a lack of information. In one case, however, where seismic information was available but no drilling had occurred they proffered estimates: Greenland.

> There are several reasons for treating the 2000 USGS report on the world's undiscovered as unreliable. . . . [T]heir estimate of the Undiscovered was based on a one-page form (seventh approximation sheet) filled by an academic geologist, without the help of oil industry expertise and knowledge, such as seismic, wells and production. The best example of this flawed approach is their estimate of 47 Gb of undiscovered oil in East Greenland, which the industry would not ignore if it had any realistic substance.
>
> Jean Laherrere[4]

Greenland's geology bears a significant resemblance to that of the North Sea, which was a major center of oil production in the 1980s and 1990s, though it is now much diminished. Continental drift has resulted in movement of Greenland away from Western Europe, as in the case of offshore Brazil and the analogous province off West Africa, so that the USGS estimated that there is potential for the area offshore Greenland to hold a significant amount of petroleum.

But, and here's the rub, they also clearly stated that the uncertainty was very high, much more so than in other areas, because of the lack of drilling. Peak oil advocates have attacked them for their supposedly unrealistic optimism, but the open discussion of the uncertainty—also apparent in the numerical estimates—refutes that: Greenland was estimated to have between 0 and 111 billion barrels (5% to 95% probabilities).[5]

REAL NUMBERS

The National Petroleum Council listed the most exhaustive compilation of URR estimates for oil and gas that has yet been published, as shown in Figure 15.1. Again, the reliability of the methodology used in any given estimate is not clear, but it is obvious that these estimates are increasing over time. Note, for example, the USGS numbers that have gone from 1719 to 3021 billion barrels from 1981 to 2000 (Table 15.1). Indeed, even the ASPO numbers have increased regularly. Campbell increased his estimates from 1575 billion barrels in 1989 to 1900 in 2005, the last estimate he published.[6] Since his estimates actually increased faster than depletion was occurring, his own numbers suggested no peak was near, contradicting his claims of scarcity

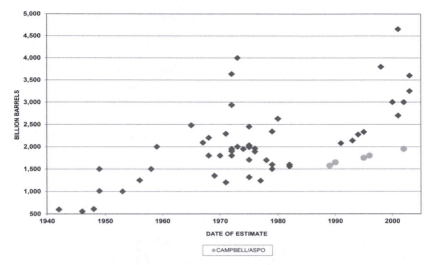

Figure 15.1   Estimates of URR Over Time (Institute for Gas Technology, "A Survey of United States and Total World Production, Proved Reserves, and Remaining Recoverable Resources of Fossil Fuels and Uranium as of December 31, 1984." Chicago, 1986 and National Petroleum Council, Hard Truths: Facing Hard Truths about Energy. Washington, D.C. 2007.)

Estimates of URR Over Time

Examining the estimates of one group allows us to control for methodology. Table 15.1 shows how the estimates of the USGS have evolved over three and a half decades. Reserve growth was only introduced in the 2000 publication, and the most recent estimates of identified reserves have been inflated by approximately 300 billion barrels due to the recent inclusion of oil sands in Canada and Orinoco heavy oil in Venezuela. What is most telling is the fact that undiscovered conventional oil in the most recent estimate is roughly the same amount as it was 30 years ago, even as over a trillion barrels have been found, or over twice the level of undiscovered oil in the 1981 estimate.

UNCERTAINTIES

[Peak oil advocates] also believe that essentially all regions of the Earth favorable for oil production have been well explored for oil, and there are few surprises left except perhaps in regions that will be nearly impossible to exploit.[7]

Table 15.1 Evolution of USGS URR Estimates

			Effective Date of Assessment			
	January 1, 1981	January 1, 1985	January 1, 1990	January 1, 1993	January 1, 1996	2010
Cumulative Production	445	524	629	699	710	1,100
Identified Reserves	724	795	1,053	1,103	891	1,354
Undiscovered Conventional Oil	550	425	489	471	732	565
Reserve growth (Conventional)					688	665
Future Resources (mode) (2+3)	1,274	1,220	1,542	1,574	2,311	2,584
Total Resources (mode)	1,719	1,744	2,171	2,273	3,021	3,684

One of the great myths about petroleum is that the world's geology is so well understood that there are no surprises remaining. Even those who don't subscribe to the Campbell notion that nearly everything has been found will still argue that no major basins are likely to be undiscovered, which proves to be mostly meaningless. Although the Middle Eastern basins are the most prolific by far, there has only been one such discovered; no other comes close. Yet the Arabian/Persian Gulf usually accounts for only one-quarter of global production and the supposedly "irrelevant" smaller basins add up to the other three-fourths.

And a noticeable portion of all articles or blogs about petroleum that discuss resources also misunderstand what is meant by resources and especially reserves. This is particularly true where the comment comes from a novice, such as a politician, or an environmentalist, but even industry members make occasional mistakes. In one unusual case, T. Boone Pickens, who considers himself a leading oil man, said in *Playboy,* that "We're halfway through all the oil in the world. We have produced about a trillion barrels, and there are probably about a trillion barrels to go."[8] The implication is that he doesn't know understand the difference between resources and reserves.

Before the 1970s, attention to energy was minimal; few knew or cared how the industry categorized petroleum resources. Occasional references were largely of interest to geologists and academics. That Hubbert's 1956 prediction of a U.S. peak received little attention in the general media shows how unconcerned the public was.

Now, however, neo-Malthusians and peak oil advocates have argued that the resource is increasingly limited, and environmentalists promote renewable energy for the same reason to where few seem to question the limits, even as the industry raises its estimates of what can be produced. Pessimism is fueled by ignorance of how the estimates are made and what they represent, including some basic mistakes about the technical jargon used.

Going Astray

We could use up all of the proven reserves of oil in the world by the end of the next decade.

Jimmy Carter (1977)[9]

First, let's examine some of the common mistakes, so that they can be avoided. These include misinterpreting technical terms like *years of production* or *resources,* but also, thinking that the recoverable resource is static.

People often become alarmed when they hear that "53 years of oil are left," in the world and only 11 for the United States (BP's estimate).[10] This measure is often used by reporters writing about the economy of a small country, such as Brunei, as they say that in 18 years, the oil will be gone and there will be no more oil revenue.

Actually, this is what the industry calls the "reserves to production" ratio, which literally means the length of time it would take to produce the amount of reserves *if* no new discoveries were made and *if* production were unchanged until the oil was gone. To show just how misleading the measure is, the world had 30 "years" of reserves in 1952, and now has 53, and the United States has not had more than 15 "years" of reserves during that entire period.

Neither assumption ever holds true in the real world, which might cause one to wonder why the measure it is used at all. In the industry, its purpose is solely to indicate the *rate* of production: a high ratio means oil is under-produced, a low one that it is produced at or near the maximum rate. The United States, where there are no government restrictions on rate of production, has a ratio that represents something close to the maximum possible: higher production would require more discoveries.

At the other end of the spectrum is Saudi Arabia, where the level has been over 100 years for a long time; they are obviously underproducing, primarily to avoid crashing the price of oil. Indeed, they have added several million barrels a day of capacity in the past five years without any significant discoveries, just by tapping existing but underutilized reserves and fields. The ratio dropped all the way to 63 years in 2014 due to increased production, but it is still five times that of the United States.

Figure 15.2 shows the global "years of production" or "reserves to production" ratio, and how it has changed over the years. Certainly, there are disputes about whether or not Canadian oil sands should be included (or how much), and the so-called spurious reserve revisions in the Arabian/Persian Gulf in the 1980s, but even if corrected for those, the world has continually found enough oil to maintain reserve levels relative to production. The claim that discoveries have not kept up with production for several decades reflects a misunderstanding of the nomenclature "discoveries," as Chapter 7 discussed.

The "years of production" measure is also useful if turned on its head: production per reserves is also known as the decline or depletion rate, and signifies the proportion of proved reserves being produced. This is useful because, given the fluid nature of the oil, capacity can be assumed to drop at a field or in a nation by the depletion rate—*if no other drilling*

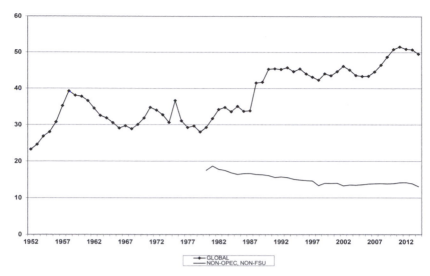

Figure 15.2   Reserves-to-Production Ratio in Years (BP Statistical Review)

occurs. Thus, the U.S. figure of 10% would imply an annual decline of 10% in production if there were no more development, although this applies more to conventional oil.

EUR and URR

One odd misstep is the use of the term *estimated ultimate resources* instead of *ultimately recoverable resources,* which is both more precise and more commonly used. The primary issue is the term *recoverable,* which some writers omit—although there are those who say Estimated Ultimate Recoverable. Few, if any, talk about the total resource (this book one exception; NPC [2007] is another) and rather consider only the portion thought to be *recoverable.*

This is an important point because, as noted below, the percentage that is recoverable depends heavily on the state of the art of geology and reservoir engineering. As geologists improve their knowledge, they have a better idea of how to get oil out of the ground, which is implemented by the engineers, who are also developing new methods and technologies as time progresses. A century ago, the recovery rate was estimated at 10%; now most analysts put it roughly at 35%.[11]

By ignoring the issue of recoverability (and scientific and technological advances), it becomes easier for neo-Malthusians to see looming scarcity. The resource base is thus treated as fixed, often at a low level. Ignoring this fact—and the reality of technological advance—are crucial to the many peak oil advocates' beliefs. In truth, the URR/EUR estimates are dynamic, with the number determined by a variety of factors most especially changing technology.

Geologists classify resources with what is known as a McKelvey Box, which includes economically unrecoverable resources as well as simply showing discovered and undiscovered recoverable resources.

Resource and Reserves: Moving among the Categories

[R]eserves represent what will be recovered in future or expected future production
[R]esource is what is in the ground; reserves are only a small part of resource.

Jean Laherrere[12]

The process by which proved reserves are increased at one end while being depleted at the other is the movement of inventory from the poorly lit warehouse of probable reserves into the well-lit shelves of proved reserves, from where it is sold and disappears.

M. A. Adelman[13]

Peak oil advocates like Colin Campbell and Jean Laherrere make much of the technical term *reserves,* which they note is not understood by many. Of course, they mistakenly think that "proved" reserves was developed by the SEC, not the engineering profession, which has had precise definitions for decades. Proved reserves, 1P or P90, are essentially oil that has been found and developed; oil that has a 90% chance of being there and available for production.

Proved plus probable, 2P or P50 reserves, which Campbell and Laherrere rely on in their analysis, represents a broader definition, and includes oil that is thought likely to be available for production with a 50% probability. In theory, estimates of proved plus probable reserves will, over time, be accurate *on average*. The reality, as Chapter 7 shows, is very different.

The term *resources* is the broadest, in part because no judgment is made of the proportion that can be produced.[14] To refer to the petroleum resource is to mean the actual amount in the ground, or "in-place" as geologists say, and few make this actual estimate, instead referring to the recoverable portion. Of course, the industry will never turn 100% of the resource into reserves but for decades to come, improvements will make a major contribution to world supply, causing URR to increase as has been happening throughout the history of the industry.

Most reports of resources are actually reserves, such as the annual *Oil & Gas Journal* survey or BP's Statistical Review of World Energy. Corporate reports only cover proved and probable reserves; oil in other categories are considered too speculative to be reported by companies whose shares are publicly traded, as they might be misinterpreted by investors.

Deductions: The Other Oil

There are enormous amounts of petroleum that are simply not counted when most professional groups publish estimates of the amount of oil that actually exists. These reflect questions of technical and/or economic recoverability, as well as information constraints. In the industry, they are generally understood to the point where they are taken for granted; many simply assume the reader is expert enough to recognize the implied constraints.

Conventional or Unconventional

Before discussing the "good stuff," conventional oil, a brief aside about unconventional oil, which is plentiful but usually expensive or even impossible to produce. At the beginning of my career in the late 1970s, we understood "conventional" to mean oil that could be produced with existing technology and at existing prices. Thus, a small, isolated oil field would be "unconventional," although it is more appropriate to say it is "unrecoverable" or uneconomic.

In recent years, following the lead of Colin Campbell, peak oil advocates have often tried to parse oil supplies as conventional or unconventional according to a number of novel definitions, most of which are neither useful nor relied on by the oil industry. Specifically, Campbell excludes from "conventional" or "regular," oil the following categories:

- Arctic oil
- Heavy oil
- Deepwater oil
- NGLs

Why Arctic oil? Because it's harder to produce, presumably. (I've never been able to find a citation to his source for data on fields by latitude.)

But why not "desert" oil or "jungle" oil? Or "oil that isn't close to a good hamburger joint?" (Okay, a bar.)

Price is also a factor, at least sometimes. Laherrere said oil in small accumulations that can't be produced for $25 a barrel ($32 in 2010 dollars) should not be considered conventional, but Campbell takes the position that price is not relevant, the oil either is there or it isn't.[15]

Many analysts did separate deepwater from other offshore fields in the early days of deepwater production (and offshore from onshore), but the meaning of "deep," is somewhat arbitrary and changes over time as rigs become increasingly sophisticated and capable of deepwater drilling. Producing oil from a field in 1,000 feet of water 30 years ago was very challenging, and some would have labeled it unconventional. Yet the processes are very similar to those of shallow water fields, and today, few label it as "unconventional."

Current Definition of Unconventional

Some categories of oil are clearly unconventional, in the sense that production is significantly different from conventional deposits. These include "continuous" deposits, which have different production requirements and profiles, and heavy oil, or tar sands, which also have different production requirements and profiles, but require some additional chemical conversion to be useful.

Continuous resources as a category refers to oil that is not found in discrete pockets, but rather large, mostly unbroken masses. The oil sands in Canada or the Orinoco tar belts are prime examples. There is no particular geological risk in these areas; nearly everywhere you drill, you will find the oil. However, the reason it is not in discrete basins is because it tends not to flow: heavy oil is highly viscous.

Tight oil, found in shales, is a similar example except that here, the problem is not the viscosity of the oil but the porosity of the shales, which is very low and inhibits the oil from flowing. As a result, there are large amounts spread very thinly throughout extremely large zones of shale.

Heavy oil, sadly, is not a precise definition but is used for oils that have a particularly low API gravity, especially if it requires special processing. In the case of Canadian oil sands or the Orinoco belt, the heavy oil is less than 10 degrees API (very close to tar), but also continuous, in the sense of being enormous and in deposits large enough to mine (if desired).

On the other hand, there are many other deposits that are heavy but would be considered conventional, some of which are very large. What they have

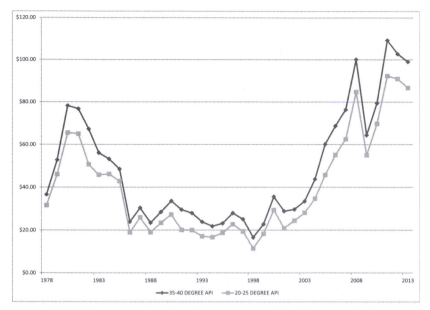

Figure 15.3   U.S. Import Prices for Light and Heavy Crude (US DOE EIA)

in common is that typically, production requires the application of steam and the produced oil usually requires some upgrading (increasing the hydrogen–carbon ratio, essentially). They do not flow as "conventional" oil does, and have much lower recovery rates. For example, the Ugnu deposit in Alaska contains an estimated 18 billion barrels of oil in place,[16] but the oil is hard to produce because it is uniquely located, in an area of permafrost that might melt if this shallow deposit were heated with steam.

For all the talk that heavy oil is "difficult," the reality is that it is not that expensive to upgrade it into the quality of conventional oil. This can be readily seen by the fact that the difference in prices rarely diverges more than about 20% (Figure 15.3) or, historically, between $5 and $10 per barrel. The cost of upgrading, which is only $8/barrel, sets the long-term margin, but in the short term, naturally, construction of refinery units does not occur and prices can vary more widely.

RECOVERY STEPS[17]

Initially, oil fields were produced using primary production techniques, that is, allowing the natural pressure in the field to be released, sending the oil and gas to the surface. This would normally allow 10–40% to be recovered, depending mainly on the chemistry/viscosity of the oil and the porosity of the rock, the low end most common.

The next step (secondary recovery) was to maintain pressure by injecting water, gas or steam into a field to increase pressure and/or reduce viscosity. Now, "enhanced oil recovery," or EOR, often called tertiary recovery, involves much more complex (and expensive) approaches, including the use of CO_2 and miscible polymer floods, and this can increase the recovery rate much further, sometimes 20–25%.

And advancing computer technology has allowed much better analysis of deposits' geology, with 3D images, so that places where pockets of oil were bypassed can be seen, allowing them to be produced, and even 4D analysis, showing the flow of oil to optimize its recovery. A variety of other methods, such as maximum reservoir contact (MRC) wells, hydraulic fracturing, and so forth, can increase recovery even more.

On a global level, however, the average recovery factor is still estimated to be about 33–35%, meaning that for every barrel in the "Reserves" classification in McKelvey Box, there are two barrels in the "Contingent Resources" category.[18] If one used the 10% recovery factor of a century ago, the implied global URR would be 1 trillion barrels, meaning oil would literally have "run out" more than a decade ago.

Needless to say, the recovery factor cannot rise above 100% and will never get very close to it; some amount will always be left behind, and some of the energy will be used to retrieve the oil. However, the current level of roughly 35% will almost certainly be improved upon over the next few decades. Chapter 7 has already shown how some areas have seen recovery factors rise well above this level, and the trend can clearly continue for decades to come.

Left behind Oil

This means that there is an enormous amount of oil that is unrecoverable— with present technology. Obviously, 100% recovery is unreachable, just as when you spill cooking oil on the floor, your cleaning efforts will leave behind ever-diminishing amounts until only CSI-Martha Stewart could detect molecules remaining. But the recovery factor is increasing, adding to the world's recoverable resources.

Some have gone so far as to suggest mining oil deposits to get every last bit of petroleum from the basin, but this would rarely be profitable, particularly if other methods have reduced the oil left behind to minimal levels, as in some North Sea fields.[19] There, a special effort by the Norwegian government has sought to raise recovery factors to 70%, while even in the

United Kingdom, numerous programs have been undertaken to improve recovery. The industry itself constantly works to apply new technologies to grow reserves. One trade journal article from consultancy Smith Rea Energy Associates described many of these and appeared simultaneously with an article by Colin Campbell.[20]

On a global level, it is hard to say precisely what proportion of oil has been left behind in conventional fields, but a typical number cited is 35% recovered.[21] Although not usually explicitly mentioned, this translates into *65% unrecovered,* or left behind oil. This probably understates the amount in a normal, non-OPEC field, as the supergiants like Ghawar have been deliberately underproduced and the Saudi goal of obtaining 75% from the Ain Dar/Shedgum/Arab D is not incorporated in the current global estimate—rather than the 60% that is now considered recoverable.[22] (A "mere" 10 billion barrel difference.)

What this means is that the amount of unrecovered, conventional oil that has already been found is enormous—as much as 7 trillion barrels, as Figure 15.4 shows. This number is rarely discussed: the USGS doesn't even mention recovery rates, for example, and Rogner (1997) doesn't raise the issue. Peak oil advocates, and indeed most others, rarely talk about this resource; a recent report by Kuuskra (2006) is a rare exception, and suggested that as much as 43 billion out of 205 billion barrels of discovered, but "left behind" oil in the United States, could be produced in the future.[23]

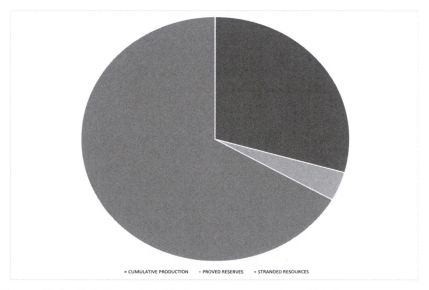

= CUMULATIVE PRODUCTION   = PROVED RESERVES   = STRANDED RESOURCES

Figure 15.4   U.S. Stranded Oil Resources (Kuuskra [2006])

THINK OUTSIDE THE McKELVEY BOX

What few people realize is that there is still more oil outside the McKelvey Box, that is, oil that is not considered part of the resource base by nearly any studies. While oil might be excluded from "proved reserves" estimates because a country has declared it off-limits and banned its development, there is also oil that, for various technical reasons, is simply not treated as part of the resource base, but sometimes moves onto it.

The most obvious example now is shale oil: despite an in-place resource of nearly 8 trillion barrels (320 times current production), until a few years ago it was not included in most resource estimates because it was simply impossible to produce it economically. Thus it was "off the radar" or "outside the McKelvey Box."

Similarly, Canadian oil sands at one time would have been omitted from a McKelvey box or resource estimate and, it could be argued, kerogen (immature shale) still is. That Canadian oil sands took roughly two decades to move from serious consideration to booming production shows how engineering or scientific advances can have huge impacts on the medium-term resource estimates and commodities market.

Long-term supply curves, as discussed in chapter 21, which show resources and their costs also tend to be incomplete. The IEA's curve, just a few years old, does not include shale oil, for example. Other resources, such as methane hydrates, are also typically omitted as they are seen as only available in the distant future. But as the case of shale oil demonstrates, that can change, unexpectedly and quickly.

Water Depth as a Microcosm

Although there were wells drilled from piers into water in the early days of the industry, it was not until after World War II that the industry began to produce serious amounts from offshore deposits, principally in the Gulf of Mexico. Steel platforms were the original method for producing an offshore field, but over time, better materials and control technologies have allowed for floating production systems (FPS) and tension leg platforms, among other things, that have meant that gradually, more and more basins can be accessed.

USGS OFFSHORE ESTIMATES

Thankfully, the U.S. government provides data on reserves by water depth. In Table 15.2, oil reserves according to water depth and (roughly)

Table 15.2   Gulf of Mexico Proved Reserves by Water Depth (Bureau of Ocean Energy Management, "Estimated Oil and Gas Reserves Gulf of Mexico OCS Region," December 31, 2013)

Year	Production System	Water Depth	Reserves MMBO
1965	Jack-up	300	11,289
1978	Compliant Tower	1,000	1,696
1990	Floating Production Systems	2,000	1,334
1995	Tension Leg Platform	5,000	4,684
To present	Subsea/Spar	Deeper	2,493

the date at which drilling technology and methods made them available are shown. Essentially, the resource has doubled since 1965, and that is *only discovered reserves*.

NOT ENOUGH STREETLIGHTS

A dog walker came across a man at a street-corner down on his hands and knees, avidly searching for something. "What are you looking for?" he asked. "I lost my keys over there," said the man, obviously drunk, gesturing towards a nearby vacant lot. "Then why are you looking over here?" the dog walker asked. "The light's better," came the response. Traditional.

The international team of scientists who discovered the 1,500 fossils said their find shows that the dark stretch in the fossil record more probably reflects an absence of preservation of fossils over the previous 25 million years.[24]

Although the conventional wisdom has often maintained that few new provinces remain to be discovered, estimates of resources have always suffered from the lack of information about many areas. This is due primarily to an inability to explore, either because governments have not allowed access or due to an absence of infrastructure that would make it easy (i.e., cheap) to drill.

This explains one reason why so much of sub-Saharan Africa's production comes from offshore: it's much easier to bring in a drillship than to move an onshore drilling rig through territory lacking good roads, implying a significant amount of oil remains to be found in many areas.

Thus, "new" resources sometimes appear, not only in new types of plays like the pre-salt offshore Brazil, but in conventional onshore plays like east

Africa, where serious drilling only began in the last decade. Now, a number of near-billion barrel oil fields have been found, suggesting that the area will make at least a modest contribution to future supply.

Although Thomas Ahlbrandt, who led the USGS 2000 World Petroleum study, tends to oppose the peak oil arguments, he must grind his teeth whenever I talk about resources. So far as I know, only Peter Odell and I have tended to be more optimistic than the USGS resource estimates over the long run, and I have tended to be explicitly critical—but primarily over interpretation, especially the conservatism of their estimates and not the quality of the work.[25]

Beginning with the USGS 2000 report, the organization has explicitly acknowledged its evolving estimates of recoverable resources, something that ASPO has usually failed to do. Needless to say, there are those who would argue that these numbers are unlikely to increase, that our understanding of the resource is so advanced that further revisions will not be significant. However, this assumes either that science or technology will stand still (see Chapter 7) or that the analysts have already incorporated future advances in their estimates. The former is absurd and the latter simply wrong: most geologists explicitly (and wisely) avoid projecting future advances, and instead qualify their estimates as representing what is available with current technology and economics. Ignoring this qualification has led many astray.

There are other reasons to believe that these numbers continue to be conservative. First and foremost, the fact that most areas are said to have little undiscovered oil despite having much less exploration than the United States is indicative of a cautious approach. Table 15.3 compares the density of drilling in the United States with other parts of the world and the percentage of the resource estimated to have been discovered.

While it is true that some areas, like the Eastern province in Saudi Arabia, will not have the typical creaming curve or distribution of field sizes, overall there is little reason to expect that other regions will see their curves cut off at much larger field sizes. Thus, simply extrapolating oil reserves and drilling effort to estimate resources in an area is simplistic and flawed, nonetheless, it can give us an idea of just how conservative some of the estimates are. As far back as the 1970s, Peter Odell was arguing that North Sea resource estimates were much too low, with minimal discoveries due primarily to minimal drilling.[26]

CURRENT RESOURCES, CONSERVATIVE AND MORE INCLUSIVE

Even the general media have become a bit more sophisticated about the difference between reserves and resources, with few thinking that the

Table 15.3   Oil Well Density by Region (Grossling, Bernardo F. and Diane Tappen Nielsen, In Search of Oil, Financial Times Business Information 1985. Updated with Annual Data from World Oil)

REGION	Prospective Area (sq km)	Well Density Total Wells	Wells per 100 sq km
ANZ	6,316,869	4,505	0.1
JPN	551,020	5,656	1.0
CAN	4,885,218	229,054	4.7
WE	3,608,900	60,173	1.7
FSU	9,009,305	572,000	6.3
USA (1985)	8,035,886	3,049,792	33.4
DC'S	32,407,198	3,921,180	12.1
ME	3,481,537	15,736	0.5
PRC	2,829,647	38,227	1.4
S/SEA	8,222,361	33,330	0.4
AFR	13,033,954	20,799	0.2
LA	12,428,357	141,527	1.1
LDC'S	39,995,856	249,619	0.6
WORLD	72,403,054	4,170,799	5.8
Non-ME LDC	36,514,319	233,883	0.6
ldc adj	26,663,904	8,000,000	30.0

United States has 10 years of oil "left." But there remains a serious failure to realize just how big oil and gas resources are, and not just because of the invalid estimates from peak oil advocates.

The USGS, for example, presents its range of estimates for recoverable resources, but makes no mention of unrecoverable resources—which is to say, currently unrecoverable resources. We've already seen that they have repeatedly revised their estimates upward, and that they appear to be extremely conservative about resources in lightly drilled areas, but this begs the question of just what is an appropriate estimate?

Hans-Holgner Rogner made an attempt in 1997 to collect estimates for various petroleum resources, including then-unrecoverable unconventional amounts, at close to 20 trillion barrels.[27] These are not higher, more optimistic estimates of the numbers that the USGS and others made, but values for resources that they did not include.

Table 15.4 shows various estimates of petroleum resources, including oil-in-place, which nearly all analysts ignore. To reiterate, not only has

Table 15.4   Petroleum Resources in Billion Barrels

Conventional Oil	URR	Oil-in-Place
ASPO	2,450	No estimate
USGS	3,700	10,571
Kerogen	103	2,826
Heavy	308	5,446
Oil Shale	345	6,900
Total (non-ASPO)	4,455	25,743

the Association for the Study of Peak Oil produced the most conservative estimate, but also the least inclusive, ignoring oil-in-place as well as most unconventional oils. The USGS is less conservative, but also not inclusive of much of the resource base; I have assumed a 35% recovery factor to estimate oil-in-place for the USGS.

The IEA provides the kerogen and oil sands/heavy oil resource estimates, and ARI's 2013 report to the Department of Energy is the source of the shale oil estimate.[28] On the bottom row, I have assumed that the actual shale oil resource is five times their estimate, inasmuch as exploration and testing of this resource has just begun. That number is probably conservative.

In all likelihood, a good portion of this resource will remain in the ground, probably by choice as new fuels and technologies slowly displace the use of petroleum. And it will be argued that the oil-in-place is misleading, as it will take many years for much of it to become accessible, if ever. But given over 100 years of conventional resources now available, and the past history of progress, that is not a serious objection.

CONCLUSIONS: ABUNDANZA

Examining the data, both historically and in-depth, demonstrates that concerns about resources were totally misplaced. Peak oil advocates were clearly off-base, as they have completely ignored reserve growth and the recovery factor. Beyond that, the continued expansion of resource estimates should also be comforting to those who worry about scarcity.

But it also puts the focus on above-ground factors, including the economics of production and the politics of access to resources. Most of the important developments in oil markets in recent decades have reflected this, and the following chapters will address them.

SIXTEEN

The Cost of Production

As the cheap oil from old mature fields is depleted, and we replace it with expensive new oil from unconventional sources, it forces the overall price of oil up. . . . Research by veteran petroleum economist Chris Skrebowski, along with analysts Steven Kopits and Robert Hirsch, details the new costs: $40–$80 a barrel for a new barrel of production capacity in some OPEC countries; $70–$90 a barrel for the Canadian tar sands and heavy oil from Venezuela's Orinoco belt; and $70–$80 a barrel for deepwater oil.[1]

Nelder (2012)

Two experts . . . calculated that the daily capital cost per barrel of new production outside OPEC jumped from $3,000 to $70,000 between 1963 and 1982. [Implying costs were twice 1985 prices.]

OECD Observer, October 1985

The use of oil production costs to predict prices has long been a challenge because in most commodity markets, prices and costs are fairly similar over the long run. Historical efforts to estimate costs have usually been very cursory and often had the effect had of supporting invalid expectations of rising prices.

Examples include the above-cited study that argued prices in 1985 were *too low* to cover costs—just before they collapsed, but also a 1989 report on OPEC production costs supported the BP CEO's belief that prices would rise in the 1990s—when they actually fell throughout. (He fell too.) Now, numerous industry pundits from the late de Margerie of Total to T. Boone Pickens insist that high costs necessitate a return to prices at $100 or more.

THE PROBLEMS

It is relatively easy to find numbers for production costs in various areas around the world, but most are rough guesstimates made by industry executives rather than detailed, precise analytical efforts. There are several very good reasons for this, unrelated to any "conspiracy of silence" in the oil industry, but instead due to the poor quality of data.

The first is our old bugaboo reserve uncertainty. If a company spends $100 million discovering a 100 million barrel oil field, then it seems pretty clear that the discovery costs were a dollar a barrel. But we've already seen in Chapter 7 that "discovery size" is a very fuzzy concept: most probably, the field will prove to be several times the original announced size. Thus, discovery costs are never very precise.

Also, accounting for dry holes is difficult to resolve. A company can show that it has spent a certain amount on exploration, and charge those costs to all its finds, but is it valid to charge a discovered field with the costs of dry holes drilled elsewhere? The government of, say, Ghana, might feel that its citizens should not have to bear the cost of a dry hole in Brazil, but to the company that drilled both, it makes sense. This is one reason that individual field costs often appear to be lower than aggregate measures, particularly when using national or company-wide expenditure data.

Fortunately, discovery costs are usually a fraction of total costs, which is perfectly logical. It takes only a few wells to find an oil field, but usually multiple wells to develop it, plus infrastructure such as gas-processing equipment, pipelines, meters; and offshore, platforms (which can be ferociously expensive). Figure 16.1 shows the breakdown between exploration, development and operating costs in Alberta, Canada (where the data are best).

Other problems are more difficult to resolve, including lack of transparency. Most companies report aggregate expenditures, usually by region (Africa, Europe, etc.) which can mean that small, expensive fields can be lumped in with large (relatively) inexpensive ones. Further, many companies are government owned, and do not release detailed information about their budgets. Some of these invest outside their own borders, further muddying the data available.

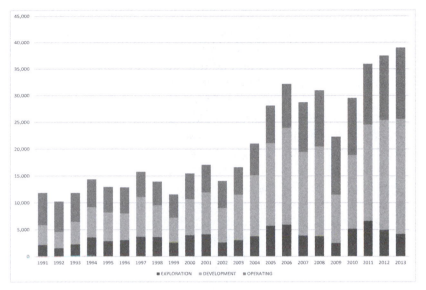

Figure 16.1   Alberta Upstream Expenditures by Sector (CAPP Statistical Handbook)

At the field level, the trade press includes many stories (based primarily on press releases from the respective companies) with numerous details of new or ongoing projects. Sometimes the cost of only part of the project, such as the offshore platform, will be reported, or long-term operating costs are included as part of "project costs" even though they would not be considered by economists to be "development" costs. With careful reading, discrepancies can be noticed and avoided, so that while it must be presumed that any given report could be inaccurate, in large enough numbers the estimates should provide a reliable estimate of costs.

Tax Man

Probably the biggest problem in trying to estimate the cost of oil production is the divergence between costs and prices, which is largely filled in by taxes. In a competitive market, the price should equal the cost of production over the long run (simplified version). But because oil markets face extensive governmental interference, there is a large gap between the long-run production cost and price. This means that costs can be anywhere from trivial (as in parts of the Middle East) to nearly the level of prices (small North Sea fields or U.S. fields onshore), and everything in between.

Thus, in North America, the cost of producing natural gas will usually be close to the price, especially if looked at over a period of time. If costs were above prices, drilling would drop (as it has recently) and production would moderate or decline, bringing prices back up, although rarely in a smooth or predictable path. Alternatively, if prices were driven up above costs, drilling would increase, production rise, and prices moderate. This is economics 101 (well, 14.01 at MIT).

The same is not true for oil, however, especially outside the United States. When prices rise, production might not increase very much because governments absorb the majority of the price increase as taxes. Most production contracts have sliding scale taxes that increase with higher prices, based on the presumption that costs are not rising with prices and the government (and people, at least in theory) deserve the "profits" from any price spike, not the oil companies (which are often foreign-based).[2]

For example, if the cost of producing oil in the fictional country of Ruritania were $25 a barrel, the fiscal contract would state that the government's share of incremental revenue would rise as prices go up. At a price of $50, the operating company might get $32 or so (cost and profits), while the government would get $18. But as the price rises, a higher percentage of the additional revenue would go to the government. At $100, the company might get $40, and the government $60. Thus, the fact that Ruritania's oil is much cheaper to produce than the current price doesn't mean that investment will become much more attractive there, nor does the global price tell us anything about the cost of producing oil in Ruritania.

This is one of the reasons that early predictions of higher production following price spikes did not prove correct. The two major studies in 1974 (described in Chapter 2) saw U.S. production soaring, using "price elasticities" wherein the supply was assumed to respond proportionally to changes in price.

COST ESTIMATES

As mentioned earlier, there are many anecdotal reports of costs provided by individuals involved in the industry, but these are frequently unreliable. Total's Marguerie commented that his company's costs were $100 a barrel, despite the fact that (a) prices have rarely been that high, implying most of his operations are unprofitable; and (b) the implication would be that his company was paying no taxes even at $100 a barrel. Alternatively (c) he was including taxes in his definition of costs, which is incorrect; they are transfer payments, not costs, and will change with prices, not provide a floor to them. Of course, he could simply have been misinformed.

Consulting firms and investment banks occasionally provide estimates of costs, but they too might also include taxes, especially if focused on the costs to companies (and the companies' stock price). Some of the work done historically has appeared to be reliable, but it is often not transparent, including only aggregate results and not the methodology used.

Analytical efforts have often suffered from lack of data. The U.S. Census Bureau used to provide aggregate data on U.S. upstream investment, but ceased publishing such long ago. However, Adelman (1991) was able to use it to show that U.S. costs had not increased from 1945 to 1986, although production had declined some, showing an increase in the cost curve.[3] Work in Canada also showed costs for the western Canadian sedimentary basin, which rose sharply after the first two oil crises in the 1970s, before coming down afterward.[4]

But factor costs changes have undercut the reliability of such studies, because they are extremely difficult to track over time. One consulting firm, IHS, has recently developed an inflation index for upstream operations, but the data only cover the past decade or so. It clearly shows that severe inflation in costs has occurred, without explaining the causation.

Size Matters

Yogi Berra once said, "If you see a fork in the road, take it." I would add to that, "If you have a numerator and denominator, divide them." Any number of analysts point to large oil projects as being supremely expensive, suggesting the higher costs for oil production due to deepwater and technically challenging oil fields such as Kashagan (Kazakhstan) or Mars (Gulf of Mexico).

But these frequently ignore the size of the fields, meaning that the cost per barrel is not estimated. For example, a shallow water field like Fionn in the British North Sea might only cost $62 million to develop compared to the Clair Ridge development West of Shetlands, which will cost about 100 times more.[5] But Clair Ridge has roughly 100 times the recoverable reserves of Fionn, meaning the actual cost per barrel is approximately the same.

Onshore/Offshore

In the past, onshore fields had one great advantage over offshore fields, namely, the fact that the drilling platform was the Earth. You can simply drive a truck with a rig nearby to drill another well, while an offshore field needed extra platforms if the deposit size was too big for it to be accessed from a single well. Thus, step-out drilling, where wells are put further out to access additional reserves, was much easier to do onshore than offshore.

This has changed for two reasons. First, horizontal drilling means that a much larger area can be accessed from a single platform than in the past. Second, subsea templates allow wells to be drilled without the installation of an entire platform, meaning that incremental supplies can be developed more cheaply than in the past. This applies to both small fields and incremental amounts in large fields.

Time Factor

Holding a barrel of oil in the ground is not the same as having a barrel of olive oil in a store. Because oil is not present in empty caverns but in porous stone, it must be produced relatively slowly. Historically, it was estimated that the "optimal" rate of production was about 6.7% per year, also known as a reserves to production ratio of 15 to 1.[6] Producing faster would mean that some oil would be left behind, wasting money. Producing slower would mean that the net present value would be reduced.

This means that the value of the oil *in the ground* is reduced depending on how long it takes to produce, but also the interest rate involved. Historically, engineers often used the 3 for 1 rule to estimate this: the value of the barrel in the ground was one-third the price of oil. Thus, if the price were $10 a barrel, a company purchasing a field with 100 million barrels of proved reserves would pay $333 million (more or less).

Such discounting for time explains why a company will spend $10 billion to develop a field with 1 billion barrels of oil and not say its costs are $10 a barrel, but closer to $30/barrel. And it means that simply looking at development costs can be an imprecise measure and lead unsophisticated investors and analysts astray.

COST COMPONENTS

The best way to break down costs is to look at exploration, development, and operating costs separately. The difference varies according to the type of field, with oil sands having high operating costs due to fuel needs, while deepwater fields involve major development costs, as the platforms can be worth billions of dollars.

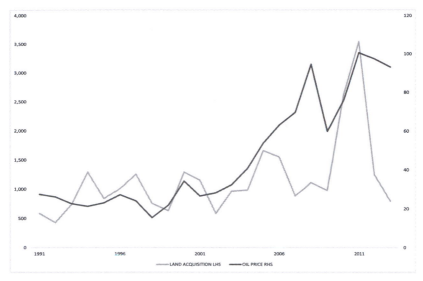

Figure 16.2   Oil Prices and Land Acquisition in Alberta (CAPP Statistical Handbook)

Payments Not Costs

Naturally, it would be foolish to say that taxes might go to zero, but the reality is that they fluctuate quite dramatically with the price of oil. In some high-cost areas, they can actually approach zero if the government is seeking to maintain employment, for example. But they certainly do not indicate scarcity, as increasing physical costs would.

And some such costs are voluntary, mostly land payments, in the sense that companies make bids in some areas, depending on the fiscal systems, to be allowed to drill. The amount of the bid is based on calculations about the presence of oil and gas in the area, and its value, which naturally fluctuates with price. Thus, as Figure 16.2 shows, there is a strong correlation with price levels and payments for leases.

Exploration Costs

The cost of exploration consists of three primary factors: lease acquisition, seismic, and drilling. Drilling costs tend to dominate, as can be seen in Figure 16.3, and they are likely to be even more important when exploring in new areas, where expensive equipment must be brought in from great distances and often over poor infrastructure. This is one reason that

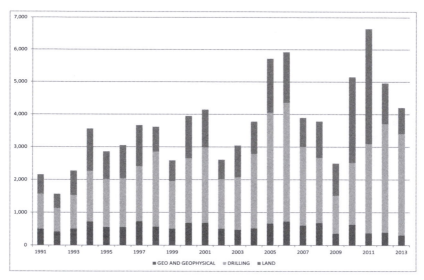

Figure 16.3   Exploration Expenditures in Alberta (CAPP Statistical Handbook)

offshore production is so prominent in many regions: ship-borne equipment needs no roads, utilities, or pipelines.

Development Costs Aka Capital Costs

Development of a field involves drilling and connecting wells, plus installation of production equipment, which can be minor (onshore conventional) or major (deepwater offshore or unconventional). Onshore is essentially pumps and gathering equipment, while offshore means a platform, sometimes several. Unconventional can include steam generation for injection and hydraulic pumping equipment. Many fields require processing equipment to separate water and other contaminants from the oil, but especially natural gas, which then usually needs further processing to extract liquids such as propane and ethane.

At a deepwater field like Mars in the Gulf of Mexico, the platform can easily cost billions of dollars. Plus the cost of drilling the wells, hooking them up, building processing plants to extract natural gas, and pipelines to connect the field to shore (or another platform, in the case of a small field). In Figure 16.4, drilling costs obviously make up more than half of development expenditures, and most of the rest is equipment.

Additional development frequently occurs during the lifetime of a field, ranging from simple infill drilling (new wells between old ones) to entire

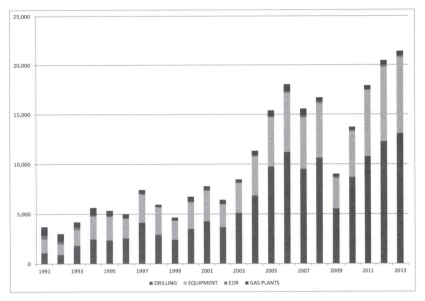

Figure 16.4   Development Expenditures in Alberta (CAPP Statistical Handbook)

new platforms. This is intended to extend the life of the field and add to the proportion of reserves that can be covered. It is treated as development costs, not operating, even though it occurs during the field's operations, because it represents new capital expenditures, which are applied to the incremental, not original, reserves.

Operating Costs

Operating costs refer to ongoing annual costs such as labor and fuel. Historically, it has been a rule of thumb that operating costs in the petroleum industry are about 5% of capital costs.[7] The petroleum industry is notoriously capital intensive, meaning labor costs are a relatively minor portion of total costs, but they can vary by type of deposit. Fuel costs also tend to be much higher when the oil is heavier and requires steam injection, with oil sands in Canada being the most extreme example.

Thus, costs are split between functions roughly 10/60/30 for exploration/development/operating, respectively (Figure 16.5), and transfer payments (land acquisition and royalties) make up 20% on top of that. Obviously, these data represent a relatively mature area, and in countries like Libya and Iraq, transfer payments would be much higher. For newer exploration zones, such as deepwater West Africa or onshore East Africa,

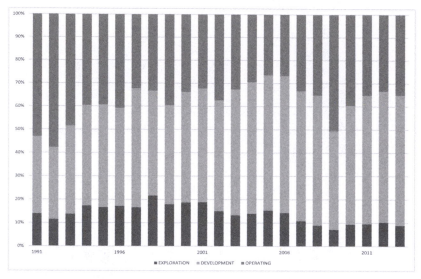

Figure 16.5   Expenditures in Alberta by Sector (CAPP Statistical Handbook)

exploration would be a relatively larger portion of the total. In remote areas, like offshore Russia's Sakhalin Island, operating costs would be elevated.

ESTIMATING COSTS

Several methods are used to estimate costs, but not all estimates are created equal. Using all of them, however, provides a certain amount of validation (assuming the results are similar).

Barrels in the Ground

A measure that is frequently used involves estimating the cost of obtaining barrels in the ground (as opposed to above-ground inventory). This is especially useful as a comparison when a company is buying reserves, and the price paid can easily be expressed in the dollars per barrel acquired. The value of a company's stock usually is dominated by its oil reserves in the ground, as well. But often, field development costs will be estimated using the total cost and the size of the field, dividing one into the other. Finally, a number of sources calculate "finding costs," which are the amount spent on exploration and development divided by the number of new reserves.

"Finding costs" are often reported for companies and groups of companies, taking exploration and development investments and dividing them
by reserves added. However, this is made imprecise by including the
poorly estimated "discovered" oil, with the more precisely measured
"developed proved reserves." Since exploration expenditures are a fraction of development expenditures, the error is not very large, and the
measure is primarily useful for showing relative changes and trends.

However, as reserves are sometimes found but not "booked" or reported
by companies for years, the exploration investments tend to occur long
before the reserves are added, and thus finding costs can fluctuate wildly,
especially for a single company. This is why some estimates aggregate
companies to average the numbers out, as DOE does in Figure 16.6, and
others used a moving average when looking at a given company.

One noted example is the survey by Scotia Howard Weil, which shows
finding and development costs for 61 companies ranging from $3.85 to
$76.34 even when averaged over five years (2010 to 2014).[8] Taking the
average of all companies still leaves significant volatility: $11.96 in
2010, $24.87 in 2011, 34.76 in 2012, $19.83 in 2013, and $22.74 in
2014. Thus, finding costs are useful as rough estimates, but great care
must be taken in interpreting results.

Project Costs

Most private oil companies, and many state oil companies, publish
moderately detailed development plans, in part as they seek contractors
to build different parts of offshore platforms. These can include field
reserves, expected peak production, and total development costs. Operating costs are rarely reported, and exploration costs are moot once the field
has been found.

Given the work of M. A. Adelman, it is possible to convert development
investment into per-barrel costs, as he did many times in his career.[9]
Using his methods, per-barrel costs can be estimated for many different fields. Note that all have been converted to 2010 dollars using
general inflation indices, not converted according to industry inflation
indexes.[10]

Costs by Nation

Historically, only the United States and Canada collected detailed
data on aggregate upstream spending, although Chase Manhattan

Bank, in the Golden Era of Energy Data, published a small booklet showing regional upstream investment. (Cost reductions and a desire to reduce the burden on industry resulted in the discontinuation of most of these.) By estimating capacity additions (gross not net) with the use of decline rates, it is possible to then show average capacity costs each year. Adelman found that U.S. costs did not rise over the long term, although production did decline after 1970, showing the effects of depletion.

In theory, the best approach would be to subdivide production by regions, trying to keep them homogenous and then build supply curves. This is roughly what some organizations do, such as the IEA's *World Energy Outlook*. However, these tend to be fairly aggregate, often by region, such as the Middle East, and limited by data availability. So Alaska might be shown separately, but not Siberia.

There are a number of shortcomings of this method, the worst of which is the static nature of the estimates. (Cost trends are discussed in the following text.) With one exception, there has been no major attempt to show how costs are changing.

LONG-TERM COST TRENDS

Diminishing returns are opposed by increasing knowledge, both of the earth's crust and of methods of extraction and use. The price of oil, like that of any mineral, is the uncertain fluctuating result of the conflict.

M. A. Adelman[11]

There are four primary factors affecting costs in the long term. In a perfect world, statistical examination would allow measurement of the different impacts, but the interference of politics—especially access for drilling—and the heterogeneous and uncertain nature of field sizes means this cannot be done. At least not for any amount of money spent to date.

The four primary variables affecting changes in long-term costs are as follows:

- Depletion
- Technology
- Infrastructure
- Regulation

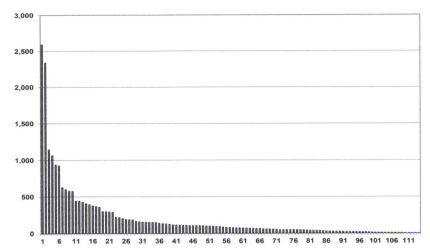

Figure 16.6   UK Oil Fields Ordered by Date of Discovery (UK Department of Trade and Industry; Series Discontinued)

Depletion can be measured by the rate at which a resource is being used up. The United States would be at one end of the spectrum, producing 8–12% of its reserves in a typical year, versus some Middle East players, such as Saudi Arabia or Iraq, which produce only a fraction of that. The challenge is to translate depletion into a cost escalator.

Measurement would be difficult, given the unreliability of both proved reserve data and ultimate recoverable resource estimates, but while neither is precise, both are useful. Other measures, such as production per well, could also indicate whether a resource base is mature or not.

Sadly, the measurement of oil fields especially is so uncertain that developing such an estimate is extremely difficult. Even so, looking at a static picture from a well-studied area, such as western Canada, Texas, or the North Sea, always shows clear evidence of field decline, even though the field sequence is hardly smooth (Figure 16.6). Actually, in Figure 16.7, which shows per well production in Texas, possibly the most open, "free" market for petroleum development, there is a very consistent pattern of decline, especially after 1981 price decontrol, with a 3.6% per year drop over the course of three decades (reversing recently with new technologies like hydraulic fracturing).

Thus, as is well known, all else being equal, depletion causes upstream costs to rise as the resource is produced.

Technology would arguably consist of both revolutionary and evolutionary technology. Evolutionary progress represents the ongoing improvement

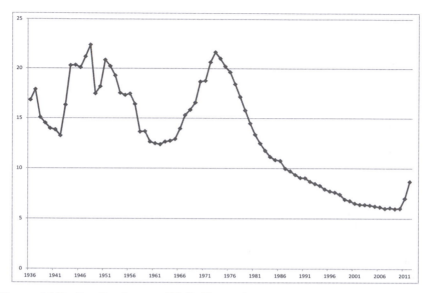

Figure 16.7   Production per Well, Texas (barrels per day) (Texas Railroad Commission, http://www.rrc.state.tx.us/oil-gas/research-and-statistics/ production-data/historical-production-data/crude-oil-production-and-well -counts-since-1935)

in every aspect of the business, such as better metallurgy and reservoir engineering. The impact of individual improvements are hard to measure, but in aggregate, it would appear to be on the order of 2–3% per year, similar to what Robert Solow found for the economy as a whole, or John Tilton reports on the hard rock mining industry.[12] Bohi reviewed a number of different indicators of productivity advance, but didn't calculate an overall effect.[13]

The number of feet drilled per rig has increased over time which should be due primarily to better equipment. However, short-term and cyclical changes in worker experience, as drilling expands or contracts, can muddy the analysis. Still, it appears that there is a long-term improvement of 1.6% per year.

Harder to estimate would be advances in geological knowledge that enable drillers to find and produce oil more efficiently. This can be reflected in new areas being explored that weren't considered earlier, improved recovery rates, fewer dry holes, or better extraction rates, but in every case, technology might also be a factor. Specific cases might be described, but it would be impossible to show the global effect.

Success rates for drillers do give some idea of how improved geological knowledge, combined with better seismic and computing equipment,

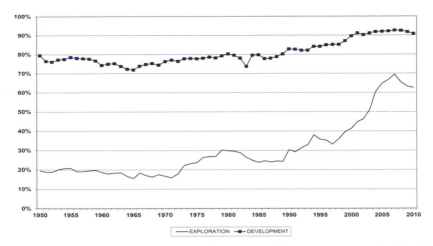

Figure 16.8   Success Rates for Development and Exploratory Wells, the United States (Energy Information Administration, US Department of Energy)

can reduce costs. In Figure 16.8, the success rates for exploratory and development wells in the United States show a clear improvement over the course of six decades.

Revolutionary progress is much more difficult to predict, but can cause major shifts in the supply curve. A sudden, major change in costs moves a resource from off the right side of the curve to the left, or viable, side. This is such as in the case of shale oil and gas, which have become much less expensive and thus viable, as will be discussed in Chapter 19.

ACCESS/INFRASTRUCTURE

Depletion and technology effects are often overwhelmed by changes in access to resources, especially as virgin areas are opened up with the development of roads and other infrastructure, but also as governments allow companies to explore in areas previously off limits. This is especially true for areas whose resources are less mature. Adelman and Shahi demonstrated this a quarter-century ago, by looking at the drilling needed to add production globally.[14] By considering only the non-OPEC Third World countries—like Argentina, Egypt, and Oman—it was possible to exclude the effects of OPEC quotas, and the results are shown in Figure 16.12.

It doesn't take much scrutiny to notice that the curves shift to the right, not left, contrary to theory. At least the simplistic version of the theory.

This reflects the fact that these areas were opening up; access to fields was increasing, more than offsetting the depletion of existing fields.

In a case of mixed effects, Adelman and Lynch (1997) found that approximately 25% of 1995 UK oil production came from small fields that had been discovered before 1980 but didn't begin production until after 1990. A combination of technological change—the use of subsea templates to develop small deposits cheaply—and infrastructure improvements—the spread of pipelines making it cheaper to connect a deposit—had made these field economic.

The biggest impact would come from improved access to OPEC resources. Iran and Venezuela are two obvious examples where greater opportunities for foreign investment would mean an addition of low-cost supplies to the resource curve. Costs in those nations wouldn't change, but the global cost of supply would. This phenomenon will be explored in Chapter 20 more fully.

REGULATION

In theory, regulation is almost always going to raise cost, and regulation will almost always increase over time, with only the occasional change to rationalize them and reduce their impact. But this is not a major factor affecting upstream costs, despite oil industry complaints. According to the American Petroleum Institute, industry environmental expenditures are $10–15 billion a year for the past two decades, rising at the rate of inflation.[15] Upstream expenditures were not much more than $5 billion of that in 2013. Compared to approximately $150 billion spent on exploration and development by the 50 largest oil companies in the United States, this amount is significant, but hardly overwhelming.[16] More important, they have not been increasing above the general level of inflation, and so cannot explain rising long-term costs.

CYCLICAL PRESSURES

Finally, the role of cyclical factors is often completely overlooked by analysts, as with the many who now insist that prices cannot decline because of high costs. High upstream activity levels drives up factor costs like labor and equipment, irrespective of geologic effort. Figure 16.9 shows how U.S. drilling cost per foot rose and fell with the late 1970s boom and subsequent bust and there is clearly an element of that occurring now.

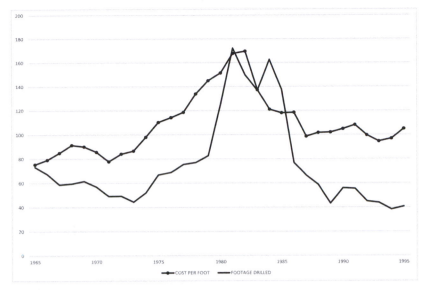

Figure 16.9   Drilling Cost Onshore United States (Energy Information Administration, US Department of Energy)

Comparison of changing costs shows that the increase in costs has not been the result of moving into more difficult environments, but rather across-the-board inflation. Figure 16.10 shows how costs shot up in just five years. Obviously, the nature of the petroleum being produced has not changed so much in such a short time: costs have not moved up because of a change in the location of production to more difficult areas, rather, costs in all areas have increased.

Another piece of evidence comes from the Canadian oil sand deposits. At the beginning of the chapter, it was said in 2012 that costs there were $70–80 per barrel, a figure that most would have agreed at the time was roughly correct. But in 2000, the IEA's *World Energy Outlook* said:

> Costs for synthetic crude production (including capital expenditures) are competitive at crude oil prices of $12 to $15 a barrel, as evidence by the growing number of new project announcements.[17] ($13.5–17 in 2010$)

Given that mined oil sands have depleted only 14% of the resource base in three and a half decades of production, and in situ only 1% in two decades, the rate of depletion is obviously very low and would not have driven up costs notably in the course of a decade.[18] Yet they are said to

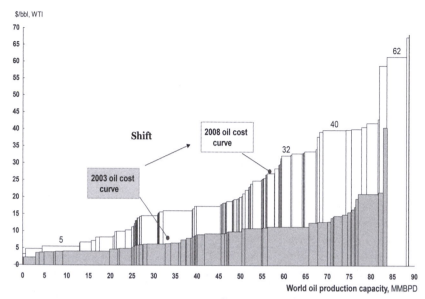

Figure 16.10   Inflation in the Global Oil Supply Curve (Courtesy of StatoilHydro)

have at the least quintupled. Clearly, this reflects the rising cost of doing business in that area, not any resource effect.

The takeaway from this is the lesson that given points on the supply curve move individually, and sometimes in both directions, including resources moving onto the current curve unpredictably. For the reserves or capacity curves, of course, movement can be in either direction, as governments can sometimes make supply off-limits.

CONCLUSION

Most of the world's oil can be produced for less than $20 a barrel, and even that has been inflated by the activity boom of the past decade. Higher estimates usually reflect a focus on the most extreme cases and outliers, which would probably not have been undertaken at lower prices. The long-term increase in costs from depletion of the better resources is generally offset by improvements in technology, although precise estimates have proved elusive. As in the case of higher prices, too many analysts fail to recognize the dominant cyclical element driving costs higher, assuming them to represent a permanent shift.

SEVENTEEN

How High Were My Prices?

Marx complained that modern economists converted the value of everything into financial terms (i.e., money), but his criticism was misguided. Although it is easy to find places where people engage in barter, the reality is that no matter how quaint or pro-community it sounds, it is extremely inefficient. (One can't help but suspect that a major portion of the attraction of barter is that it does not leave a financial trail that tax agents can track.)

And there have been schools of thought that argued for a labor theory of value or a land theory of value, the former being especially popular among Marxists and labor economists, who argued that other inputs simply existed and should be ignored. Rather, the thinking went, the only real input of value was the effort undertaken by workers, and as such, the amount of labor was the only important measure of a good's value.

Once I found myself on an airplane next to an MIT professor, who told me that he began his first class on energy by telling students that bottled water was far more expensive than oil, therefore oil was cheap. I was tempted to remark that for economists, this was considered evidence of economics ignorance, but since I had to spend four hours next to him, erred in the favor of discretion. (He was an engineer, of course, who promoted a particularly expensive alternative energy technology, which he insisted was great and didn't understand why people wouldn't pay far above market prices for its output.)

Many have attempted to come up with alternative means of measuring the value of oil, such as comparing the price of other liquid commodities like bottled water, Coca-Cola, or even perfume to that of crude oil or gasoline. This is usually done by people who are trying to show that oil is cheap, not that bottled water is expensive. Simmons argued that oil was not expensive—if priced by the cup. I'm not sure if he tried buying caviar by the egg, but that would be an apt comparison.

My favorite refutation of this was a radio opinion piece by a Boston disk jockey in the 1990s, who noted that if we were invading the Middle East because the price of oil was so high, the bulk of our army should be in Lynchburg, Tennessee, protecting our precious stocks of Jack Daniels (then worth about $5,000 a barrel, retail; some would say it's priceless).

Similarly, the argument might be made that water, being an essential ingredient of life, should be far more valuable than petroleum, which was more or less a luxury. But while some pay outrageous amounts for a commodity plus a brand name, oil has a global market price and no one pays a higher amount.[1]

This raises the question of why oil is so much cheaper than Coca-Cola, if it's so much more valuable. Or has been cheaper than water on occasion, usually to the industry's consternation.

It is worth noting that nobody who *buys* oil argues that it is underpriced. And I have suggested to those who think it is too cheap that I am willing to sell them oil for a price that is equal to, say, Perrier ($250/barrel) or, even better, Jack Daniels. They usually appear confused and reply that they don't want to buy it at prices above what's offered in the market, they only want to sell it that way.

ENERGY RETURN ON ENERGY INVESTMENT

A number of experts have adopted a "biophysical" approach, wherein certain other measures of the value of oil are substituted for prices and/or costs, and as with peak oil, *Scientific American* heralded this method.[2] The one that has particularly been embraced by peak oil advocates is known as the "energy return on energy investment" or EROEI. This describes the net energy generated by exploitation of fossil fuels or any other type of energy, including ethanol, which some estimates have said uses more energy in its production than it contains.[3]

This approach stands out in some cases, such as oil sands production, heavy oil production, and upgrading, gas-to-liquids and even liquefied natural gas, where the fuel used is quite large, often a significant fraction of the energy content of the final product.

However, peak oil advocates—and others doing the research—argue that this is also true for conventional oil. The seminal work was done by Cleveland and Hall in 1981, and updated recently by Guilford, working with Cleveland and others, and their basic point conforms, in many ways, to resource economics theory: increasing depletion of the resource means that greater effort—and energy—is necessary over time.[4]

The actual measurement is fraught with problems. For one thing, estimates made by Guilford show several distinct trends. Indeed, the initial 1981 research by Cleveland and Hall predicted that U.S. conventional oil production would become "energy-negative" before 2000, implying onshore drilling would cease.

A more viable interpretation is that the high drilling rate in the late 1970s caused efficiency of drilling to plummet, and it then recovered, dropping again with higher prices and drilling in the last few years. Under this view, the difference between the value in 1960, about 17, and 2003, about 15, is negligible and it appears that the primary decline in EREOI is from short-term effects.

And two prominent errors can be seen. One is the use of a single estimate for the amount of discovered oil, ignoring the growth that occurs. Hall uses data from the API Blue Books to show the amount of oil discovered in a given year, apparently unaware that these numbers were revised annually. Even in the course of a few years, the amounts were increased meaning that, first, the energy recovered was significantly understated, but also that the declining trend is probably exaggerated by the lower revisions in younger fields.

Another, rather novice mistake, is that Hall and Klitgaard assume that the difference between produced and marketed gas, which is roughly 25%, represents gas consumed in the production process, rather than (mostly) shrinkage from the removal of liquids.[5] It's a little odd that no one has read his stuff and corrected him, which suggests that perhaps his audience does not include anyone familiar with the industry. The correction is actually noted in Guilford et al.[6]

There's one other major problem. The implication of this research is that nearly the entire expenditure for producing oil will soon be taken up by the energy costs. This would seem rather unlikely, given that even a small, onshore conventional oil well requires a large amount of equipment, such as a drilling rig, portable power generator, thousands of feet of drill-pipe, several skilled workers, often a pump, and sometimes a baseplate.

This is a case where the practitioners of this particular methodology are primarily environmentalists: I have seen nothing in the petroleum industry literature that discusses this problem, and telling an oil driller that he was using too much energy to continue on would produce a hearty guffaw. And most economists regard it as a distraction: if something is profitable but not energy-positive, who would care? Indeed, it doesn't even inform you about the relative emissions, since it turns all energy—solar, coal, nuclear—into basic units.

IMPERFECT

The actual cost of producing oil is discussed in Chapter 16, but it is worth reiterating that even these most efficient and effective of economic measurements are imprecise in many ways. Accounting difficulties, uncertain reserve estimates, and cyclical cost changes make not just the production costs difficult to estimate, but further, the future trend is hard to predict. Even so, it remains by far the best indicator of whether prices are near the long-run market clearing price, or inflated by a variety of factors.

Market fundamentals can be said to be a good long-term indicator of prices, but there are a variety of factors that would seem to have changed the relationship. But, these data only cover the period from January 1995 to January 2004. After that time, the relationship is reversed, as higher prices coincide with higher inventories.

But, these data only cover the period from January 1995 to January 2004. After that time, the relationship is reversed, as higher prices coincide with higher inventories. As Figure 17.1 shows, with data from January 2004 to March 2013, the correlation is not as strong as before 2004, but still significant and *reversed*. Higher inventories have meant higher prices—the opposite of how economics tells us they should behave.

The point is that economic fundamentals haven't changed; rather there is a third factor affecting both price and inventories. What it is has been the subject of much debate, but three arguments have been made recently.

QE(x)

The unique position oil and gas occupy in our nation's and the world's economy make it one of the most desirable and sound investments available. The fact that there is no suitable substitute for the role oil and gas play in driving industry and creating wealth

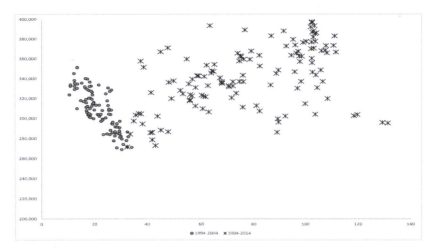

Figure 17.1   U.S. Crude Inventories Related to Price, 2004–2013 (US EIA)

has economists referring to oil as the "new gold" and a safe haven commodity in inflationary times.[7]

During the great recession of 2008, the U.S. Federal Reserve Bank unleashed a mighty weapon to fight the collapse of what was belatedly recognized as a housing bubble (especially by the Fed): quantitative easing (QE). This involved, essentially, pumping tens of billions of dollars into the economy in an attempt to encourage spending and investment, and thus resuscitate economic growth.

One side effect was that the surplus of funds needed somewhere to go, and this apparently was one cause of the run-up in asset prices from stocks to gold to oil. For some, oil had become a "safe haven" and so was thought to be a good place to store money. But this had the side effect of raising oil prices relative to market fundamentals.

This means that an end to quantitative easing should see some downward pressure on oil prices. Sadly, this is all but impossible to quantify, given the intermix of different variables, and the timing depends on the U.S. economic situation and *how it is perceived by the Fed's governors*. Historically, this group has done an excellent job (I think, though others disagree), and it seems highly likely that they will reduce the quantitative easing sometime soon. In mid-February 2013, talk of this saw gold and oil prices drop sharply for a couple of days, but the longer term effect on prices isn't clear. At any rate, the QE program appears nearing an end, meaning that whatever positive effect it had on oil prices is not long for this world.

SPECULATORS

A decade and a half ago, a number of investment banks (and others) created what are known as commodity index funds. This enabled buyers to speculate directly on the price of a given commodity, or a group of them, where previously it had been necessary to buy future contracts. Because of the nature of commodity futures, many funds (such as pension funds) were prohibited from investing in them, but the new index funds made them acceptable and so resulted in a flood of money into commodities.

This, some argued, meant an increase in prices relative to the market fundamentals and the historical price, as these funds were buyers, never sellers. A hedge fund, on the contrary, could buy or sell, often in the same day, but commodity funds were only asset holders, that is, buyers. Arguably, such an increase in pure buyers would require that prices rise to offset. Some believe this explains a part of the price increase during the past decade.

Again, it is nearly impossible to prove this with the existing evidence because of the many intangible factors that affect the price of oil, but there is likely to be some truth to this. Over the past few years, market observers have often suggested that there is a speculative premium of $10–20 per barrel, but this reflects a rank guess.

Run for the Door

This raises the issue of whether or not such investors will "run for the door" at some point, that is, perceive that oil, or commodity indices more generally, no longer offers attractive returns and seek to move out of them en masse. Sharply dropping prices would encourage others to pull out as well.

This raises the question of investor psychology, much debated among market watchers. Sellers of commodity index funds argue that it is best to hold them as a balancing factor in a large portfolio, implying that most would not sell off just because of a bad quarter. There is definitely a lot to this argument; few large funds shift enormously between sectors in the short term. On the other hand, at the margin, there might be enough speculators who leave to encourage the large funds to reduce their holdings—at least for a time.

Examining the annual change in the price of oil from 2003 to 2011, every year but one saw strong gains, so that holding oil or a commodity index fund weighted for oil, would perform very well. Since then, the returns have been poor to horrible, which might move many to abandon the funds.

GEOPOLITICS—BLACK SWANS AND GREY SWANS

The 1990s were a period of extraordinary political calm in the oil industry.[8] Partly, this reflected the much weaker market, with huge amounts of unused production capacity available to offset any disruption of supplies. But also, the opening up of the FSU and the *apertura* in Venezuela, among others, generated a sense that the industry was no longer as politicized as before, and that oil had become just another commodity.

This was to change, however, very much for the worse after 2001, when an extraordinary number of events combined to raise the political risk far beyond what was experienced at any time since, at least, 1979, and arguably ever.

Lost Barrels

The first military attack on oil supplies occurred in 1863, when General William "Grumble" Jones's Virginia cavalry attacked what later became Burning Springs, W. Virginia.[9] Needless to say, it didn't accomplish much: presumably General Grant had to rely on candles to read his maps (and do his drinking) for a while, but histories of the war note no greater impact.

Since that time, oil has often been subject to a variety of attacks that have led to the actual loss of supply, although many of these "attacks" have been fiscal or regulatory in nature. High taxes in Mexico in the 1920s strangled investment in oil production long before the famed 1937 nationalization, and Libya in the 1970s restricted production levels for conservation purposes (allegedly).

Other problems arose with the 1951 nationalization of BP holdings in Iran; the company used legal sanctions to prevent the sale of "its" oil; the 1956 Suez Crisis shut the canal and delayed delivery of Gulf oil to Europe by several weeks; a 1971 accident cut the TransArabian pipeline from Saudi Arabia to the Mediterranean, having a similar effect, but on a much smaller volume.

Still, these were nothing compared to the 1973/74 Second Arab oil embargo and, especially, the accompanying production cutbacks, coming at a time of soaring oil demand and tight markets. This took nearly half a

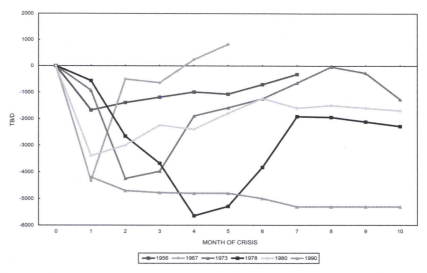

Figure 17.2   Losses in Major Oil Supply Disruptions (pre-2000) (US Dept. of Energy)

billion barrels off the market over seven months (roughly; see Figure 17.2), greater in absolute terms than earlier disruptions. And the 1978/79 Iranian oil crisis had a longer impact, in terms of both lost supply and, more important, the *perceived threat* to other countries' political stability, as the new Iranian leaders talked of exporting revolution. Fears peaked when a group of radicals seized the Grand Mosque of Mecca in November 1979, making the stability of the Saudi regime seem at question.

The 1980–1988 Iran-Iraq War took a large amount of oil off the market, but did so at a time when there was a large amount of surplus capacity, and so prices rarely moved even as massive slaughter occurred on the battlefield and oil tankers came under attack. After peace, Saddam Hussein turned on Kuwait in 1990, and again markets did not respond more than slightly.

This ushered in a period of calmness. In the 1990s, the biggest problem with physical supply was in Colombia, where rebels occasionally bombed the pipeline from Canos Limon, shutting off supplies repeatedly but briefly. Iraq, though under sanctions, was able to maintain more or less stable supply, and even the Iran–Libya Sanctions Act had no significant impact on either country's ability to maintain production, although they might have increased output without the sanctions, at least in theory.

Figure 17.3   Lost Venezuelan Production (EIA [The Production Stability Reflects Poor Data Updates from the Venezuela Government])

NEW ERA

The new millennium saw all this change (probably not for Biblical reasons, or even Mayan). There were two big downward hits on demand: the terrorist attacks of 9/11 resulted in lower air travel and depressed economic activity, and the SARS epidemic did the same shortly afterward.

But these were quickly overshadowed by two large, near concurrent, disruptions to supply, which had major, long-lasting effects. Although the run-up to the invasion of Iraq began shortly after 9/11, it was preceded by a strike at Petroleos de Venezuela, which stopped all exports for several months (Figure 17.3). Subsequently, the firing of a large number of employees meant that production recovered only slowly, aided in part by the startup of a number of heavy oil projects.

The invasion of Iraq had a more lasting effect, as the lengthy political turmoil meant that decisions about restructuring took years to make. As Figure 17.4 shows, production had largely recovered by early 2008 (a mere five years!).

In these two cases, there is the problem of estimating the amount of production that would have occurred without the problems described earlier. Would Iraqi production have risen earlier, if not for the change of government? Or would it simply have been stagnant with the continued reign of Hussein and economic sanctions? In the first case, the implication

Figure 17.4a   Lost Iraqi Production (EIA)

is that as much as 1.5 billion barrels of oil supply were lost in the 2000s but, in the second, only about 1.2 billion barrels.

Delta Blues

Another somewhat uncertain case occurs in Nigeria, where, beginning in 2006, rebels in the Delta area began to attack oil installations. Sometimes, workers were kidnapped; other times, equipment and/or wells damaged, but the primary intent was to get money from the oil companies operating in the area. The grievance centered on the perception that most of the oil revenue was not only being lost to corruption, but the corruption was benefiting people outside the producing area.

Over time, as much as 600 thousand barrels per day of production was shut in, partly replaced by new fields in the deepwater areas. Despite a cease-fire and efforts at reconciliation, production continues to be hampered by, increasingly, theft from pipelines as well as the occasional rebel who refuses to cease their activity. In March 2013, Shell again had to declare force majeure on 100 thousand barrels per day of production, claiming rampant theft was causing it to shut down a major pipeline.

Nibbled to Death by Ducks

The past three years has seen physical losses of production, which, individually, are fairly small but in aggregate total about 700 thousand barrels

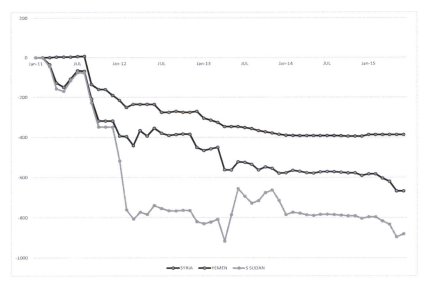

Figure 17.4b   Oil Supply Losses from Small Producers (EIA)

per day, hardly earth-shattering but still enough to matter in a market that remains somewhat tight. Figure 17.4b shows the rough amounts lost from the Sudan (now South Sudanese) production, in a dispute over pipeline tariffs, Yemen, where the overthrow of the government has created enough political instability to affect production, and Syria, where the civil war has stopped most activity.

PERCEIVED RISK

Unlike other elements affecting prices, political risk includes a heavy intangible effect, one that would be quantifiable if there were no other intangible factors. As my old friend Geoff Ward used to say, you have more equations than variables, so you cannot solve for x. Several years ago, I had to explain to a reporter that the various estimates quoted for the "security premium," that amount of price that is supposedly due to the perceived risk of disrupted supply, were simply guesses. No doubt, some clever grad student can come up with an algorithm that can provide some precise-seeming numbers, but the fact remains that the role of perceptions is to economics like daylight to vampires.

And again, it is the perception of oil traders (or speculators, gamblers, investors, etc.) that matter, not that of geopolitical risk analysts. This can explain why the 2003 arrest of Mikhail Khordokosky, CEO of Russian oil company Yukos, appeared to elevate prices for a time, even though it had no particular impact on Russian oil production. To traders, oil company problems of whatever nature in a place like Russia would translate into a "buy" signal, without a complex process of estimating the precise risk to any given amount of supply.

Of course, there is the potential for "crisis fatigue," where traders finally become accustomed to some particular concern and begin to ignore it. Iran has made many threats to shipping in the Straits of Hormuz (or, more accurately, various Iranian officials have done so), and in all likelihood, the price impact has declined over time as traders slowly realize that they are unlikely to be followed by action.

Table 17.1 is an attempt to quantify the risk, with the understanding that this is a case of "factiness" as Steven Colbert would say. It is impossible to say how much more serious sanctions against Libya are than, say, legal

Table 17.1   Oil Supply at Risk (mb/d)

1980 Country	Problem (Threat)	Production Capacity	% World
Iran	War	**6**	9.1
Iraq	War	**3.5**	5.3
Kuwait	Feared unrest (low)	2.5	3.8
Saudi Arabia	Feared unrest (high)	11	16.7
UAE	Feared unrest (low)	2	3.0
Lost		9.5	14.4
@ High Threat		11	16.7

2015 Country	Threat	Production Capacity	World
Iran	Sanctions	4	4.5
Iraq	Insurgency	3.5	3.9
Venezuela	Civil unrest (High)	3	3.4
Libya	Civil War	**2**	2.2
Nigeria	Civil Unrest (Localized)	2.4	2.7
Syria, Yemen, South Sudan	War	**0.8**	0.9
Saudi Arabia	Feared unrest (low)	12	13.5
Kuwait	Feared unrest (low)	**3.2**	3.6
Algeria	Feared unrest (low)	1.5	1.7
Lost		4.4	4.9
@ High Threat		3	3.4

troubles for oil companies in Russia, or how they are perceived to affect prices by traders, but the figure gives the sense that the levels were not insignificant in the past decade, and often increased.

A MIGHTY WIND

A peak oil advocate once chided me for suggesting that Katrina was an outlier, an unusual or transient event, not so much a black swan as an albino python—rare but hardly unheard of. The reality is that while hurricanes are hardly unusual, the one–two punch that was Katrina and Rita was exceptional in the manner that they blanketed the entire Gulf of Mexico oil industry.

Many have heard about platforms knocked off their bases or drill ships torn from their moorings, but the damage to the offshore service industry received less attention. Every manned platform, and there are hundreds of them, required deliveries of workers, supplies, and equipment on a regular basis, using a mix of ships and helicopters, based on the coast that was hit by the two storms, leaving many of these destroyed or badly damaged.

Not only that but the services *for* the service workers were badly damaged. Housing, power, and water supplies were frequently damaged or unavailable and usually received first priority from workers who might otherwise have been restoring oil production. And those restoring damaged oil production capacity were not adding new capacity. As Figure 17.5 shows, production did not recover after Katrina the way it did after Ike or previous storms.

BEEN UP SO LONG IT LOOKS LIKE DOWN TO ME

When prices dropped in 2008 after the outrageous spike, few took the subsequent price of $37 as indicative of the likely long-term price, but amazingly, many have taken both the $140 peak in July 2008 as close to "normal" and the more recent $100 a barrel three-year plateau as a new floor. However, this appears to be a repeat of the 1970s misinterpretation of short-term or transient events as representing long-term or physical effects. There seems to be little question but the overwhelming cause of high prices was the many political disruptions of supply.

This is not to say that other factors, such as the Chinese economic boom, renewed resource nationalism, and/or speculator expectations of "peak oil," were not important. If production is restored in Libya, Iran,

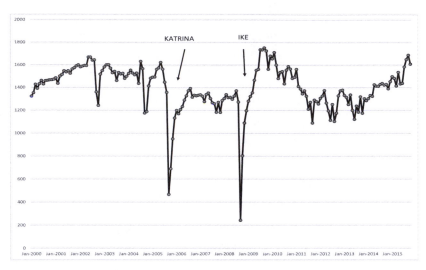

Figure 17.5   Gulf of Mexico Production (DOE)

and Nigeria and prices remain below $60 a barrel, then the case for a geo-politics will be made. If instead, those countries become stable producers and prices remain near $100 a barrel, arguments that "the cheap oil is gone" will be bolstered. As following chapters will show, I can be placed firmly in the prior camp.

EIGHTEEN

Tipping Point for Prices

(This chapter was written when the price of oil was $100 and nearly every forecast called for prices to rise over the long run. It is, as of this writing, $50, but it is not yet clear if those levels will persist, with numerous forecasters arguing "Current prices are unsustainable. It's hard not to see oil hitting $100 a barrel at some point in the next five years."[1] The focus of this chapter is what will determine a new equilibrium.)

In Figure 18.1, which shows the average monthly import price in the United States (a good but not perfect analog for global prices), the pre-1986 period stands out sharply. At that time, OPEC's market share was very high (Figure 18.2), and the group had no trouble announcing a price and having that stand. People either bought or didn't, but OPEC was, literally, the price maker then.

Unlike the Texas Railroad Commission, which heavily influenced U.S. prices during the period 1932 to 1970, in part by using military power, OPEC relied on market power that declined precipitously during the 1980s.[2] In late 1985, Saudi Arabia decided it would no longer serve as the "swing producer," absorbing fluctuations in demand, and forced the organization to switch to setting production quotas, which were intended to influence but not set prices.

At the same time, the first post-1973 price collapse occurred, with prices going from $45/barrel to $20/barrel (adjusted for inflation to 2010 dollars), recovering to a new equilibrium level of $24, which persisted

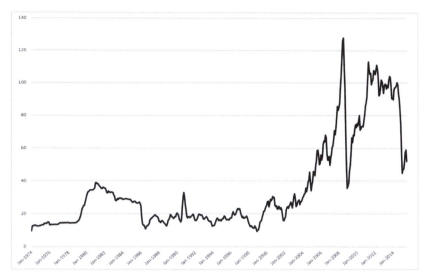

Figure 18.1   U.S. Imported Price ($/barrel, Not Adjusted for Inflation) (EIA)

for roughly 15 years. This was thought to represent the sustainable price, and it proved so for quite a while.

There have been a number of theories as to how this happened, including the conspiracy theory that President Reagan sought to have the Saudis crash the price of oil to hurt the Soviet Union economically, bringing

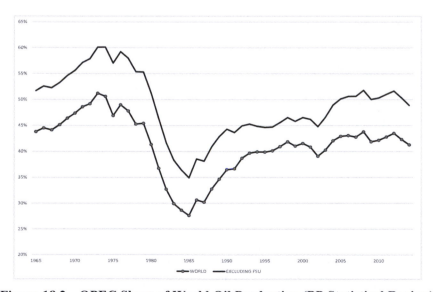

Figure 18.2   OPEC Share of World Oil Production (BP Statistical Review)

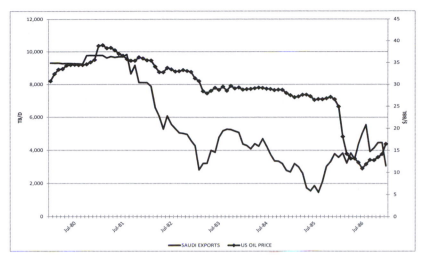

Figure 18.3   Saudi Export Decline in the 1980s (EIA)

about its downfall. This was propagated in Schweitzer's *Victory* and has been embraced by those who think that the price was too low during the 1990s, including some in the industry.[3] (Colin Campbell, among other peak oil advocates, has also adopted it, possibly to rationalize the 1986 price drop as not being due to fundamentals.[4])

But while it is undoubtedly true that, like many other U.S. presidents, Reagan urged the Saudis to work for lower prices, and the Soviets certainly suffered when they came about, this theory largely ignores the underlying reality of what was going on in the market at that time. As Figure 18.3 shows, Saudi oil exports were sliding downward rapidly from the peak in 1980, when they had increased to replace lost Iranian (and later Iraqi) supply and were trying to calm markets.

This was entirely due to the fact that the post-Iranian Revolution oil prices were so high that oil users were switching to other fuels and conserving, and new supplies like Prudhoe Bay and the North Sea were ramping up. OPEC production dropped by 50%, far beyond what the experts predicted. Indeed, at a conference in Stanford in 1980, comparisons of OPEC production were all flat, except for the model designed by Nazli Choucri of MIT (with some assistance from your humble narrator, among others), which predicted a 10 million barrels per day decrease by 1985.[5]

Again, this reflected the new consensus among experts that non-OPEC had peaked and demand was only marginally price responsive, with any weakness in oil markets due to temporary economic conditions. Even after the 1980 recession ended and demand failed to recover, most continued to expect an imminent turn around, which helps explain why OPEC (and the Saudis) did not reduce prices earlier. A review by Chase Manhattan Bank in the summer of 1985 found that the oil industry was expecting continuous gradual price increases, for example.[6]

But when Saudi production dropped below 2.5 million barrels per day by late 1985, of which a third was used for consumption and much of the rest to pay for barter deals, such as for airliners, the country reached the point where they would have had to *buy oil* to support prices. Regardless of any other political considerations, continuing their price support policy of acting as swing producer was completely untenable (or unsustainable, as the kids would say).

If something can't go on, it won't.

Herbert Stein[7]

The funny thing was that the year before, a cold winter had caused heating oil prices to spike in the United States, and prices had risen sharply. The relatively new futures market saw some serious money made by those who had been long before the winter, and so, in the fall of 1985, traders were buying into the heating oil market again, driving up the price of WTI to a relatively new high of $30 per barrel late in 1985.

At the same time, the Saudis announced, quite openly, that they would no longer act as the swing producer, cease defending the price of oil, and take steps to increase their exports by offering buyers contracts that essentially guaranteed them a certain profit level. The idea was to preempt any price-cutting by other OPEC members, which worked extremely well.

As late as early December, there were comments about the fact that Saudi oil did not seem to be showing up in markets. A story in *Petroleum Intelligence Weekly,* the leading industry newsletter, said, "The surge in spot prices is persisting as refiners search but fail to find any signs of excess oil in the market, despite repeated warnings of an imminent flood of supply."[8] Remember that after 1979, the Western oil companies' transport of oil had dropped sharply, so information on how much oil was in transit to importing nations had become much more uncertain.[9]

Again, although some had been bearish about oil prices, no one to my knowledge had predicted the exact timing of the market shift, even though many now seem to claim to have done so. Figure 18.4, which shows domestic U.S. drilling for oil and gas compared to the price, indicates that

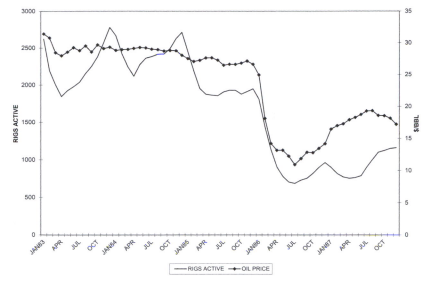

Figure 18.4   U.S. Oil Price and Active Drilling Rigs (EIA)

the companies (mostly independents) that were investing in the U.S. upstream had no clue as to what would happen.

Mea culpa. I did not predict it either, as I was working on global natural gas markets at the time. Good excuse, eh? But in February 1986, when most thought a recovery in prices would occur, I gave talks subsequently published as "Structural Changes in World Oil Markets and Their Impact on Market Behavior," at MIT, and argued that changes in the nature of the market meant that OPEC had much less influence, that prices were likely to remain low, but also that they would remain volatile.[10]

INTERLUDE

The U.S. oil and gas industry has been dramatically weakened by the recent oil price collapse. Domestic drilling activity reached a new post-World War II low during the summer of 1986. Given a weak, unstable oil price outlook, U.S. capability will continue to deteriorate. ... These facts, coupled with the nation's generally short-term orientation, suggest a strong likelihood of a new U.S. energy crisis in the early to middle 1990s.[11]

Thus began what many considered an "interlude" of low and volatile oil prices, with repeated predictions that they would recover sometime soon.

Most forecasters, as we saw in Chapter 2, thought that the trend was for an increase above inflation of 3% per year, and in 1989, the typical forecast for 2000 was $50 per barrel (in 2010$). Indeed, the *Petroleum Economist* correctly referred to my "heretical" view that prices would not keep up with inflation.[12] (This actually proved to be the best forecast, to my knowledge.)

This was when the first of the new-wave peak oil publications appeared, Colin Campbell's "Oil Price Leap in the 1990s," which predicted a near-tripling of prices in the early years of the decade.[13] But more commonly, Bob Horton, CEO of BP, remarked on the inability of OPEC to raise sufficient funds to add enough capacity to meet expected demand, meaning prices would rise soon.[14] This seemed likely to be borne out when Iraq invaded Kuwait, triggering the first Gulf War and seeing oil prices soar to $50 a barrel—for all of three months!

And despite the anguish of industry insiders like Donald Hodel and Robert Dietz, co-authors of *Crisis in the Oil Patch,* production outside of the United States largely recovered.[15] OPEC's market share stopped growing, and areas like the North Sea experienced a surprising increase in production. At the same time, political stability in most OPEC nations (Iraq the exception) resulted in the occurrences of few or no disruptions of supply.

DROWNING IN OIL: 1998

No doubt many will remember the 1998 oil price collapse and (in the United States) gasoline prices threatening to reach $1 a gallon, and casual readers will think of this increase as a replay of the 1986 price collapse. However, the resemblance is not very close.

The oil market in 1998 was in a very different situation from that of 1986. Demand for OPEC oil was strong; Saudi production was nearly 9 million barrels per day, which is to say fairly robust; but there were pressures. First and foremost was the expansion of production in Venezuela, which had defied nearly everyone's expectations. Adding insult to injury, Venezuela oil officials not only exceeded their quotas (Figure 18.5), but openly defied the other OPEC members to retaliate. Some officials even assured me they could win a price war with Saudi Arabia, but fighting an oil price war with Saudi Arabia is rather like the old admonition against fighting a newspaper: don't get into an argument with someone who buys ink by the barrel.

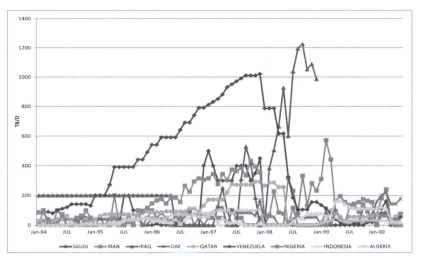

Figure 18.5   Production over Quota Late 1990s (Data from Energy Intelligence Group)

Considering OPEC behavior as a case of game theory, Saudi Arabia would be the "enforcer" of discipline. As Figure 18.5 shows, the egregious behavior of Venezuela gave other members cover for their own quota-busting, to where nearly every country was producing at capacity except for the Saudis. At the quota meeting in November 1997, the Saudis pushed for a quota increase that would allow them to raise production without breaking their quota. Actual production barely moved for the other members, who were already producing at their maximum levels.

Unfortunately, this occurred at a time when inventories had been increasing and the infamous Southeast Asian recession had just begun, so that by the second quarter of 1998, storage tanks worldwide were said to be overflowing and prices were bottoming out. But intransigence among the members meant that a quick solution was not forthcoming.

In part, this was because some in OPEC thought that the previous price level, roughly $20 a barrel, might be too high. Shocking as it seems, there were suggestions at the time that non-OPEC oil supply would continue to rise at prices near $20 a barrel. North Sea production had boomed in the mid-1990s, nearly every other producer was, at the least, flat, and three "new" areas of unconventional oil were expected to make major contributions. Venezuela's Orinoco heavy oil belt was attracting foreign investment capital, which would add up to 600 thousand barrels per day, and oil sands in the Athabasca area of western Canada were increasing production by 60 to 80 thousand barrels per day per year. Both were said to be profitable at $20 a barrel ($26 in 2010$), and seemed to have little political risk.

The third area, gas-to-liquids, was somewhat more problematical. The company Syntroleum announced that it had reduced the cost for converting natural gas (methane) to light, superclean, petroleum products, namely, diesel fuel and naphtha (which can be made into gasoline).[16] At the time, most of the major oil companies began considering the construction of commercial scale plants, although it is now obvious in hindsight that investment in this method would progress only slowly.

Although it's not clear how widespread fears of low prices were, there were cases of companies, expecting further drops, locking in the low price, and the *Economist* magazine published the infamous "Drowning in Oil" issue, which warned that prices would remain depressed, and possibly drop lower.

BLOWBACK

In a development that contains an important lesson for future market developments, the recession created in Venezuela by plummeting oil income in 1998 was a major factor in the election as president of Hugo Chavez, a populist and socialist who ran effectively against the existing political elites and their management of the country. (This is a gross oversimplification of the Venezuelan political situation, but sufficient for this venue.) He reversed the country's aggressive oil policy and embraced cooperation with other OPEC members to rein in production and stabilize and increase oil prices.

Unlike 1986, prices returned to their precollapse level and, ultimately, much further. In hindsight, this might seem obvious, but at the time, the prices in 2000 were a surprise to nearly everyone (certainly me). In large part, this occurred because the price collapse in 1986 reduced costs that had been high due to massive upstream investment; in 1998, the price dropped well below the long-run marginal cost of nearly all oil, meaning investment collapsed in many areas, setting the stage for the subsequent high prices.

FORWARD INTO THE PAST[17]

These two price collapses are clearly very different animals: one occurred because prices were too high to be sustainable in the long run, the other mostly because of aggressive production by one member into a relatively weak market. In the first case, the solution was lower prices, in the latter, better quota adherence. Both required a price war to accomplish, however.

What would cause the price to drop for a lengthy period? It could actually be a combination of the factors that led to the two previous collapses. As this book argues, $100 a barrel is too high to be sustainable and eventually will need to be lowered significantly. The drop in 2014 is still, to many minds, not permanent. OPEC market share has remained relatively stable for the better part of two decades, with production growing with demand, and it could be that the group is satisfied with such a state of affairs.

However, if, as seems likely, in the next few years the demand for OPEC oil is roughly flat, then rising capacity within OPEC could mean a growing perception that the price is too high. Should demand for OPEC oil fall while some members are expanding production, as I expect, the price pressure would be more acute.

Most notably, the potential for rising Iraqi production to pressure the cartel is much greater than Venezuelan behavior in the 1990s. The goal of 12 million barrels per day by 2020 has long been abandoned (and was always seen by most as unrealistic); they nonetheless are almost certain to add between 300 and 500 thousand barrels per day for the next eight to ten years. If the past is any indication of future behavior, the other OPEC members will expect Saudi Arabia to make the necessary cutbacks to make room for the extra Iraqi oil. This could easily lead to the Saudis reaching a point where they insist that the other members contribute production restraint, which could lead to a 1998-style price war.

Beyond Iraq, there is a distinct possibility that both or either Iran and Venezuela will have policy changes that would allow for a renewal of outside investment and noticeable recovery in production. Both countries have a desperate need for hard currency, and both have new governments that might find it easier to attract foreign companies.

Iran, in particular, has shifted political course, with new president Rouhani taking small steps to liberalize life for his people, and what appears to be a successful nuclear agreement with the P5+1 powers. As a result, there is every likelihood that, first, production will grow by nearly 1 million barrels per day in 2015/2016, as fields shut in by sanctions resume operation. Longer-term, foreign investment (and technology) will mean a lot more supply.

The story is different in Venezuela, where the passing of President Chavez brought his associate, Nicolas Maduro, to power. He is a much less charismatic leader, albeit with similar misconceptions about economic policy. In mid-2015, the economy shows signs of nearing collapse, with widespread shortages of basic goods and ongoing protests against government incompetence. Maduro has talked about seeking new upstream investment, but hasn't proceeded very far with the idea, although it seems all but certain that either he, or his replacement, will feel the necessity of finding new upstream partners.

All in all, it appears that in the next five- to ten-year period, there will be at the very least some pressure on OPEC's market share, and possibly actual volume of sales, while inside the organization, Saudi sales will be under serious threat from other members. This should result in prices remaining in the $50/barrel range, if not lower, but the path is uncertain.

PRICE COLLAPSE SCENARIOS

There are several different paths to a price collapse, depending on a combination of political developments and policy choices, particularly among OPEC members. Relying on past behavior, these can be described well enough that observers might anticipate them. How well the recent price collapse fits into one of these models is an important question.

First would be the developing surplus scenario, in which supply more or less continually grows faster than demand. This would be, as described in Chapter 20, the result of rising conventional oil production in a variety of places, along with spreading shale oil production both in and outside of the United States. Ultimately, demand for OPEC oil would fall, forcing the Saudis especially to cut back, and creating the perception that the price was unsustainably high. Traders would then recognize that $100 was not a sustainable long-term price, and prices would drop.

The second scenario, Internal Stress, involves the resolution of some, if not most, of the political disruptions affecting primarily OPEC oil supply. Stability in Libya and/or a diplomatic resolution of the conflict over Iran's nuclear program could return 1–2 million barrels per day to the market fairly quickly. But more important, Iran and Venezuela might either or both have a resumption of upstream investment, including by foreign companies, which would provide them with steady production growth. This would threaten both a supply surplus but also reduce the Saudi share of OPEC oil sales.

The very real possibility that Kurdistan, Iran, and Venezuela could, between them, contribute a half a million barrels a day of production growth per year, even as non-Kurdish Iraq adds about the same amount, would mean that only abnormally high demand growth could allow the other OPEC members to maintain their market share. Such would necessitate prices closer to $50 than $100 a barrel, and this would become quickly obvious to markets.

The third scenario involves a more strategic consideration by the Saudis that the price has or will become unsustainable because of threats from other supply, including shale oil but also gas-to-liquids and liquefaction of kerogen, both of which have received increasing consideration at $100 a barrel. This is a tricky proposition, since it is all too easy to get it wrong: reacting too early, when it's not clear if the synfuels will proceed, or waiting too long, and having the capital in place, as was the case with the Canadian oil sands.

There's a fourth scenario, that could be called the doomsday scenario, which is a combination of the three above, in all or part. It is entirely possible that rising non-OPEC production will occur at the same time that some of the OPEC countries resume currently disrupted production, putting huge pressure on markets. And, if prices were to return to $100 as many expect, the development of synthetic fuels seems likely to proceed, or as least to appear so.

Is there a fifth scenario that calls for rising prices? Nothing is impossible, but market balances that would support long-term prices above $100 a barrel would be difficult to attain without some major supply problems. Demand growth above 2 million barrels per day per year seems impossible to attain at these prices, unless some truly extraordinary economic growth occurs globally. And only two likely events could prevent supply growth from levels described above: the cessation of hydraulic fracturing of shales, presumably due to political opposition reflecting environmental fears, or a major political crisis in the Middle East, such as widespread religious war or a lengthy uprising in Saudi Arabia. Either is possible; neither is likely.

The price collapse in 2014 does not fit precisely into any of these cases, although the second scenario would be closest. It is important to realize that neither OPEC nor the Saudis took any action to bring prices down; rather they simply declined to step in when they were falling. While gas-to-liquids and kerogen were not threatening their market share, the greater-than-expected shale oil boom in the United States and rising Iraqi production, after years of delay, was clearly a concern.

Unless there is a significant change in non-OPEC production trends, and especially for shale oil, it is highly likely that we have now entered a new, lengthy period of lower prices (i.e., $50 or so).

TRADERS

> Successful investing is anticipating the anticipations of others.
>
> John Maynard Keynes[18]

As Keynes figured out, the price is set in the shortrun by what traders think it should be, so it is important to consider not just the intersection of OPEC production, world demand, and non-OPEC supply, but *expectations* of trends in those variables. These can be almost as hard to change as the actual market fundamentals, partly because it is not always clear if those fundamentals are changing or just experiencing a temporary deviation from the trend.

Over the past decade, the primary expectations driving traders' market-clearing price perceptions have been booming global oil demand, and especially in Asia, without regard to price, and anemic non-OPEC oil supply. Given the current economic situation, weak oil demand in Asia will hardly be seized upon immediately as a sign of longterm, price-induced conservation. Already, nine of the last ten years have seen average oil demand growth below 2%, without altering the view that economic recovery will lead to tighter markets. Every piece of good economic news that came out in early 2013 moved prices higher, although recent slowing in China has moderated this.

Expectations for non-OPEC supply are also slow to change. There are few old-timers trading oil who remember the period before 1998 when non-OPEC regularly confounded pessimistic forecasts. For every Yoda, there seems to be a dozen Luke Skywalkers, "Always with you it cannot be done." Even if conventional oil production grows in the United States, Colombia, and various African nations at a robust pace, it is far from certain that traders will feel this represents a change in the recent tendency for anemic behavior.

And while tight/shale oil production is likely to spread, it might be three to five years before countries like Argentina, Australia, Colombia, and France, for example, show that their shales are going to be as productive as the ones now in the early phases of exploitation in the United States.

The big question now is: will lower prices seem to be curtailing global energy investment? Already, the U.S. rig count is down sharply, and some Canadian oil sands investment has been deferred, but the longer-term impact is not clear. If shale oil production only grows modestly, say, 300 thousand barrels per day or so per year, then traders are likely to feel that $50 is unsustainable and bid prices back up. But if it remains robust, and most currently planned deepwater projects proceed, then prices will persist in the vicinity of $50.

MARGINAL COSTS DO NOT MAKE THE FLOOR PRICE

Asked by the BBC if oil could go back "well-above" $100 a barrel, Mr Margerie said "Yes," adding: "The problem is when, the problem is to anticipate this, not to send [a] message to scare people but to send a message ... it's important to invest now." He said prices of $60–$70 a barrel was not enough to protect our long-term investment.[19]

Confusing short-run and long-run marginal costs is also a potential analytical trap. Supply curves showing costs represent long-run marginal costs, meaning that they are not current marginal costs, but rather full costs (capital and operating). In the short run, marginal costs are operating costs that are typically a fraction of full costs, rarely more than half.

That is why the price can drop below the supposed level of marginal costs without much of an effect on short-term oil supply. As the late Total CEO Margerie indicated, the response would be of lower drilling and development, not operations, meaning that the supply effect would be delayed and somewhat uncertain. But there would be much less shut-in of supply than would be suggested by the long-run marginal cost supply curves in Chapter 22.

And remember that the right-hand point on most supply curves is actually an *outlier* rather than a true representation of the long-run marginal cost. There are many ways in which an unusually expensive supply can be developed and enter the database, that is, the supply curve. Companies might, for example, make unprofitable investments expecting some non financial benefit, such as good will. This was apparently the case when some companies invested in Saudi Arabian natural gas exploration, despite the low fixed domestic gas price.

Thus, in Figure 18.6, a price drop of 40% would lead to a loss of approximately 10% of production (A–D versus A–C). In the real world, this means the small number of very expensive projects would be cancelled, and *if and only if* they needed to be replaced by projects with similar costs, then the price would be forced to recover to the earlier "marginal cost" level. But even this proposition will be questioned in Chapter 21.

The empirically driven supply curves in Chapter 16 seems to confirm this view; most showed a large amount of cheap oil and a small "tail" of

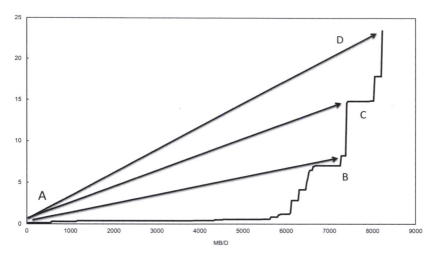

Figure 18.6   Impact of Lower Prices on Supply (From data in Adelman and Shahi [1989])

high-cost oil. Implying that the highest cost oil on the supply curve is the marginal cost is generally misguided, as discussed next.

IF WISHES WERE PRICES. . .

The marginal cost of production is the ultimate floor in the oil market. In the North Sea it can be $80 to $100 dollars

But the real marginal cost of production also includes social costs that some big oil producers need to pay. When you add social costs in Russia and Saudi Arabia, it means that the effective floor on Brent is around $100 a barrel.[20]

Andrew Moorfield from Scotiabank.

Overemphasis of oil producers' financial needs is another common mistake made by market analysts. A number of organizations have produced estimates of what various producing countries require to balance their budgets and some go so far as to assert that this will decide OPEC's target price, but they are confusing wants and abilities. While many governments consider their revenue needs in thinking about quota decisions, few make explicit commitments on that basis.

Partly they recognize that every country has different needs, making negotiations to set a price accordingly all but impossible, but also because those needs are constantly changing.

And if determining the likely price were that simple, how to explain the period from 1985 to 2000, when prices were much lower? Revenue needs should be considered only as a minor and short-term influence on prices, and not a significant factor.

CONCLUSION: ARE WE THERE YET?

The price drop of 2014 has aspects of both the 1986 and 1998 price collapses, making it hard to know whether it will persist or prices recover, at least for a time. The primary indicator—pressure on demand for Saudi oil—is not apparent at this point. However, given the arguments of this book and the perceptions of some analysts, the rise of shale oil production suggests that within a few years, the Saudis would face serious problems and need to move prices lower.

Similarly, soaring supply from Iraq and the possibility that Iran will soon have a more open oil industry implies that new quota wars lie within the near future, even if demand for OPEC oil is not falling. And the longer-term prospect of reform in Mexico and Venezuela and a return to political stability in Libya means that large amounts of relatively cheap oil will be hitting the market.

The possibility of a recovery in oil prices in the near term cannot be ruled out, especially with new supply disruptions as Chapter 19 discusses, but the long-term indicators all appear to suggest that without major new restrictions of access to petroleum resources, the long-term price is unlikely to be above $60/barrel for any length of time.

NINETEEN

The Shale Revolution

The late George Mitchell is a hero to many in the petroleum business, and deservedly so. In fact, he did far more to reduce greenhouse gas emissions than the winners of the 2007 Nobel Peace Prize, Al Gore and the IPCC, even if only the United States is considered. Since 2009, U.S. CO_2 emissions have declined because cheap natural gas has displaced nearly 300 million tons of coal in power generation. This is roughly the same reduction in CO_2 that solar and wind combined have accomplished, and not only without the massive subsidies but providing tax revenue.

In many ways, Mitchell is the epitome of the American entrepreneur: a stubborn visionary who persisted until he accomplished the breakthrough that has created a revolution in the energy industry. For years, most dismissed his efforts to improve recovery rates from the Barnett shale, thinking it could not be done economically. He succeeded beyond the wildest dreams of virtually everyone, to the point where now, approximately one-third of U.S. marketed gas production comes from shale gas wells.

In consequence, U.S. natural gas prices have dropped two-thirds or more from their highs of a decade ago, and even global LNG prices are showing signs of weakness. Plans to export LNG from the United States have proceeded fitfully, stalling as lower oil prices help to bring down gas prices in Europe and Asia. All this while a variety of pundits insist that most shale gas is uneconomic and that high well decline rates mean a near-term peak.

And with the burgeoning shale oil production—growth of 2 million barrels per day in two years—and the drop in prices in 2014, the implication is that the $100/barrel price is not sustainable, and many assume that the Saudis wanted prices to drop in order to restrain investment in this new resource. If significant amounts of shale oil can be produced for $50/barrel, that will become a ceiling on prices.

PERTINENCE

All of which is of great scientific curiosity, but what does this mean for the oil industry and market? Several things stand out:

- Shale basins are enormous.
- Shales are ubiquitous.
- They contain massive amounts of oil and gas.
- This requires significant effort to produce.

The near omnipresence of shales is a reflection of their origin not in the carcasses of large animals over millions of years but in the deposition of a myriad of often-microscopic organisms. Compare the size of the human population: roughly 335 million tonnes at the present time. The weight of the world's population of prokaryotes (bacteria and other unicellular organisms) is roughly 400,000 million tonnes, and they have been present on earth about 4 billion years, versus 170 million for the dinosaurs, so clearly they dominate the organic material available for conversion to fossil fuel resources.[1]

The impact: while peak oil advocates talk about 2 trillion barrels of recoverable conventional oil, and the USGS estimates that there are 3.5 trillion barrels of recoverable oil, the additional known amount of shale oil is estimated at 8 trillion barrels in place, of which 345 billion barrels is considered technically recoverable, and this is extremely conservative, as we shall see.

Although there are some incredible deserts on the face of the Earth, it is often referred to as a water planet for good reasons. Oceans cover three-fourths of the face of the planet, and much of the land surface is well-watered. In addition, areas that might be dry now were often covered in water, oceans or fresh, including from the Gulf of Mexico to Canada in the center of the United States and similarly, Central Australia, both during the Cretaceous period.

Table 19.1   Conventional and Shale Fields (Prudhoe Bay Fact Sheet and Intek (2011), in http://eaglefordshale.com/companies/)

		Prudhoe Bay	Eagle Ford
Size	Sq Miles	333	3,423
Thickness	Feet	600	200
Reserves			
Oil	Bln bbls	13	3.4
Gas	Tcf	26	20.8
In-place resource			
Oil	Bln bbls	25	300
Gas	Tcf	46	100

The best analogy of shale to conventional deposits is of a farm stand to the orchard: conventional deposits represent localized occurrences where the oil has been gathered and stored, while the shales are the origin of that resource. The East Texas field was the source of wealth for its developers, the Hunt Brothers, but the Eagle Ford shale has 200 companies now active in exploiting oil from it.[2] Table 19.1 compares the physical characteristics of Prudhoe Bay, the largest conventional field in the United States with the Eagle Ford shale in Texas, and the difference in size and in-place resource is astounding.

Thus, huge amounts of the Earth's surface have had the opportunity to generate shales, and geological studies confirm this, as Figure 19.1 shows, and new shales and layers are being constantly discovered. Any map even five years old is horribly outdated: Jarvie's 2010 map, for example, did not show the Utica shale, Granite Wash, Rogerson, Mississippian Lime, Permian, or the SCOOP play in Oklahoma.[3] The continued discovery of new shales is occurring despite the extensive knowledge of U.S. geology resulting from a century and a half of oil drilling.

The importance of this is crucial: it means that there is unlikely to ever be a large concentration of the shale resource in any given country or region, the opposite of conventional oil's situation, where just five countries in the Middle East possess half of the world's conventional oil reserves.[4]

This is of vital importance for the gas industry, whose conventional resource is widely scattered but often far from major markets, necessitating massive investment in transportation. The development of shale gas in the United States is less important than its possible development in central Europe and east and south Asia.

THE RESOURCE BASE

As mentioned in Chapter 15, no estimates of the shale oil resource base were made until a few years ago, instead, kerogen shale has been studied intensively for many years. The most comprehensive study to date was by Advanced Resources International for the U.S. Department of Energy, covering 137 formations in 43 countries.[5] Table 19.2 lists the countries with the most significant resources, as well as the total global estimate. Although the ARI report does not break down recovery factors by country, they have assumed a 3–6% rate overall, and the right hand column of Table 19.2 shows the implied in-place resource assuming 5% recovery rate.

These estimates are necessarily very conservative. Two data series illustrates this. Estimates for the Bakken reserves have increased from 151 million barrels, made before hydraulic fracturing had been applied to shales, to 3 billion after fracking began in the deposit, to 7 billion now, with the Three Forks shale included. This demonstrates the fact that both improved recovery and knowledge of the expanded resource contributes to the growth in recoverable resources.

On a global level, between the 2011 and 2013 reports performed for the Department of Energy, the estimated non-U.S. shale technically

Table 19.2 Technically Recoverable Shale Oil and Gas Resource (Advanced Resources International [2013]. Recovery Factor for Shale Oil Assumed to be 5%. Conventional Estimate from NPC [2007])

	Gas TRR	Oil, Bln bbls	
	Tcf	TRR	In-Place
France	137	5	94
Poland	148	3	66
Russia	287	76	1,516
Canada	573	9	176
Mexico	545	13	262
US	567	58	1,162
Australia	437	18	350
China	1,115	32	644
Indonesia	46	8	158
India	96	8	158
Total Non-U.S.	6,634	345	6,900
US	567	58	1,160
Total	7,201	403	8,060
Conventional	15,400	3,400	10,000

Table 19.3   Potential Recoverable Resources at 10% Recovery Factor (Calculated from data from Robert Kleinberg, Schlumberger, http://www.fool.com/investing/general/2014/07/30/we-could-unlock-more-than -267-billion-barrels-of-o.aspx)

Shale Play	Est Recoverable Oil	Current Oil Recovery Factor (%)	Implied Oil in Place	Recoverable at 10% Recovery Factor
Permian Basin	75	3.50	2,143	214
Bakken/Three Forks	31.5	3.50	900	90
Eagle Ford	26	6	433	43
Niobrara	7	1.4	500	50

recoverable resource has grown by a factor of 10, from 32 to 345 billion barrels, in part reflecting greater coverage but also better knowledge of both the various shales and production methods. Table 19.3 shows a recent estimate for the primary U.S. shales and how the recoverable resource changes when a 10% recovery factor is assumed. The total amount, 397 billion barrels, is greater than the earlier estimate for non-U.S. basins, reflecting better knowledge not a larger resource.

There are, of course, adjustments downward as well, particularly as drilling provides additional information about resource density and recoverability. The best case would be that of the Monterey shale in California, which in 2011 was estimated as the largest U.S. shale oil resource, but is now thought to be too expensive to produce.[6]

Industrial Progression

In many ways, then, shale development is to deepwater fields what gas turbines are to nuclear power plants. The risks are much lower because a small investment will quickly yield revenue, whereas the deepwater field usually requires at best a very expensive well and installation. A company can spend as much or little as it wants on shale development, stopping in uncertain times or when it has low cash flow and be profitable. But if progress is halted on a deepwater field, the initial investment is lost.

The analogy is hardly perfect: natural gas turbines face minimal opposition because there is little danger to the public and environmental impact is minimal. This does not mean they never suffer from the NIMBY problem: even during California's 2000 power shortage, wealthier parts of the state opposed construction of such plants.

Fracking has much more environmental impact, particularly if done wrong, but the small scale of production means that the worst possible oil spill is trivial compared to, say, an oil tanker accident or offshore drilling disaster. The Macondo disaster released an estimated 5 million barrels over 83 days, or 60 thousand barrels per day, but the typical shale well produces less than a million barrels, and rarely more than 1 thousand barrels per day.[7]

And the geographical impact tends to be much more minimal. Although opponents claim a huge impact on the local environment, most of that occurs during the fracking operations, as water is brought to the site. Afterward, the site is largely self-contained equipment, pumps, and pipes, providing only visual pollution.

Unlike the drillers in Backyard Oil, on the Discovery Channel, who rent a rig out by the day and send a relatively unskilled laborer to watch over it, fracking is a multimillion dollar operations that requires a much more specialized set of equipment.[8] If governments like Russia or India liberalized their industries and allowed individuals to drill and develop oil and gas, a significant amount of production could come from small producers, but the source would be conventional deposits. Initially, at least, few would have the knowledge or expertise to drill and fracture shale wells.

This could actually spur changes in those countries' policies, especially in China and India, which are energy poor and would greatly benefit from such entrepreneurial activity. The former is much less likely to undertake any sweeping change, but it is not inconceivable that India, particularly under the new, more commercially inclined leader P. M. Modi, might move in that direction.

Outside the United States, there are ongoing efforts to exploit shale in a number of countries, including Argentina, Australia, Britain, Canada, China, Colombia, and Russia. Argentina and Canada are the furthest along and should see significant production beginning in the next year or two, although the operations, especially in Argentina, are still on a much smaller scale than in the United States. Countries like Poland and the Ukraine have tried to develop their shale deposits to reduce reliance on Russian gas, but so far success has been elusive. In China, the government is enthusiastic about developing its shales, but has made only slow progress to date. Some foreign companies, notably Shell, are assisting in the effort.

CHALLENGES

Quite a number of pessimists have pointed to the various challenges that face the shale oil and gas industry, and most are debatable. The most obvious is the question of the economics of production, which some say is abysmal. Others point to the high decline rates that shale wells typically experience as evidence that production will never be very robust. And there is a significant amount of public opposition, which often affects regulation and, in some places, constrains development.

Decline Rates

Pessimists insist that decline rates for shale wells are so high that it will be impossible for them to achieve or at least maintain levels of production that will be meaningful. Most famous of these is Arthur Berman, who made a name for himself by analyzing well data and made some relatively apocalyptic predictions about future production. Most notably, he has argued that 22–23 Tcf/year of shale gas capacity needs to be replaced annually, and for shale oil, he says "the party is over."[9]

The issue resembles that for conventional oil, namely, decline rates are very high and require constant effort to replace production (Chapter 14). As discussed later, the high decline rates for shale gas did not prevent it from growing at phenomenal rates throughout the past decade, even when drilling dropped sharply. Whether this will be reproduced in the case of shale oil is an empirical question that has not yet been answered but will be discussed below in detail. The important thing is to avoid treating "decline rates" as a silver bullet that will slay the shale revolution.

Economics

The economic analysis of shale oil and gas production has been severely deficient mainly because of the static nature of the estimates in rapidly evolving production methods. Even so, observers have been baffled by the industry's ability to increase production of shale gas even as prices collapsed to half the supposed "breakeven" point. One writer noted in 2009 that the *optimistic* estimate of breakeven costs was $5–6/Mcf, that most producers were losing money (the price was then $3.80/Mcf) and called optimists "clueless."[10] Since then, the price has averaged $3.8/Mcf while shale production has quadrupled!

One problem has been the use of average results for shales that are somewhat heterogeneous and where the operators are very heterogeneous, as well as the nature of play development. Since each shale is somewhat different, it is necessary to do a significant amount of testing to discover the optimal production method: pressure, length of horizontal laterals, number of fractures, and chemical mix. Early wells are often vertical and designed to allow the study of the shale's geology, and as such are low producers. Averaging their cost and productivity with latter wells yields an inaccurate view of the actual costs. In addition, some companies are very diligent about developing the best production methods, while others are less knowledgeable and/or experienced and concentrate on putting holes in the ground.

Play Evolution

Because production techniques for shale gas are different from conventional deposits and because each shale has different characteristics, there is a much greater tendency for costs to evolve and drop as companies move into and evaluate a play.

Jeff Wojahn (2010) of Encana described the process as follows:

1. Identify prospect
   a. Basin history, rock properties
   b. Hands-on rock work
   c. Experience!
2. Map gas in place
   a. Correlations between logs, cuttings
   b. Decide net play, play limits, drivers
3. Define prospect
   a. Model type curve, costs
   b. Divide play into economic buckets
4. Simulate development
   a. Monte-Carlo prediction of outcomes
   b. Both production and cash flows

Thus, as well costs are compiled it is important to recognize that data for any given well might not be representative of the play as a whole, because of heterogeneity. Beyond that, costs should be dropping sharply from the early wells to the later wells, but existing analyses don't seem to address these issues.

This also highlights the rapid evolution of methods and costs. Aside from a drop in cyclically inflated costs due to high activity rates, companies have reported numerous improvements in methods that have brought continuing savings.

Innovations include:

- *Longer laterals:* The horizontal wells are now up to 6,000 feet, double or triple early wells.
- *More fractures:* The number of spots per well where the shale is fractured is increasing.
- *Stacking wells:* Given multiple shale layers in some places, multiple wells can be drilled from the same drilling pad.
- *Walking rigs:* Because numerous wells are drilled in one area, rigs have been designed to move from one site to the next on their own, cutting down idle time.
- *Zipper fracs:* Fracking one well while another is being drilled at the same time cuts down on completion time.

All of these and many others have resulted in cost reductions of 50% for some places and many companies, and additional savings of 10–15% are being projected for 2016 by most drillers because the lower investment levels are seeing service costs drop. This explains, in part, the great success of shale gas despite lower prices and suggests the same might happen with shale oil.

SHALE AND ITS DISCONTENTS

The shale revolution has its share of naysayers, with three primary objections. Some complain that the resource is being overestimated and that high decline rates will cause production to peak and decline soon; some argue that non-U.S. shales are unlikely to be exploited, at least not until far into the future, and there are a significant number who oppose the use of hydraulic fracturing in shales.

Political Opposition to Fracking

As discussed in Chapter 13, political opposition to operations is a growing problem for the oil industry, and particularly those engaged in fracking. In a variety of countries around the world, public opposition has been strong, and resulted in fracking bans in countries like France, as well as parts of the United States. Well-known environmental activists like Alec Baldwin and Yoko Ono have weighed in, arguing for greater use of renewables.

The industry in this case was helped by the overreaction of many opponents who have often resorted to apocalyptic proclamations. For example, David Letterman described the Delaware River basin as "destroyed," Yoko Ono, in a full page ad, said, "President Obama, you have two beautiful daughters. Do you want their health, environment and futures to be irreversibly destroyed by fracking, like the suffering children of Pennsylvania, West Virginia and Ohio?"[11] Any number of articles on the Internet correlate fracking with health problems, while ignoring other prominent pollution sources or treating normal occurrences, such as natural gas in tap water, as startling events that are claimed to be due to fracking.[12]

The long history of fracking and its widespread nature provides the strongest evidence against the idea that it is dangerous to the environment or health, rather as claims that cell phones had carcinogenic effects became widely dismissed as it was obvious that the omnipresence of the phones was not causing an increase in brain cancers.

Two serious concerns have proved to be manageable. Although methane's greenhouse effect is a problem, a recent study by the Environmental Defense Fund found that while methane emissions by the industry was higher than believed, it occurred primarily in the processing and distribution, not the wellhead production.[13] And the EPA found that contamination of water supplies was rare, and usually from surface activities, not the drilling and fracking procedures.[14]

That said, fracking has been slowed or even banned in a number of places, such as New York state and France. The more interesting question is whether the continued success of fracking in the United States, with its minimal side effects, will encourage others to follow suit. Already, Britain has argued that the practice is not dangerous, and liberal governors in California and Colorado have supported it, the latter engaging industry and environmentalists to write regulations that both found acceptable.

Only in America?

D. Nathan Meehan of Baker Hughes published a nice summary of the differences between U.S. shales and those overseas, which I paraphrase here:

- Land access
- Difficult terrain and other operability issues
- Service company capabilities

- Well costs
- Few risk-taking firms
- Social and environmental issues
- Poor quality resource (low well productivity)
- Low gas prices
- Inadequate infrastructure
- Well stimulation technology needs to be optimized for local conditions
- Data availability (geological and production)[15]

Obviously these apply in different amounts in different countries, but many of them apply in most countries. A lot of them are technical in nature, such as more difficult shales (e.g., softer and thus harder to fracture); however, it is not clear that they cannot be overcome. And especially, pessimists seem to always assume that not only are technical challenges insurmountable, but as with peak oil advocates, they often mistake political problems for physical ones.

Development of shales in areas like Brazil, Siberia, and the Caspian will probably progress slowly, as each has other more attractive investment targets, especially conventional oil. However, to the extent that the shales are profitable (i.e., oil producing), then governments might allow companies to develop them. No company avoids an attractive investment just because another project nearby is more attractive, if that project is not available for them to invest in.

SHALE GAS AND WORLD OIL PRICES

Natural gas remains abundant, despite some attempts to make an argument for "peak gas" and shale gas has only magnified that abundance. Even if shale gas is slow to develop in the eastern hemisphere, conventional gas fields continue to be found around the world, including supergiants in the Mediterranean and offshore east Africa. Robert Hefner's argument that "the age of gas" has come looks very solid.[16]

In the early 1980s, an enormous amount of residual fuel oil was replaced in power generation and industrial use by coal and natural gas and that transformation cannot be repeated. However, there is still a significant amount of fuel oil being used that could be displaced, but that requires a combination of gas development in oil-rich nations, and lower prices to importing countries.

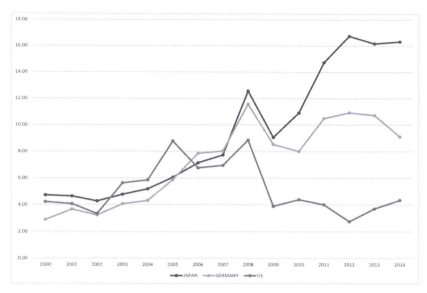

Figure 19.1   World Gas Prices (BP Statistical Review)

Price Dilemma Resurrected

For gas importers in Europe and Asia, price setting is a problem, as there is not a functioning market that sets international gas prices. Starting with the use of natural gas to replace crude oil in power generation in Japan four decades ago and bolstered by the 1970s hysteria for avoiding insecure crude imports, consumers in importing nations agreed to pay the same price for imported natural gas as for crude oil imports, converted on a heat-equivalent basis. As Figure 19.1 shows, this has generally meant that natural gas prices in most of the world have been well above those in North America, the only major market were prices are set by supply and demand. This practice of equating oil and gas prices makes as much sense as pricing tea sales based on coffee prices, adjusted for caffeine content, but it has persisted because of the market power of natural gas exporters.

Enter Shale gas

To date, shale gas has made an impact only in the United States, but expectations of soaring LNG imports in North America have now been reversed, possibly to the point of the United States (and Canada, and conceivably Mexico someday) becoming serious gas exporters. Projects in

Qatar and Russia that had been aimed at the U.S. market now find themselves either delayed or casting about for new buyers, and other projects could soon experience the same trouble. What happens if this trend spreads to other importers (current or potential)?

Most notably, the LNG industry has great hopes for China as a market. But China has an estimated 1275 Tcf of shale gas,[17] and while it has only recently begun to learn hydraulic fracturing techniques, its companies are certainly capable of large scale, repetitive drilling operations. Given that natural gas demand has been increasing by about 500 Bcf/yr in the past five years, shale gas production should soon reach levels that satisfy most demand growth. It took the United States about eight years to reach that level of incremental production, and while China might take a little longer to accomplish that, it doesn't face the same kinds of skepticism and resistance to the viability of shale that the early U.S. drillers did.

Europe is a different story, with much greater—and more effective—resistance from the public to large scale drilling and fracking. Restrictions and/or outright bans on hydraulic fracturing have been problems in mostly western European countries, but Poland and England appear to be moving ahead with at least some development. Supply growth does not look likely to be rapid, but European natural gas demand will be anemic for several years, at least, and supplies from North Africa will make up much of their incremental needs. Gazprom will be especially hard hit by weak demand in markets east and west.

Impact of Increased Shale Gas Production

Because shale gas wells are expensive and less productive than wells in areas like East Africa offshore, eastern Siberia, or Australia, the implication is that shale gas will prosper when it is found near consuming areas, such as the United States or China, as transport costs will be minimal. (The differential is heightened by price-setting clauses described earlier.) Thus, investment will be done locally, and reduce imports in many nations, as already in the United States.

The outlook, then, is for much slower growth in the international trade of natural gas, especially LNG. Major exporters like Russia and Australia will find that markets are largely closed to them, and that they must compete fiercely for small volumes of new sales. And with the growing number of players in the market, it seems highly likely that the oligopoly will break down and some exporters will try to expand market share by discounting prices relative to crude oil equivalence, taking the industry back to the 1960s, when the Dutch created a European market by pricing Groningen gas to compete with coal and residual fuel oil in the power sector.

Of course, there will be strong resistance from current exporters who will hardly welcome revisions to existing, high-priced contracts, and only when there is enough pressure will the practice break down. The tipping point will be when shale gas is recognized as ubiquitous and cheap enough to compete with oil price-linked imports, and exporters scramble to lock in contracts and avoid being frozen out of markets. This could be similar to 1986, when gas exporters like Norway decided to abandon efforts to achieve ever-higher prices so as to cope with lower-price competition.

Already, there are some signs that price indexes are changing, as it is reported that Kansai Electric in Japan has signed a contract tied to Henry Hub prices with BP (although it is not clear what multiplier is used) and Statoil is offering a contract to Wintershall with prices linked to European spot prices.[18] These amounts are small, but could be just the beginning of a broader shift.

An important implication is that exporters will probably be forced to choose between price-cutting and having their gas "stranded" especially if oil prices retreat from currently elevated levels. One option that would then present itself is to utilize it domestically, essentially onsite, where the low production costs could be translated into low fuel or feedstock prices. Increasing the use of gas for power generation should certainly be pursued in many of these places, but one interesting alternative is as feedstock for gas-to-liquids (GTL) plants.

Oil Market Impacts

A combination of lower gas prices and higher production should be expected, the best of both worlds for some, but a challenge for others. Lower gas prices would mean more fuel substitution in the short run, and greater creation of pipeline systems in the long run, with Latin America, West Africa, and South Asia all candidates for extensive pipeline networks. Power generation would become cheaper and more reliable in many areas, as small gas turbines would be preferable in most instances to renewable installations.

There is also a distinct possibility that NGL production will increase significantly, as is already being seen in parts of the United States. Prices for propane, for example, have become depressed by about 40% compared to historical levels vis-à-vis oil prices, even more relative to Brent, and are likely to remain so as infrastructure struggles to keep up with rising supply. This raises the potential for propane markets to expand into new areas, especially where people lack access to commercial cooking fuels. While some displacement of kerosene or heating oil could occur in countries like India and Japan, it would be hoped that people now using biomass will become the new consumers, rather as whale oil was replaced by kerosene for lighting in the 19th century, reducing deforestation and habitat loss.

The losers would be those LNG exporters who hold out for high prices and find there is no market for them. This could particularly affect high-cost supplies in places like eastern Siberia, the American Arctic, and the Shtokman deposit offshore northwest Russia. Others, such as new discoveries in east Africa and Australia, with lower production costs but no market, might attempt to develop local consumers, such as in power generation, petrochemical plants, or conversion to liquids through gas-to-liquids (GTL). Recall, after all, that the petrochemical industry has generally migrated to areas with surplus natural gas, beginning with the 19th century in Appalachia, then the Gulf Coast, and now the Middle East.

Of course, aside from the question of oil equivalence in pricing, the possibility that oil prices will fall could affect natural gas trade as well. If oil prices remain high, then discounted LNG prices could lead to greater market share for gas in many Asian countries, especially the historical LNG importers, which still use significant amounts of residual fuel oil for power generation. Also, some oil producers like Mexico and Saudi Arabia that still use lots of oil for power could cut back as well (Table 19.4).

Table 19.4   Residual Fuel Use in Selected Countries (EIA)

Oil Producers	Thousand barrels per day
Mexico	213
Russia	408
Iran	421
Iraq	123
Kuwait	160
Saudi Arabia	323
Egypt	210
Total	1,858
Gas Importers	
China	551
Hong Kong	150
Japan	550
Korea	269
Singapore	797
Total	2,317

Lower oil prices, say $60 a barrel, could discourage high-cost LNG producers, such as the United States, from exporting at a discount, meaning lower cost producers like Australia and Russia would again have an advantage.

Thus, the oil market over the next decade faces a triple storm: higher conventional oil production, competition from shale gas and its attendant NGLs, and rising shale or tight oil supply. This will almost certainly lead to lower prices, but hopefully it will also mean cleaner energy: more natural gas instead of coal for power generation, and commercial energy like propane for the energy impoverished.

On the other hand, there is a significant probability for a major expansion of natural gas use, including substitution for coal in South Asia and China and heavy fuel oil in East Asia, as well as the creation of major systems supplying utilities in other less developed countries, notably in Africa. The rise in NGL production would allow LPG markets to grow serving those not on any pipeline system, and improve the lot of the energy impoverished substantially.

Timing of the spread of the shale gas revolution might be uncertain, but there is no doubt it will prove transformative to world energy systems, and the industry should be prepared to cope with these changes.

SHALE OIL

The U.S. shale oil revolution has not only made an enormous impact on world oil markets, but astonished nearly every one with its speed and success. Despite the shale gas revolution, which was well under way by 2005, many believed that the same could not be accomplished for shale oil, primarily because petroleum hydrocarbon molecules are much larger than methane and thus less likely to flow through tight formations. Even when oil production in the Bakken shale in North Dakota began to grow strongly in 2009, there were arguments that the process could not be replicated in other locations because the Bakken shale characteristics were uniquely favorable.

Obviously, this has not proved to be the case, but there remain numerous questions about the economics and production of shale oil. The oil price collapse of 2014 has caused many argue that shale oil production will at least stop growing as rapidly as before, and possibly began to decline.

Shale Oil Resources

Estimates of shale oil resources must be considered highly preliminary and evolving, with revisions both positive and negative as more information comes in. Negative revisions usually come as the result of drilling efforts, which find productivity less than expected, while positive often reflects studies of new areas or extensions of existing areas. To date, recovery factors have not changed very much, but in the future are likely to be important.

In the United States, the country is heavily explored and there are many less blank spaces on the map than in Russia or Africa, for example. But even here estimates of resources are continually evolving. Initial estimates of recoverable Bakken oil resources were only on the order of 151 million barrels, which was increased to 3.65 billion in a 2008 analysis by the USGS, and more recently, to 7.4 billion barrels, when the Three Forks layer of shale was included. This makes the Bakken/Three Forks about half the size of Prudhoe Bay, the largest U.S. oil field.

On the other hand, optimism has sometimes turned out to be excessive. The 2011 Intek estimate of resources at the Monterey shale in California have been downgraded to zero, on the evidence of the difficulty of producing oil from the basin.[19] And expectations of 100 billion barrels of oil in place (not recoverable) in southern England have been clarified as still quite preliminary. The Monterey downgrade in particular serves as a cautionary tale regarding assessments made on the basis on minimal drilling.

To date, the biggest problem remains the lack of information about the recoverability of shale oil in many basins as well as the heterogeneity of the deposits. Table 19.5 shows three estimates made for different basins, but the most reliable, those of the USGS, are often seriously out of date. Still, claims from oil companies, such as Continental's argument that the Bakken has 24 billion barrels of recoverable oil, need to be treated with a degree of caution.[20]

Table 19.5   Recoverable Resource Estimates for Various Oil Shales (Intek [2011], USGS [2012] and Kleinberg [2014])

Play	EIA/Intek	USGS Date of Study	Schlumberger
	2011	2012	2014
Eagle Ford	3	0.76	26
Bakken	4	7.38	31.5
Monterey/Santos	15		
Niobrara			7
Permian			75
Avalon & Bone Springs	2		

Recovery Factors

The biggest element of uncertainty regards the fraction of oil, which can be recovered from any given shale deposit. The USGS has assumed a level close to 1% for the Bakken, while others have argued 3–5% is more reasonable, and at least one company claimed that 10% could be achieved. (An early study used 50%, but this was completely conjectural.) To date, there has been relatively little broad, statistical analysis to demonstrate recovery rates across different basins.

Given the relatively recent advent of shale production, this is not surprising. Advanced Resources International has described the potential recovery factors and what would determine them:

- *Favorable oil recovery:* A 6% recovery efficiency factor of the oil in-place is used for shale oil basins and formations that have low clay content, low to moderate geologic complexity and favorable reservoir properties such as an over pressured shale formation and high oil-filled porosity.

- *Average oil recovery:* A 4% to 5% recovery efficiency factor of the oil in-place is used for shale gas basins and formations that have a medium clay content, moderate geologic complexity and average reservoir pressure and other properties.

- *Less favorable gas recovery:* A 3% recovery efficiency factor of the oil in-place is used for shale gas basins and formations that have medium to high clay content, moderate to high geologic complexity and below average reservoir pressure and other properties.[21]

There have been more recent claims about much higher recovery rates, such as Whiting Petroleum's estimate of 15–20% in the Niobrara, and UK Oil and Gas that it can get 5–15% recovery rates in the Weald Basin in south England (although they haven't yet begun testing).[22]

Only as various basins are more intensively drilled and studied will it become clear what recovery factor is likely in each basin, but over time, they should be expected to increase. In Table 19.3, the main four shale oil basins are shown with the current estimates of recoverable oil, the recovery factor used, and the amount that could be recovered if the recovery factor increased to 10%. Reserves would go from 140 to 400 billion barrels. Even if the improvements take place over two decades, it will still mean that the amounts of shale oil are technically recoverable in the United States are similar to Saudi reserves of conventional oil.

Notice that the Permian basin is estimated by Kleinberg to have the largest recoverable resource by far, although the amount of fracking there is still relatively small. The basin has a long history of conventional oil production, but the application of fracking led to a near doubling of production in three years. This is because the basin includes numerous shale levels, and in many places, multiple completions can be made from the same drilling platform into different layers.

Primary sources of production in the Permian are the Bone Spring, Wolfcamp, and Spraberry shales, and there is some overlap among them and some of the smaller shales. Much of the Wolfcamp has four "benches" or levels that can be produced, greatly increasing the potential productivity from a site.[23] Given the large areal extent of the basin as a whole (86,000 square miles), and the large size of the sub-basins (Wolfcamp is 4,000 square miles), there is clearly ample opportunity to develop the region. As seismic is performed and test wells are drilled, it should become increasingly clear how much oil is likely to be recovered and from which plays, but the reserve estimates will probably increase sharply.

PROSPECTS

Table 19.6 provides a list of various forecasts for 2020 only. Clearly, the earliest ones were too pessimistic, and the later projections have increased significantly. There is still substantial divergence among them, and it remains unclear what prices are being assumed in each.

The reliability of these forecasts is questionable, in part because oil price forecasts remain highly uncertain and production is very sensitive to the

Table 19.6   Projections of 2020 U.S. Tight Oil Production by Date (Citicorp, "Energy 2020: North America, the New Middle East?" March 2012; International Energy Agency, World Energy Outlook, 2014; OPEC, World Oil Outlook 2014; Rystad Energy, "Shale Companies Have Proven More Efficient but not More Productive," July 2014; BP Energy Outlook 2035, February 2015)

Forecaster	Date	Barrels per Day
Citicorp	2012	3.5
Rystad	2014	5.5
OPEC	2014	4
IEA	2014	4.2
BP	2015	4.5
EIA	2015	5.6

world oil price. But also, most official forecasters have been too conservative in the past, raising the possibility that they will be so again.

ECONOMICS OF PRODUCTION

Some investment banks made statements about breakeven prices as the oil prices were dropping in 2014, including the following:

- Bernstein Research said one-third of U.S. shale oil would be uneconomical below $80.
- KLR Group estimated that a WTI price of $90 was needed to maintain 1500 rigs operating (the level of mid-2014).
- Morningstar, Inc., said the average breakeven was $70 a barrel.
- Stifel, Nicolaus & Co., Inc., says "weaker portions" of shale plays require prices above $80 a barrel to yield a return of 20%.
- Continental CEO Hamm says, "We can live with $65 and hopefully that will grow to $75."[24]
- Jan Stuart of Credit Suisse said, "Fringe shale rocks and fringe conventional production will struggle at $70/b WTI, let alone at $50."[25]

Of course, these are averages for all shales in production and/or the most expensive parts, which isn't very useful in describing how lower prices will affect operations. In Table 19.7, several groups' estimates of the breakeven by different shales are shown, and the variability of costs across shales is quite evident.

It is noteworthy that Goldman Sachs is far above the others, and indeed, their cost estimate suggests that drilling should have begun declining fairly quickly before prices had reached $80. Yet as the graph in Figure 19.2 shows, until prices hit $60, drilling was very steady. There are a number of reasons why drilling might lag prices, but the complete lack of response before that point is very telling about the price at which operators can continue.

The interesting effect is that the drilling in the various basins has declined by very similar amounts (Table 19.8). Despite the apparent disparity in breakeven price by basin, there's no sign that drilling is shifting from low-quality to high-quality plays. More likely, rigs are being redeployed within plays, as many of the smaller companies would not have operations in varied basins that would allow them to deploy to different basins instead of within them. The three primary producing shales,

Table 19.7   Estimates of Breakeven Costs by Shale Basin (Reuters 10/23/2014)

Shale	UBS	Goldman Sachs	Credit Suisse	Baird
Eagle Ford	43.34	$80–90	46.05	53
North Wolfcamp	52.56		53.92	57
South Wolfcamp	62.74			75
Bone Spring	64.67			
Bakken	65.06	$70–80	64.74	61
Niobrara	72.75		46.1	68
Delaware Wolfcamp	74.86		68.54	
Mississippi Lime	85.54		64.05	84
Avalon	85.87			
Wolfberry			64.63	
Granite Wash		73.1		

the Bakken, Eagle Ford, and Permian, do show the smallest drop, but still are sharply down.

Of course, there are a number of factors that would explain changes in drilling, besides profitability, first and foremost, cash flow. Smaller companies are particularly prone to relying on cash flow to finance operations,

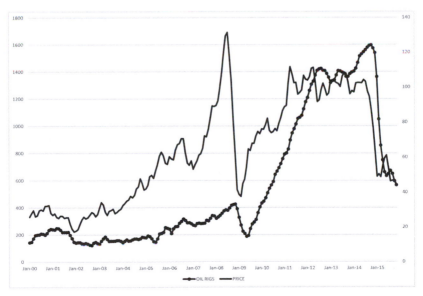

Figure 19.2   Drilling and Oil Price Levels (EIA)

Table 19.8   Change in Oil Rigs Active (Baker-Hughes, http://phx .corporate-ir.net/phoenix.zhtml?c=79687&p=irol-reportsother)

	Rigs Active	Y-on-y Change
Bakken	80	−56.0%
Cana Woodford	35	−67.0%
Eagle Ford	86	−59.0%
Niobrara	30	−43.0%
Granite Wash	14	−72.0%
Mississippian	23	−67.0%
Permian	236	−56.0%

although the shale industry has been said to have raised enormous amounts of capital from outside sources, one reason analysts like David Einhorn worry, since companies have to pay interest on all their loans, even if the money was borrowed for assets not yet producing.

On the other hand, it has been said that there is still capital flowing into the sector in large amounts, including for new startups but also for debt consolidation, which can be attractive given current low interest rates.[26]

Resumption of Drilling

With a partial recovery in the price of oil in the second quarter, 2015, some have been talking about the price point where they would resume drilling. For example:

- EOG says it would resume growth at $65/barrel.
- "We cannot grow US supply at $60/bbl," stated Scott Sheffield, chief executive officer of Pioneer Natural Resources Co. The United States can add $10/bbl to the oil price to reach $70/bbl by 2016 by lifting the export ban, he said, adding that U.S. producers need $70–80/bbl to increase production by 500,000 barrels per day.[27]
- "We're planning to ramp up" once ConocoPhillips executives believe oil prices are back into the $60–70/bbl range for a sustainable period, [CEO Ryan] Lance said.[28]

Early in 2015, the brief recovery of WTI prices to roughly $60 resulted in drilling increasing slightly, but the most recent price of $45/bbl (October 2015) has led to the number of rigs active resume their decline.

Coping with Low Prices

Whether or not shale oil producers can respond as well to low oil prices as shale gas producers did to low shale gas prices cannot yet be determined, but there are already signs that they have made progress. First, anecdotal evidence suggests that there have been significant reductions in costs, including:

- Hess says it has seen 15–30% lower costs in the Bakken.[29]
- Continental Resources' costs have dropped 20–25% in four and a half months.[30]
- Occidental CEO Steven Chazen says service companies have offered moderate reductions in prices, but "they still do not reflect the current reality."[31]
- Halcon Resources says Bakken costs are down 30%.[32]

Recent government data show the fall in drilling costs has approached 20% in the past year, while service costs are largely unchanged. This could represent inertia in contracts, and future data should be monitored closely.

Another indicator that suggests low prices have not made shale oil uneconomic comes from corporate reports of their internal rates of return (IRR) in different shales. Companies like Pioneer and EOG have reported making large profits even at $45–55/bbl.[33]

Still, official forecasts tend to be fairly pessimistic. The U.S. Department of Energy expects an increase in U.S. production of 500 thousand barrels per day in 2015, but flat levels in 2016, despite a number of large fields coming on line in the Gulf of Mexico. Shale production, they predict, will peak in the second quarter of 2015, and recover slightly in the second half of 2016, as prices are projected to increase. The IEA, relying heavily on DOE's estimates, concurs with their conclusions, and OPEC provides a similar (but less precise) estimate, again, apparently relying on the DOE work.

Others who think production will be declining by yearend 2015 include Paul Horsnell of Standard Chartered, Deutsche Bank, and ConocoPhillips Chief Executive Officer Ryan Lance, who he sees U.S. production falling in the second half of 2015, helping boost prices in the next three years as high as $80. "There is a supply response happening. You don't see it in the first half of the year because of the investments that we made over the last two years."[34]

High Grading

Major potential could come from Spraberry Wolfcamp, although this is being held up by higher horsepower rigs being used in the Haynesville shale gas play in East Texas; if drilling there falls with continued low natural gas prices, there could be greater development in this area.[35]

No one was really doing great work. Everyone was just doing a lot of work.[36]

As the section on shale gas showed, redeploying gas rigs can improve results by high-grading, or seeking the best opportunities from among the possible drill sites. The bulk of drilling in the Bakken is already focused on counties with the lowest production cost; while in Table 19.7, there was a noticeable difference in estimated production cost for different shales, but the Eagle Ford and Permian basins appear to be the cheapest, and others, like the Mississippi Lime, are not attracting much investment already. This was confirmed by Occidental E.V.P. Vicki Hollub, announcing that it was selling off its Bakken acreage to concentrate on the Permian basin.[37]

Factors Affecting Costs other than the Geology

A number of things determine costs in a given play (or portion of a play) that change over time in various ways. These can mislead observers as to the potential for future production.

First, infrastructure costs has two parts, social and industry. Isolated, lightly populated areas like North Dakota where the Bakken is located have much higher costs for everything—food, water, housing. This especially drives up labor costs but can also raise the cost of bringing equipment into an area. On the other hand, places like Pennsylvania and West Virginia have abundant housing and roads, but lack the density of oil field service providers found in areas like Texas, as well as gas processing and pipeline capacity. This can not only raise costs of producing from shale, but also result in discounts on the price of the produced oil and gas to cover additional transportation expenses. The situation is more dramatic in Europe, where a well is thought to cost 75–100% more in Poland than an equivalent well in the United States.[38]

Social costs are typically highest when a play is first developed and should grow more slowly as incremental activity becomes relatively smaller. Costs will also tend to plateau as activity plateaus, since, for example, new housing is no longer needed. Industry costs should also drop as activity becomes denser and service companies establish operations in an area. Also, once activity slows, there is less need to import equipment from distant locations, so costs should drop.

Second, there is a steep learning curve for shale oil and gas production. Initial wells are usually vertical test wells with low yields, but intended to allow geologists to examine the deposit properties. After that, a variety of wells are drilled that test different lengths, fractures, and proppants. Usually only after eight to ten wells does a company achieve efficient results, and afterward, progress appears to continue at a rate of 8–15% per year, either in reducing well costs or improving yield.

Third, cyclical elements are clearly important, especially in new plays where the petroleum industry is not particularly active, such as the Bakken and Marcellus, but to a lesser extent even in the Permian. A large increase in activity brings higher costs for equipment and personnel, which drop with activity levels. This year, numerous companies have reported that they expect a 10 to 15% drop in the costs of services (rig rentals, fracturing, etc.), because of lower oil prices. (See above on breakeven prices.)

ESTIMATE OF ECONOMIC PARAMETERS

As with shale gas, it is possible to examine several factors to determine the relative economics of production in different basins. Again, the most important are initial production and decline rate, which allow us, given price assumptions, to describe discounted cash flow and compare that to the cost of drilling and completion wells in a given area.

Oil well costs vary quite a bit by company and location; the two primary factors determining them are the relative isolation of the site and the depth of the shale. However, a review of company reports suggests that $6–7 million is typical for the great majority of wells.

Initial production rates in early 2015 were frequently observed at 800–1,000 barrels per day, significantly higher than amounts observed earlier in the Bakken, which tended to be on the order of 400–600 barrels per day, implying that there is continuing progress in raising recovery rates, or at least initial production rates. Claims by Halcon that it has recently been achieving 3 thousand barrels of oil equivalent per day and even 5,000 barrels of oil equivalent per day are intriguing but it would help to have more data.[39]

Table 19.9   Shale Oil Delivered Cost

| | Above Ground | | | Transportation | Delivered |
	$/bbl	Tax	Royalties	Differential	Cost
Bakken	$43.04	$3.00	$7.50	$8	$61.54
Eagle Ford	$8.83		$7.50		$16.33
Niobrara	$23.37	$3.00	$7.50		$33.87
Permian					
Bone Springs	$12.14		$7.50		$19.64
Leonard	$10.74		$7.50		$18.24
Wolfcamp	$12.53		$7.50		$20.03
SCOOP	$23.54	$4.20	$7.50		$35.24
Woodford					

The USGS has estimated the flow rate in Bakken/Three Forks wells based on existing data, and the implied monthly production decline rate for the first five years is a little less than 4%.[40] Given that most production to date has been from the Bakken and the Eagle Ford, which is a mixed oil and gas play, it is hard to find good evidence of decline rates elsewhere. To be conservative, we shall test decline rates of 4–7%, and add one percent per month to represent the interest rate, thus giving us decline in cash flow of 5–8%.

Table 19.9 suggests that only the Bakken has costs that are too high to allow production to expand at $50/barrel, and even that is probably an overestimate based on earlier costs and production data. High-grading, longer laterals and more fractures, stacked wells and continuing progress in production methods should translate into significant expansion of production after the current fiscal weakness in the industry.

EXPECTATIONS

The next year will be very telling for shale oil production. Optimists like myself expect a resumption of production growth after the smaller, weaker companies' situations are resolved, whether through refinancing, bankruptcy or acquisition. Exploitation of oil shales globally should accelerate, making a significant contribution to the market by 2020 or thereabouts. There is certainly the possibility that shale oil can meet the bulk of incremental oil demand in the long run, but over the next five to ten years, it will be enough to prevent prices from going up.

TWENTY

In Vino Veritas: Conventional Oil Production Outlook

A common misconception about supply is that production normally rises or falls across the board globally, the only difference being the split between OPEC and non-OPEC trends. Pessimists will often downplay the significance of periods when supply was growing robustly, as in the early 1990s, by saying the progress was concentrated in one region. In the case of the early 1990s, this region was the North Sea. (Shale was dealt with in Chapter 19.)

But this pattern is actually quite typical of historical patterns of production. As Table 20.1 shows, in the early 1990s, Norway and the United Kingdom did, in fact, account for one-third of the gross increase in non-OPEC production, that is, excluding the countries that declined. In the early 2000s, Russia, Azerbaijan, and Kazakhstan were responsible for 60% of the gross increase. And more recently, the United States and Canada were responsible for two-thirds of the gross increase in non-OPEC production.

Table 20.2 summarizes the trends, including the total change in OPEC production and the net change in non-OPEC production. (The amounts excluded in line 3 as "special cases" represent the decline in the Russian Federation production during the 1990–1998 period and the disrupted supply from Yemen, Syria, and South Sudan in the 2008–2014 period.)

But row 5 of Table 20.2 is crucial. From 1990 to 1998, when prices were supposedly "low," only the United States experienced a decline of

Table 20.1   Change in non-OPEC Oil Production (Thousand Barrels Per Day)

Major Changes	1990–1998	1998–2008		2008–2014	
Norway	1,422	Russian Federation	3,840	United States	4,860
United Kingdom	895	Kazakhstan	926	Canada	1,085
Canada	705	Brazil	896	Russian Federation	887
Mexico	558	Azerbaijan	664	Brazil	447
China	439	China	597	China	432
Argentina	373	Canada	535	Colombia	402
Brazil	353	Sudan	445	Sudan	−348
Colombia	329	Mexico	−333	Syria	−373
United States	−903	Indonesia	−514	Mexico	−381
Russian Federation	−4,232	Norway	−672	Norway	−572
		United States	−1,227	United Kingdom	−705
		United Kingdom	−1,273		

more than 100 thousand barrels per day (if Russian data is excluded). Small producers were either increasing production or showed only minimal declines, meaning that the increase in production in areas like the North Sea and Latin American had a much greater impact on the net increase in non-OPEC production, rather than merely offsetting decreases elsewhere.

This changed after 1998, when declining non-OPEC producers experienced a total loss of over 5 million barrels per day in production, heavily concentrated in Norway, the United Kingdom, and the United States (a decline of 3.2 million barrels per day). At the same time, seven other producers saw production drop by more than 100,000 barrels per day, so that the large increase from the former Soviet Union (5.4 million barrels per day) was largely offset. Pessimists might say this shows the geological maturity of many areas and their inability to maintain production, but it is worth noting that the shift from robust (1990–1998) to weak (1998–2008) non-OPEC production coincides with the 1998 oil price collapse.

After 2008, the trend was partly reversed, with North Sea losses continuing, FSU production growing more slowly, and the United States and

Table 20.2   Trends in Non-OPEC Production by Period (Thousand Barrels Per Day)

Change	1990–1998	1998–2008	2008–2014
1. Net global production	8,072	9,390	5,828
2. Net non-OPEC production	1,549	3,860	5,511
3. Excluding special cases	5,782		6,402
4. Amount from countries increasing	7,046	9,168	9,096
5. Amount from countries decreasing	−1,264	−5,308	−2,707

Canada accounting for much of the gross increase. (The loss of nearly 1 million barrels per day from South Sudan, Syria, and Yemen certainly contributes to the lower supply although it is excluded from row 5 as being due to external factors, i.e., war.)

Clearly, increasing production is important, but whether or not the amount of supply lost from countries with decreasing production offsets the increases plays a major role in determining the trend in overall production. And it would seem that low prices are the primary reason for widespread production declines.

Two questions need answering going forward: Is there a major play or plays that can provide serious production growth over the next decade? And also, will production outside the major areas be weak or strong, where strong could just mean no major decline that would offset growth elsewhere, as in the 1998–2008 period? This chapter argues that the oil market will see not just old wine in new bottles but also old wine in old bottles and new wine in new bottles.

NEW WINE IN OLD BOTTLES: CONVENTIONAL OIL FROM TRADITIONAL PRODUCING AREAS

Part of the ongoing evolution of the oil market is the resurgence in production from areas that had been in decline, especially after the 1998 oil price collapse. This includes the OECD producers, particularly the North Sea, but also some of the smaller producing countries like Oman, as well as Mexico, where reform could mean substantial new amounts of supply.

"Dogs that didn't Bark"

Negative information is so valuable that nearly everyone has heard the phrase (from Sherlock Holmes) "the dog that didn't bark," referring to a

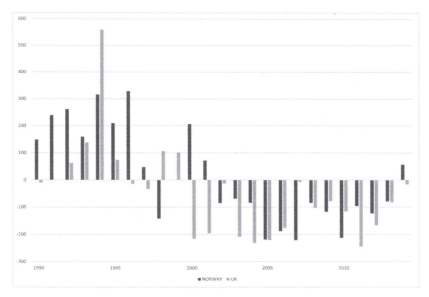

Figure 20.1   Trends in OECD Oil Production (BP Statistical Review)

dog's failure to bark as evidence that it recognized the person passing by in the dark. Others joke about starting a "Journal of Negative Results," to accommodate the many experiments that fail to prove a thesis, but by doing so provide useful information.

In the oil market, this has an analog in the oil production that didn't decline. As seen in Table 20.2 earlier, a major factor in the tightening of the market was the decline in production in countries like those in North America and the North Sea. Few remarks on the fact that, contrary to predictions from peak oil advocates, they have ceased to decline, even ignoring U.S. shale oil production. Figure 20.1 shows the annual change in Norway and the United Kingdom; these countries helped to weaken the oil market in the 1990s, but saw production going from annual growth of 500 thousand barrels per day to an annual decline of 200 thousand barrels per day. The gross drop was small, but the net change, of –700 thousand barrels per day, played a major role in keeping oil prices high in the 2000s.

This trend has reversed, with little notice. Production in Norway and the United Kingdom is dropping only slowly, and new finds offshore Norway, such as Johann Sverdup, scheduled to add 300 thousand barrels per day of production, should see at least some gains over the next decade. The Norwegian Petroleum Directorate, usually producing conservative forecasts, projects the country's output to remain constant for the next five years.[1]

The United Kingdom has not seen any dramatic discoveries like Norway, but the West of Shetlands area is showing signs of life. Development has been slow and technically challenging, but new fields like Lancaster, which is a fractured basement reservoir, possibly holding 200 million barrels, should make a strong contribution to offsetting decline in older fields.[2] Also, a variety of developments in the traditional producing areas should see production flat and possibly increasing over the next decade, including heavy oil fields like Bressay and Mariner and enhanced oil recovery projects like Chevron's Captain.[3]

It is possible that Canada will move in the opposite direction. Canadian production has grown steadily for years, primarily of oil sands. Accelerating in recent years from 50 thousand barrels per day per year in the 1990s to 125 thousand barrels per day per year since 2008, prospects for the future are not as bright, especially given low oil prices. Even before the prices dropped, there were concerns about high costs, much of which reflected the isolated location and underdeveloped infrastructure. That, plus growing local opposition to further development and political pressure to reduce greenhouse gas emissions, was already threatening the industry's expansion plans.

Now, the Canadian Association of Petroleum Producers has lowered its forecast for 2030 production by 1.1 million barrels per day, reflecting the impact of lower prices. Most of the slowdown would come after about five years as a lot of investment is already underway, and it is possible that slower investment rates will lead to a deceleration in costs, but this is one area where pessimism is warranted.

Investment in Canada's east coast appears to be picking up, despite the area's high-cost environment (the nearest oil industry center is Scotland). Fields like Statoil's Bay du Nord in the deepwater Flemish Pass basin off Newfoundland suggest there is significant promise, but the contribution will be to offset the decline of conventional oil production in Alberta.[4]

Nibbled to Death by Ducks

The oil industry's penchant for seeking so-called elephants, giant or supergiant oil fields, disguises the fact that a lot of the world's oil comes from smaller fields—and producing countries—that are often under the radar of analysts. However, they do make a difference to the oil market balance, especially in aggregate.

In the early 1990s, Argentina, Colombia, and Oman added 900 thousand barrels per day in production, not an earthshaking amount but significant nonetheless. After the oil price collapse of 1998, production for the three dropped by 450 thousand barrels per day over the next decade, a net change of 1,350 thousand barrels per day in market trends. The primary factor seems to have been the cutback in drilling following the drop in prices, plus political inertia that failed to adjust fiscal regimes and release new areas for drilling. Colombia demonstrates this beautifully, as production not only recovered after hitting a low in 2005, but reaching new highs, passing 1 million barrels per day in 2013, almost double the low. Considering Colombia was described as "post-peak" by various peak oil advocates, this dramatically refutes their belief.

MEXICO RETURNS

Mexico provides a strong cautionary tale for oil production forecasters. In the 1920s, governments had repeatedly taken steps against the interests of foreign (primarily American) oil companies as part of the domestic political struggle. Even conservative rulers felt it necessary to raise taxes on oil production, to the point that production had peaked in 1922 and declined afterward. The 1937 nationalization of foreign assets actually had little impact on production.

After decades of minimal operations, the 1970s, when oil prices were soaring, saw increased effort and the discovery of some significant resources, including offshore fields near Tabasco and the supergiant Chicontepec field, thought to contain as much as 100 billion barrels in place (not recoverable). This led to great enthusiasm in many corners, with some in the United States seeing Mexico as a savior for the oil market, knocking down the market power of OPEC. The 1982 *World Energy Outlook* produced by the International Energy Agency actually saw Mexico as the only country outside the Persian Gulf that had the ability to increase production.

There was at least one cautionary note: Thomas Stewart-Gordon, writing in *World Oil*, noted that despite the hype about Chicontepec, Pemex was apparently planning little or no investment in the field. Subsequent developments proved him prescient, as the field remains largely undeveloped (see below), but other fields, such as Cantarell, allowed Pemex to raise production sharply.

For over three decades, Mexico has remained one of the world's largest producers, exporting mostly heavy oil to the nearby U.S. market. However, after 2003, Cantarell entered a sharp decline, with production dropping from 2 million barrels per day to 350,000 barrels per day last year. Peak oil advocates seized on this decrease as evidence of impending global decline and certainly the end of Mexican oil abundance. More recently, Pemex announced that it would invest $6 billion to stabilize production in the field, adding 100 thousand barrels per day per year of new capacity to offset decline.[5]

Reality, needless to say, is more complicated than the inevitable decline of a supergiant field and the irreversible decline of natural oil production. Pemex, as a national oil company, does not respond to price signals the way that privately owned companies do. Its budget must be allocated by the government and often depends more on the nation's fiscal situation than on its opportunity for profit.[6]

The negative impact of this was obvious in the last decade. As oil prices rose after 2000, so did drilling in the Gulf of Mexico and the cost of renting drilling rigs. Pemex, with a budget fixed by the government, found itself unable to maintain exploration levels, and drilling fell off slightly at a time when increased drilling was needed to offset the Cantarell decline.

Since then, the government has relented and increased funds, which has meant more drilling and renewed discoveries. This has also resulted in an end to the decline; after dropping by 800 thousand barrels per day from 2004 to 2008, production has largely stabilized. This means that one of the major factors causing non-OPEC production to decline during the 2000s has disappeared.

The primary uncertainty is the direction of conventional production, although three sources of "unconventional" supply could play a major role. Like Venezuela, Mexico has a number of fields where production has declined sharply. These could be leased out as the Venezuelans did with their "marginal fields," which added nearly 1 million barrels per day of production in a short period of time, before Chavez's counterrevolution stalled investment.

Shallow water offshore should also make a significant contribution: Pemex recently announced the discovery of four new fields containing 350 million barrels of oil.[7] These are expected to produce 200,000 barrels per day and are far larger than fields being found in the U.S. shallow-water Gulf of Mexico or most parts of the North Sea. UK production peaked when new discoveries were averaging about 10 million barrels, for example, implying the Mexican shallow water is still relatively immature.

Further, the heavy oil field Ayatsil, with an estimated 750 million barrels of oil equivalent of 3P reserves, is planned for production of 150,000 barrels per day.[8]

Chicontepec

The Chicontepec field could be considered a supergiant field because of its in-place resources, but the recoverable amount remains uncertain and is officially considered to be small. Highly fractured, development has been ignored for several decades because of its higher costs, especially since other cheaper fields were available to be developed.

Given that it has an estimated resource of 139 billion barrels, comparable to some shale oil deposits in the United States, the potential exists for as much as 1 million barrels per day of production from this field alone.[9] The current industry reform, and the offering of 90 different areas of Chicontepec to private operators, should see an era of experimentation, almost certainly followed rising production. The use of a combination of better seismic and other techniques like horizontal drilling and hydraulic fracturing could see significant supply from this field in the next decade.

Deepwater

Until recently, Mexico has not exploited its deepwater areas because it lacked the technology and funding. However, there was great excitement when in 2012 President Calderon announced a major oil find from the first deepwater well, which he described as containing 400 million barrels of oil in an area he claimed might hold 4–10 billion barrels.[10] Sadly, the field proved to be gas. Even so, Pemex estimates that the deepwater contains 30 billion barrels of oil equivalent, and so far, four oil and three gas discoveries have been made, the oil fields holding an estimated 300 million barrels.[11]

Shales

At present, Mexico has made no major effort to exploit its shale resources, which ARI estimated at 13 billion barrels in 2013, some of it an extension of the highly productive Eagle Ford shale from across the U.S. border.[12] Since the Eagle Ford is now well understood, exploitation of the Mexican shale should be technically simple; the political effort might be more difficult, but even without reform, Pemex can be expected to advance in this direction.

Mexico has now begun a process of major reform that resembles Venezuela's 1990s *apertura* and should see similar results. The first round of leases offered received few bids, but observers believe that the major companies are waiting for more prospective acreage. All in all, it appears that Mexican reform will proceed, if slowly, and production will rise, probably surpassing its earlier peak of 4 million barrels per day. If various parts of the country's resources are developed simultaneously, including margin fields, deepwater, tight oil, and shale oil, production could grow robustly.

DEEPWATER: NEW WINE FROM NEW BOTTLES

The industry has gradually expanded its ability to produce oil from greater water depths, considerably increasing the world's oil resources and accounting for increasing oil production. Until recently, most of the activity was in the U.S. Gulf of Mexico and the Campos basin off Brazil, but the presalt basin discovered off South America and West Africa is showing great progress. Exploration in deepwater areas in Southeast Asia is nascent, but has shown signs of at least significant gas resources.

Brazil

In one of the more amazing incidents of serendipity, Brazil's Petrobras discovered the supergiant oil field Majnoon in Iraq during the 1970s. This led to extensive criticism, given the national oil company's inability to find significant reserves at home—the critics ignoring the role of geology. It has still found only meager onshore reserves. However, the stroke of luck came when the Iraqi government nationalized the field and reimbursed Petrobras with crude oil, luck because before the field could be developed, it became part of the front lines during the 1980s Iran-Iraq War and remained undeveloped until recently. Petrobras was far better off with a small payment in crude oil than if it had retained ownership.

But the company was hardly idle at home and became a leader in deepwater exploration, largely out of necessity. By 1974, it had made the first discovery in the Campos basin, now containing seven giant fields like Marlim and Roncador, with approximately 12 billion barrels of reserves plus historical production.[13] This enabled the country to pursue a goal of oil independence, although rising domestic consumption has made that difficult to achieve. The fields range from shallow to very deep, which necessitated developing cutting-edge production technology.

Then, in a classic case of technological advance making uneconomic resources into reserves, the company drilled into the presalt area of the Santos basin in 2006, finding the (now) Lula field, which is estimated to contain 5–8 billion barrels of proved reserves, putting it on a par with Prudhoe Bay as one of the largest Western Hemisphere fields found in the last half-century.

By early 2014, production from this region had surpassed 500 thousand barrels a day, a remarkable achievement given the operating difficulties inherent in both the very deep water depths (over a mile for Lula) and the presalt geology. Wells produce an average of 20,000 barrels a day each, which is an indication of just how attractive the fields are. Some fields, like Lula, might produce as much as 1 million barrels per day ultimately.

Presalt geology is difficult, as the salt layer can deform drilling pipe and the depth at which the oil is found means the oil is under high pressure. That, combined with the water depth, will probably result in some technical problems and project delays. However, the industry has worked in such conditions before, and these challenges will not be insurmountable. Funding, on the other hand, is a more serious challenge. Petrobras now intends to spend $130 billion through 2019 to pay for its development plan, a reduction of 37% from earlier plans.[14]

The recent "Carwash" scandal involving oil industry contractors and service companies has dealt a major blow to the country's plans to raise production, but this should be temporary. By requiring that foreign oil companies rely on domestic companies to provide as much of the equipment and services as possible, the government inadvertently created the opportunity for large-scale corruption. The revelation of this means that many projects will be delayed, as firms involved are investigated and some contracts either renegotiated or canceled. However, progress should continue, albeit at a slower pace than the government had originally projected; the updated figure for 2020 is 2.8 million barrels per day, a modest increase from the current 2.5 million barrels per day.[15]

And although reserve estimates remain uncertain, the Iara, Libra, and Guara as well as Lulafields are all thought to contain more than 1 billion barrels, with a total, for the four fields, said to be 13–28 billion barrels. Numerous other discoveries have not yet been sufficiently evaluated to allow for reserve estimates to be announced, but many of them appear to be of a similar scale. Overall, the presalt Santos basin should be on the order of the North Sea in terms of ultimately recoverable resources, and possibly more.

West African Deepwater/Other Presalt

For decades, if not centuries, school children looked at the maps of Africa and South America and noticed that they seem to fit together. Numerous scientists had also wondered about this over the centuries, but the first major theoretical work on the subject was done in 1912 by Alfred Wegener, whose arguments were rejected by mainstream geologists for decades. Now that continental drift is an established fact, it has proved important for petroleum because the area offshore Brazil that is now proving so prolific might have a counterpart offshore West Africa.

To date, drilling in the West African deepwater has been focused mainly in Angola and Nigeria, where shallow water oil production has a long history, but countries like Gabon, Namibia, Ghana, and the Congo have already found large fields. Tullow's $4.9 billion Tweneboa-Enyenra-Ntomme (TEN) development project offshore Ghana has an estimated peak production of 80,000 barrels a day;[16] ENI has found 4 billion barrels of oil equivalent in place between Congo and Gabon in presalt plays, including the Nene Marine field with an estimated 1.2 billion barrels of oil in place.[17] Although it is early to say, the belief is that the West African presalt might contain a number of billion barrel fields.

In other parts of South America, explorationists have begun to see if the presalt extends beyond Brazil. So far, Exxon has found a field estimated at 700 million barrels offshore Guyana, although Venezuela is disputing the border demarcation.[18] Exploration in this area is still in its infancy, and it could prove to be prolific.

THE RUSSIAN BEAR CHARGES AHEAD

Russia was one of the early centers of oil industry, with the Nobels producing in the Caspian area around Baku in the tsarist era, resulting in the foundation of Royal Dutch, later merged with Shell Trading and Transport, to create one of the Seven Sisters. At various times over the decades, surging Russian oil exports threatened to collapse oil markets, partly because the Soviet Union's industry was not transparent. Export increases could come at a complete surprise. However, supply disruptions due to war and political upheavals have also been a boon to Russia's competitors.

But this provides an excellent test case of peak oil advocates' ability to understand the factors that actually drive supply. After the dissolution of the Soviet Union, oil production fell off a cliff, but due primarily to the political turmoil and the collapse in domestic oil consumption. Oil exports did not decline at all; they would have increased except that there was no spare pipeline capacity to accommodate the surplus.

As a result, tens of thousands of wells were shut down. The drop in production was entirely driven by a lack of demand, either domestic or foreign—due to the inability to actually export the surplus. But at the same time, the industry was beginning to open up to outsiders, from consulting engineers to major oil companies seeking to invest in the upstream. The poor technical situation in the Siberian oil fields became well known, and the lack of the most modern technology was seen as a huge opportunity to resurrect its aging fields. As discussed in Chapter 7, Samotlor production plummeted from over 3 million barrels per day in the early 1980s to about 0.5 million barrels per day before recovering slightly.

From that point, many of the outside forecasters expected that production would recover and soar in the former Soviet Union. My own analysis in 1997 found that the unexploited reserves were very large, and production costs were quite low, implying that reform of the sector would lead to large supply increases.[19] If anything, the optimists were not optimistic enough, and even after the resurgence slowed, the country has continued to post production increases, albeit small. Moves into Arctic waters, which should be delayed by the Ukraine-related sanctions, might add some new supply, and shale as well, but neither seems set to make a contribution in the next few years. Production might not increase much in the near future, but a decline does not seem to be in the cards. East Siberian oil reserves might become prominent in the future, but probably not very much in the next decade.

KAZAKHSTAN: NEW WINE IN A LATE BOTTLE

The supergiant Kashagan field has been delayed so frequently and for so long that many seem to have forgotten about it. Projected capacity is 1.2 million barrels per day, an amount that should hit the market quickly as it is dependent on the completion (and now repair) of the export pipeline. At present, startup is anticipated for 2017, and it will probably not be delayed very much beyond that.

Other fields are able to boost production, especially as the Caspian Pipeline Consortium has just boosted capacity by 200 thousand barrels per day, to 780 thousand barrels per day. Another expansion to 1.4 million barrels per day has been planned, but lower oil prices appear likely to delay that, as the government believes a $40 price will see production stagnate (excluding Kashagan).[20]

LONG-RANGE PROSPECTS FOR NON-OPEC PRODUCTION

Again, despite the constant refrain that there are no major basins waiting discovery, various areas of the world have either shown potential already or are thought to have significant resources. The Falkland Islands will start oil production soon, several decades after the war between Argentina and England, which some characterized as an oil war (although the area is also a major source of squid). Greenland has seen only a handful of wells and has huge amounts of prospective areas that haven't been examined as yet. East Siberia is almost completely unexplored, but billion

Table 20.3   Expectations for Conventional Non-OPEC Oil

	2015	2020	2025
United States/Canada	17,000	−250	−250
Mexico	2,590	250	750
Brazil	2,520	200	750
Africa	2,280	220	500
S. America	2,000	125	125
Asia	8,000	0	−100
North Sea	3,400	250	100
FSU	13,900	1,500	250
Total		2,295	2,125

barrel oil fields have been found there. And places like the Faeroe Islands will likely see exploration soon.

However, for the purposes of this analysis, it will be assumed that these frontier areas will contribute only marginally over the next decade. Table 20.3 shows my expectations for non-OPEC conventional oil (excluding biofuels, oil sands, and shale oil). Growth of 400 thousand barrels a day/d is not exorbitant, and indeed has been reached many times in the past, but combined with shale oil and the possibility for rapid expansion in Iran and Iraq (see below), it implies that strong demand growth is needed to keep prices from being below $50 for an extended period. If oil prices were to return to $80–100 in the next few years and remain there, the numbers below would be even larger, implying that only major disruptions to OPEC production would make those price levels sustainable.

OPEC PRODUCTION POSSIBILITIES

The projection of oil supply from OPEC has always been problematical and typically involved treating it as a residual of other variables, that is, the amount needed as a result of world oil demand minus non-OPEC supply. In effect, OPEC supply was assumed equal to the demand for OPEC oil. Few attempted to predict which OPEC countries would produce how much, with the Department of Energy being the prime exception by forecasting future capacity by country. Production in OPEC, much more than elsewhere, is the result of not only economic but political considerations, and the decisions of some countries, like Saudi Arabia, have a major impact on price. Thus, anticipating fiscal, investment, and production policies over the long term is fraught with difficulty.

OPEC itself has projected an increase in OPEC supply from 2015 to 2025 of 3 million barrels per day, but they do not undertake detailed analysis—or at least don't report it, possibly given the sensitive implications of production levels for the politics of quota allocation.[21]

The pessimism that infected non-OPEC supply projections after the Iranian Oil Crisis also influenced some aspects of expectations of supply from many OPEC countries. Typically, it was assumed that outside the Arabian/Persian Gulf, none of the OPEC members would be able to increase production. Algeria, Ecuador, Gabon, Indonesia, Libya, Nigeria, and Venezuela were all thought to have peaked in the 1970s, and their production was predicted to decrease in the future. At the same time, as Chapter 8 showed, DOE repeatedly projected ever-rising Saudi production, needed to meet the gap between demand and supply.

Most of these countries have not confounded expectations, and Indonesia has actually dropped out of OPEC, due to stagnant production and rising consumption. But this is due more to various countries' political and/or economic situations than any lack of resources. Iran, Iraq, Kuwait, Libya, Nigeria, and Venezuela have all faced problems that have resulted in under-exploitation of their petroleum sector or have simply sought to conserve their resources. There is a very real chance that these constraints will ease in many of these countries over the medium-term future, given the current low price of oil and their dire economic situations.

At the same time, the short-term projections of many of the countries themselves have not proved reliable, in some cases because weak markets discouraged the addition of capacity that would go unused, in other cases, budgetary, political, or other difficulties kept nations from achieving announced goals.

That said, the question remains, first, can OPEC nations add enough capacity to meet demand, and, secondly, will they? In Table 20.4, official targets are shown for different OPEC nations along with the IEA forecast and this author's guess as to what is likely. Official goals are particularly soft and are often subject to change. Iraq is most notorious, having set a target of 6 million barrels a day repeatedly under Saddam Hussein, only to see it fail because of war or sanctions. The post-Hussein government similarly set a very ambitious target of 12 million barrels per day by 2020, but this was clearly unrealistic and has slipped substantially.

So the primary question involves the extent to which various countries are willing or able to expand their production. Iran, Iraq, and Venezuela all have substantial potential for significant production growth, but at this writing, only Iraq seems on target to achieve this.

Table 20.4   Expectations for OPEC Production (IEA Medium Term Oil Market Report 2015, and National Reports)

	2015	IEA Forecast	Official Goal	Likely 2020	2025
Algeria	1.14	0.95	1.68	1.2	1.2
Angola	1.8	1.86		2	2.4
Ecuador	0.57	0.59		0.5	0.55
Iran	3.6	3.6		4	4.8
Iraq	3.9	4.73	6	5.2	6
Kuwait	2.82	2.76	4	3.5	3.5
Libya	0.5	0.98		1	1.2
Nigeria	1.92	1.89		2	2.5
Qatar	0.7	0.73	0.8	0.8	0.8
Saudi Arabia	12.34	12.39	15	12	12
UAE	2.94	3.21	3.6	3.6	3.6
Venezuela	2.49	2.56	4.4	3	3.5
Total	34.72	36.25		38.8	42.05
Increase				4.08	3.25

Iran

Iran is often considered the poster child for the failure of a disrupted oil sector to recover to pre-revolutionary levels. Although the Khomeini government had attempted to prove it could match the 6 million barrels per day level of the Shah's regime, it managed to get only a little above 4 million barrels per day. This was due to policy decisions, not resources, as Iranian proved reserves have long been adequate to support the 6 million barrels per day production level of the 1970s. While it might be that the peak oil advocates are correct in questioning the 1986 reserves' revision to 90 billion barrels, the pre-revision level of 60 billion barrels, at least 150% of what the United States had when it reached 10 million barrels per day, certainly would have supported 6 million barrels per day for a lengthy period.

And recent multibillion discoveries like Azadegan and Yadaravan imply that there is plenty of room for continued expansion of both reserves and production. Until recently, hardliners had insisted on terms that few foreign companies would accept, and then sanctions imposed because of the disagreement over Iran's nuclear research program prevented most investment. Now, the new government is desperate for revenue and more ideologically disposed toward foreign investment.

Officials have begun courting foreign investment and described 45 oil and gas development projects and expect to raise capacity to 5.7 million barrels per day, although the timing of when they will be implemented is unclear.[22]

Kuwait

In the 1960s, Kuwait was a much bigger player in the oil market but instituted a policy of conservation that, combined with weak markets in the 1980s, saw its production drop by two-thirds in the early 1980s. The destruction wreaked on the oil fields by Saddam Hussein during the first Gulf War was actually overcome relatively quickly, in only 18 months, demonstrating just how prolific the country's fields are.

But because of its small population, the country has much lower revenue needs than other nations and, as a result, is under less pressure to increase production. Plans to add 400 thousand barrels per day of heavy oil production have been stalled for two decades because of a political conflict between the parliament and the regime. Potentially, the country could break the deadlock and allow the investment, but that seems improbable in the near future.

United Arab Emirates

As with Iran and Kuwait, the UAE has large reserves, possibly overstated, but still easily capable of further production increases. At present, plans are to add roughly 1 million barrels per day of capacity by 2020.[23]

Iraq

Iraq's petroleum resources are widely agreed to be extremely abundant, even by peak oil advocates. The geological potential is only outweighed by the political incapability. Nearly every decision seems to be made with glacial slowness—but they do get made. The outlook for Iraqi production is extremely bullish, with 2015 showing an increase of over 500 thousand barrels per day after a couple of years of stagnation.

The country's history does give cause for caution. As bizarre as it might seem now, the Iraqi government nationalized the foreign operators in 1970 because of their *under* investment. While countries like Kuwait and Libya were trying to conserve their oil, the Iraqis were complaining that the consortium had been restraining production there in favor of their other holdings in Iran, Kuwait, and Saudi Arabia. Ever since, the government has been

Table 20.5   Main Iraqi Oil Fields (Center for Global Energy Policy, "Issue Brief: Iraq's Oil Sector." New York: Columbia University, June 24, 2014)

Field	Region	Reserves
West Qurna	South	43
Rumaila	South	17
Majnoon	South	12
Zubair	South	8
Nahr Umr	South	6
East Baghdad	Central	8
Kirkuk	North	9

consistently bullish about its prospects, but these plans have repeatedly gone unrealized.

First, the decision to invade Iran in 1980 and the subsequent eight-year conflict put the existing capacity expansion plans on hold, and when the end of the war in 1988 saw the plans revived, they were once again shelved by the regime's invasion of Kuwait in 1990, leading to 13 years of sanctions followed by the second Gulf War and overthrow of the regime.

After the new government was established and the decision-making infrastructure put in place, a target of 12 million barrels per day of production for 2020 was announced and greeted with the appropriate. Although the engineers of INOC had done a masterful job of maintaining production during the lengthy sanctions period, it was widely acknowledged that the infrastructure needed serious repair and reconstruction.

At the same time, the presence of a number of supergiant fields (Table 20.5) whose production had declined over the years served as a signpost to the country's potential. No country outside the Middle East has fields with remaining reserves on this scale, and arguably, the Iraqi conventional oil resource base is second largest in the world after Saudi Arabia.

After much political debate and lengthy negotiations, the ministry signed contracts with a number of foreign companies to rehabilitate and expand production in known fields. Although the political situation remains turbulent, the biggest problems are in the Sunni West region, and the primary oil reserves are in the South, which is more stable.

Lower oil prices have led the government to ask operators to slow investment, which should see a pause in the growing oil production, but this is unlikely to last long. There is nothing else the government can

Table 20.6   Recent Kurdish Oil Discoveries (Oryx Petroleum August 2014. Bijeel Field Shows Oil-in-Place, Others Reserves)

Field	Mln barrels	Operator
Shaikan	2,700	Gulf Keystone
Bijeel	2,400	MOL
Tawke	771	DNO/Genel
Taq Taq	647	Genel/Addax
Attrush	647	Taqa
Chia Surkh	306	Genel

spend money on that will bring the economic returns that oil will, and at least moderate growth should be expected for quite some time to come.

Kurdistan

In a curious historical footnote, the northeastern part of Iraq is nearly completely autonomous and on its way to becoming a moderate oil producer. The Kirkuk supergiant field, which is on the border of the autonomous region, remains in dispute between the Kurds and the central government, but there have been a number of major discoveries by a variety of small companies (Table 20.6), which means that the Kurdish government's goal of reaching 2 million barrels per day of production by 2019 is feasible, if ambitious.[24]

Venezuela

As one of the early oil producers and a founding member of OPEC, Venezuela has often been considered a mature producer. The U.S. Department of Energy regularly forecast its production to decline, especially after the steep drop in production during the 1980s. Several reorganizations of the national oil company, PDVSA, did little to stem this, but under Luis Giusti, a policy of *apertura*, or opening, several steps sent production soaring.

First was the development of Orimulsion, a water-oil mix using the heavy oil from the Orinoco belt as a boiler fuel, roughly similar to residual or heavy fuel oil and priced to be competitive. This, and extensive testing of production techniques, led to a second step, four major agreements to develop production and upgrading projects with foreign majors; these projects have largely displaced efforts to sell Orinoco oil without upgrading. An additional 2 million barrels per day of Orinoco developments have been announced with a variety of partners, mostly from Asia and Russia, but most of these have not progressed very far.[25]

The third step was the leasing of marginal fields, older fields whose production had declined to very low levels, but which could be rehabilitated with new technologies and capital. By 1998, these had added roughly 600,000 barrels per day of new production (a classic case of new wine from old bottles).

The Chavez administration's program of reasserting control meant that it took over most of the marginal fields and majority ownership of the Orinoco projects. Two companies pulled out, but the others remained. However, the relatively xenophobic attitudes of the ruling party have generally made it difficult to attract new investment, and payment disputes with some contractors have made it harder to hire service companies to replace the fired PDVSA employees. As a result, the company has struggled to maintain production, despite startup of the heavy oil projects.

At present, the country is facing economic collapse, with increasing shortages of basic goods. In response, the government is trying to use high-tech methods to ration them, but this is a Band-Aid on a severed arm. Public unrest has grown, and even unions are unhappy at both the conditions and the inability of the government to keep its promises. Although much of the anger is focused on President Maduro, there is also recognition in many quarters that the policies inherited from Chavez are the root cause.

Two possibilities for regime change exist: a military coup or someone in the ruling United Socialist Party of Venezuela replacing Maduro. In almost any case, there is likely to be serious reform of the petroleum sector; it is not clear how many former PDVSA executives or technical staff will return, but replacement with foreign operators and/or contractors would ease the situation.

Can Venezuela return to the growth of the 1990s? Yes. Will they? This is not clear. However, there are growing odds that the economic crisis will force the government's hand—or other actors to replace the government—and one of the few things likely to provide short-term economic relief is reform of the oil industry. In such a case, Venezuela is likely to see significant growth in production commence within several years.

Angola

Angola's production has followed a rather unusual trajectory, where numerous giant oil field discoveries in the past two decades have primarily served to offset high depletion rates and maintenance problems at existing fields after the 2008 peak in production. The country began producing in the 1960s, when it was still under Portuguese rule, from shallow water fields, and the move into the deepwater shelf in the 1990s saw production rise sharply, challenging Nigeria as the leading West African producer.

Since that time, production has paused, but there is every reason to expect strong growth in coming years. Nearly a million barrels a day of new projects are planned, although the timing is not certain on many of them, and exploration continues to be robust.[26] Most exciting, the country's offshore is geologically adjacent to the Brazilian presalt area, and there are strong indications that Angola's presalt might yield similar billion barrel discoveries. The only find to date in that area is the Cameia field, which is estimated to have 400–700 million barrels, and there have been some disappointing results.[27]

Libya

Without a doubt, the greatest uncertainty in the short- and medium-term involves the possible recovery of Libyan oil production. The political situation, really a civil war, shows no signs of being resolved, but there is a good chance that production in the East of the country could recover within months if the security environment improves. Recent negotiations between the main warring factions have made progress, but there is no sign when an agreement will occur, and whether that will end the principal fighting.

Still, the potential is for an increase of 0.5–1.0 million barrels per day of production in the next 5–10 years, basically through the restoration of existing fields' operations. Further expansion could occur, but seems unlikely in that time frame

Nigeria

Nigerian production has stagnated in recent years, as new deepwater supply has only offset losses in the shallow waters and especially the Niger Delta, where rebels and criminals have disrupted more than 0.5 million barrels per day of production. Additionally, corruption has hampered the operations of the national oil company, and it has struggled to meet its portion of investment for fields. (The company normally takes a share of each development project.)

In Table 20.4, announced targets for production in some OPEC nations were listed, but the final column includes a more likely sequence of

Table 20.7   Market Balance for the Next Decade (Actual for 2015 Estimated from IEA Oil Market Report. Demand and Other Liquids from EIA Annual Energy Outlook 2015. Other Supply from Author's Estimates)

	2015	Change 2015–2020	Change 2020–2025
Non-OPEC conventional	58,300	2,295	2,125
Other liquids		1,686	1,200
World demand	94,500	5,500	4,500
Need for OPEC oil	36,200	1,519	1,175
Expected OPEC oil	37,500	4,800	3,250
Market balance	1,300	3,281	2,075

events, given fiscal and political problems in some countries. Compared to the expected need for OPEC oil, the potential for a serious glut looms. the final row of Table 20.7 shows the difference between what the oil market needs and what this study suggests is OPEC's likely capacity change without considering shale oil production, which will be more than enough to keep the market balanced. Indeed, spare capacity will probably increase.

The forecast for global demand growth is probably too low, especially if prices remain at $50 a barrel; however, adding the minimum likely 500,000 barrels per day of shale oil production growth each year will mean that even robust demand growth of 1.5 million barrels per day will be easily met. Quite possibly, every forecasted number will prove to be incorrect, but the evidence suggests that the errors should be roughly off-setting or at least more so than for more forecasters, who are typically too pessimistic about oil supply potential when the effects of politically inspired supply disruptions are excluded. And of course, the peak oil forecasters have usually been so biased as to be too low for nearly every prediction.

RESOURCE NATIONALISM OR RESOURCE RATIONALITY?

The biggest uncertainty regarding conventional oil supply has to do with decisions by a number of governments as to whether to promote exploration and production or not. Iran, Mexico, and Venezuela are potentially the biggest contributors to new supply in the next decade whose outlook remains uncertain, but appears positive. Only Venezuela still seems committed to a policy (more de facto than de jure) of antagonism to foreign investment, and that does seem to be changing in response to their economic woes.

Other countries, from Angola to Russia, are unlikely to revert to resource nationalism in a time of low oil prices. Prices much lower than $50 would be needed to discourage investment, whether from private companies cutting capital expenditures or by governments refusing to lower their taxes to minimal levels. The most likely outcome for at least the next decade involves shale oil production keeping a ceiling on long-term prices at $50, meaning governments do not have the luxury of holding resources off the market to raise prices. Lower production will mean lower revenue not higher prices.

Of course, high-cost areas like oil sands and offshore Arctic will be replaced by the new "easy" oil from Iran, Iraq, Mexico, and Venezuela, among others, which translates into lower capital expenditures for the industry, a minor but welcome contribution to global financial markets (through lower interest rates). The wild card, as always, remains political risk: in the early 1990s, it appeared as if the Washington Consensus or Thatcherism of conservative economic policies would persist, but corruption and politics in Russia, Venezuela, and other countries caused them to turn back to more nationalistic and statist policies. My bias is toward optimism, but forecasting long-term international politics remains astronomically complex.

TWENTY ONE

The Design and Interpretation of Oil Supply Curves

To many, the resource estimates in Chapter 15 and the cost numbers in Chapter 16 might seem hard to interpret, and economists often turn to the use of graphics in the form of supply curves to show costs and volumes in a graphical manner. Figure 16.10 was a typical example, but this chapter will go beyond what is usually done to show a dynamic version of supply curves and how they illuminate future developments.

THE TYPICAL SUPPLY CURVE

Basic microeconomic theory relies heavily on supply and demand curves and every textbook includes various examples, but primarily using theoretical curves, usually hyperbolic: higher prices means higher supply and vice versa. A notable exception was the attempt in the 1980s to describe petroleum, especially from OPEC members, as making up a "backward-bending supply curve," but this proved of dubious value.[1]

Some groups produce supply curves for petroleum, but many of them are relying on company-specific data, such as when producing cost estimates for specific regions or developments, and others produce very aggregate estimates, covering mostly national or supra-national regions.

In theory, the best approach would be to subdivide production by regions, trying to keep them homogenous and then build supply curves.

This is roughly what some organizations do, such as the IEA's annual World Energy Outlook (Figure 16.10) or (what is now) IHS CERA. However, these still tend to be fairly aggregate, often by region, such as the Middle East, and limited by data availability. So, Alaska might be shown separately, but not Siberia, let alone western and eastern Siberia.

Alternatively, a supply curve that is resource based, showing costs for billions of barrels of resources, which provides some insight into long-term market behavior. Still, these curves remain static and provide no mechanism for the transformation of resources into reserves, nor do they tend to include all resources, a limitation that has become obvious recently.

There are a number of shortcomings of these methods, the worst of which is the static nature of the estimates. Additionally, many people tend to confuse short-run and long-run marginal costs and to misinterpret the implications of the shape of the curve, as discussed in Chapter 18. One basic mistake is to think that the short-run marginal cost includes capital costs, which is important in terms of short-term price setting. Another is to misinterpret the rightward movement as temporal in nature.

This chapter will attempt to describe how these shortcomings might be addressed, resulting in a new formulation for supply.

EARLY WORK

Initial studies were largely point estimates of costs and usually contained no significant detail. Not a few analysts simply described a single cost for the production of conventional oil, particularly before 2000, after which there was a transition to exotic oil resources, including coal liquefaction, oil sands, deepwater, and shale oil. Most have forgotten that in early days of post-1973 energy economics, much of the work similarly provided a simple division of oil production costs into "conventional" and "unconventional" or backstop resource, such as kerogen or coal-to-liquids conversion, arguing that prices should rise exponentially from the lower cost to the higher (Figure 21.1).

Even studies with a focus on production costs sometimes only included one or two categories, rather as a step function and not unlike current arguments that the "cheap" or "easy" oil is gone, and the world has moved to a new era of "expensive" oil. Most such assertions are not backed up by serious analysis.[2]

There have also been numerous cases of costs described with minimal interpretation, including the claim that non-OPEC capacity costs were

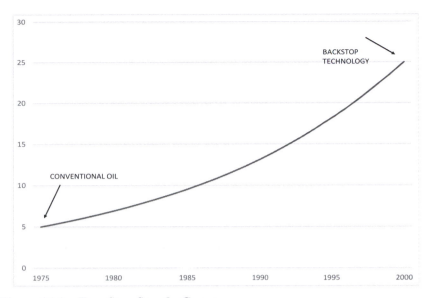

Figure 21.1   Two-Step Supply Curve

$70,000/daily barrel from 1980 to 1985, the late 1980s claim that OPEC would need to spend $60 billion over 5 years to add 6 million barrels per day of capacity; and reports in the late 1980s that Siberian costs were 12,000 rubles/daily barrel. The former two seem exaggerated and the latter is not in a form that can be readily interpreted in relation to oil prices.[3]

There are numerous recent supply curves created by a variety of sources with varying degrees of detail and reliability. The IEA's supply curves which are interesting but do not provide much detail. The similarity between them suggests that possibly many are relying on the same source, perhaps with slight modifications, rather than redoing the work, although they might simply be employing similar methodologies and data. Since many curves exist only in the form of online slideshows or are drawn from proprietary work from consulting firms, it is difficult to judge how rigorous or valid they are.

Empirical Work

The unexpected oil price collapse in 1986 raised anew the question of the price necessary to allow expansion of production. Many like Hodel (1986) decried the supposed "low" prices (really a return to historical norms) and predicted that non-OPEC supply, at least, would shrink, but

most of this writing was rhetorical in nature and not based on a true understanding of the marginal cost of production.

Two early reports were from Les Deman (1986), in the *New York Times*,
and M. A. Adelman, in the *Energy Journal*. The former, appearing as a
short note in the press, did not include notes on methodology or data sources; the latter relied on drilling and discovery data, and compared them
using U.S. cost factors.

Later studies included one by the Canadian Energy Research Institute
and another by Marie Fagan (1989), later of Cambridge Energy Research
Associates (now IHS CERA). Adelman (1989), with the assistance of
Manoj Shahi, produced a detailed examination of costs in many countries
over three decades, the years 1955, 1965, 1975, and 1985, using operating
data from published sources as in his 1986 work. This was particularly
useful because it showed the evolution of costs over several decades, from
which a supply curve based on OPEC capacity and OPEC reserves could
be derived.

Figure 21.2 compiles these estimates, along with one by Stauffer (1993),
who used a method similar to Adelman's, Francis Harper of BP, and one
anonymous estimate from an international oil company (IOC) all done
before either the price collapse of 1998 or the high prices of the 2000s.[4]

Many tend to think of long-term supply curves as resembling Figure
21.1, where costs tend to rise more or less exponentially. Examining
the three curves in Figure 21.2, which turn sharply upward at roughly
45–50 million barrels per day of supply, this view seems to be vindicated.
This is a misleading interpretation, however; note that three curves are
long and flat, especially when OPEC supply is included.

The right-hand point on most supply curves is actually an *outlier* rather
than a true representation of trend in long-run marginal costs. There are
many ways in which an unusually expensive supply can be developed
and enter the database, that is, the supply curve. Companies might, for
example, make unprofitable investments expecting some nonfinancial
benefits, such as good will. This was apparently why some companies
invested in Saudi Arabian natural gas exploration, despite the low fixed
domestic gas price.

Another case comes from the work of Adelman and Shahi (1989),
where the right-hand point was supply from Argentina. This skews the
results because the Argentinian national oil company, YPF, was grossly
overstaffed and investing in areas that did not yield good results, but
whose politicians had the political clout to direct investment to their
locales. Supposedly, much the same was true in the Soviet Union.

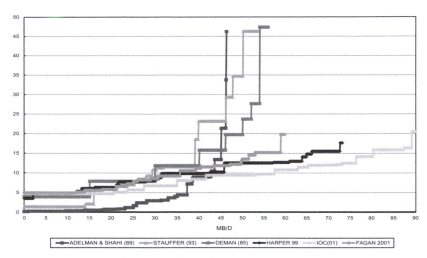

MB/D

ADELMAN & SHAHI (89) —— STAUFFER (93) —— DEMAN (85) —— HARPER 99 —— IOC(01) —— FAGAN 2001

Figure 21.2   Empirically Based Oil Supply Curves (Petak, Keven, "Funda-mentals Point to Demand Growth, Stronger Prices in the Long Term." In: American Oil and Gas Reporter, October 2010)

More recently, there can be instances where unusual factors cause costs to be excessive, as in the case of the supergiant Kashagan field in Kazakhstan. The cause of the cost overruns, about 100%, are not clear but quite possibly the field would not have been developed if the operators and investors had known in advance how much the costs would be. Examples of uneconomic fields lying fallow, like the supergiant West Sak field in Alaska, abound but are rarely discussed openly.

In response to this, it would seem best to throw out the most costly supply points as outliers, not representative of global supply generally, and treat the supply curve not as line A-C in Figure 21.3, but line A-B. In the latter case, the slope of the curve is $3 million barrels per day of supply, whereas the former would suggest something on the order of $10 million barrels per day of supply.

Costs on supply curves are perceived as rising, but this is an artifact of the data presentation, which is ordered on the x-axis from lowest to highest cost deposit or region. Casual observation suggests that costs are rising over time, when actually there is *no time effect shown*. The apparent trend, curve A-B, is illusory and meaningless as to the future direction of costs. A is not t_1 and neither B nor C is t_2. The movement from left to right is from cheapest to most expensive, deliberately ordered that way, and says nothing about the supply curve in the future.

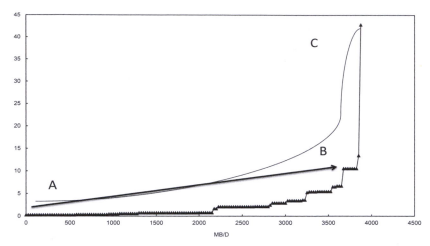

Figure 21.3   Slope of the Supply Curve (Adelman and Shahi [1989])

MOVEMENT IN AND OF SUPPLY CURVES

Whether considering resource-based or capacity-based supply curves, movement is both frequent and important, albeit not always unpredictable. The variables that affect supply curves help to explain possible movement as well as predictability, but differ according to time period. In the very long term, on the order of decades, the primary factors are depletion and technological change. In the short term, cyclical elements dominate, along with political changes in access to resources.

Understanding how and why the long-term supply curve moves is fundamental to having a picture of the long-term oil market, but as mentioned repeatedly, nearly every analysis of costs and supply is static in nature. Ask most analysts what costs will be like in the future and they will only insist that they go up, as reflected in the typical price forecasts seen in Chapter 2. (Although those are also influenced by access to resources.)

In the short term, on the order of a day to a year or two, the factors affecting supply curves are (a) infrastructure, (b) investment, and (c) policy. These are mainly important for commodity markets and short-term economic impacts, but usually not much beyond that.

Infrastructure, such as the construction or destruction of a pipeline, tends to be sudden and lumpy, but also usually not involving large increments. Out of 95 million barrels per day of global oil supply, the largest pipeline is rarely more than 1 million barrels per day and few are on that order, nor do they change often. Probably the biggest effects occur due to war or natural disaster, although war-related destruction can be considered a policy effect. Hurricane Katrina and the Iraqi destruction of the Kuwaiti oil fields in 1991 are among the largest examples, and they mattered but were important only on the order of months.

Policy changes include embargoes but, more frequently, decisions by oil-exporting nations about production meant to influence oil prices. The former are not common and usually of minimal importance, with the UN embargo against Iraq and more recent sanctions against Iran the most prominent examples.

Production decisions related to oil prices are primarily made by OPEC nations, but other countries like Norway and Oman have also participated, and some OPEC members often take no action in support of the organization's decisions. A decision by OPEC to reduce or increase production quotas (even assuming it was obeyed) does not change a capacity-based curve, but does move the production-based curve. Again, this is a short-term effect and will not be discussed in detail here.

In the medium-term of 3 to 5 or even 10 years, supply curve movements are almost entirely driven by cyclical cost changes and changes in infrastructure and/or access. Cyclical cost changes are the most prominent recently and have certainly had a large impact in the past decade, and with lower oil prices, a reversal of that trend is likely in the next few years.

Figure 16.10 compares two supply curves done five years apart by a consulting firm, and the impact of rising costs is clear. The short period involved makes it clear that this is not due to moving from the "easy oil," as does the fact that costs rise across the board. And an examination of older analyses, such as the IEA's *World Energy Outlook* 2001, shows that costs at that time were estimated to be much lower, with most supply under $20 a barrel. The implication is that costs are higher, but not permanently so.

The Effect of Changing Access

Additionally, and more important volumetrically, countries have often changed access to their resources, which moves the supply curve in the medium term. (A nation like Mexico, which for decades has not allowed foreign investment, accomplishes the same with changes in the national oil company's investment budget.) This can move both ways, reflecting most especially political decisions, but economic sanctions can have a similar effect.

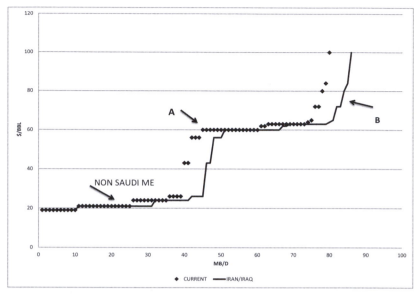

Figure 21.4   Medium-Term Supply Curve Movement (StatoilHydro)

When a number of oil-producing countries decided, in the 1970s, to emphasize resource conservation, the supply curve shifted strongly to the left, as the pieces representing Kuwait and Libya, for example, shrank. Now, with the opening up of Iraq and Mexico, probable end to sanctions in Iran, and possible reversal of resource nationalism in Venezuela, the supply curve should shift sharply to the right over the next 6–8 years. Figure 21.5 shows how this would look, assuming no other changes to the supply curve. In a sense, this is rather like saying, "The easy oil went on vacation, but now some of it is back."

AN EMPIRICAL CASE STUDY

Changes

In the case of access, this is a piece of empirical work that demonstrates very nicely the non-geological movement of supply curves: Adelman and Shahi's 1989 estimate of the evolution of costs, using drilling and capacity data from published sources. Figure 21.6 shows how the curve shifted to the right, as more and more supplies were added without costs rising, the opposite of the theoretical leftward shift that depletion should be imposing on the curves.

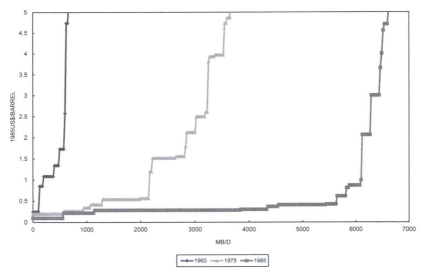

Figure 21.5   New Low Cost Supply Shifts Curve to the Right (Adapted from Adelman, M. A. with Manoj Shahi, "Oil-Development-Operating Cost Estimates 1955-1985." In: Energy Economics January 1989)

Of course, because the analysis covers the less-mature areas of the non-OPEC Third World, the shift to the right reflects primarily the opening of new territories for exploration, rather than advances in production technology. The Third World nations often released areas for investment slowly, and some areas became accessible due to improvements in infrastructure, such as the construction of roads. Similar movement is not seen in more advanced areas, such as the United States, where factors such as depletion and technology are the primary movers of the supply curves.

The takeaway from this is the lesson that given points on the supply curve move individually, and sometimes in both directions, including resources moving onto the current curve unpredictably. For the reserves or capacity curves, of course, movement can be in either direction, as governments can sometimes make supply off-limits or return them to the market.

Long-Term Supply Curves

Diminishing returns are opposed by increasing knowledge, both of the earth's crust and of methods of extraction and use. The price of oil, like that of any mineral, is the uncertain fluctuating result of the conflict.

Adelman, 1986[5]

In Chapter 16, the four primary factors affecting costs in the long term were discussed, including depletion, scientific/technological advance, access/infrastructure changes, and regulation. Too many analysts and policy-makers focus only on the first, and thus think of resources as becoming more scarce and higher in cost, the equivalent of the supply curve moving up and to the left. Regulation should have a similar effect, but it tends to be relatively small.

Changes in access and infrastructure, ignoring those that technological advance brings, are most obvious in the medium term, as described above, but should have a moderating effect on costs, particularly as an area first opens up to economic development and/or petroleum exploitation. The more dramatic changes are discussed below.

Scientific/technological advances, although obvious on the small scale level, have an impact that is hard to estimate, at least in an evolutionary sense. Again, there is a sense that gradual improvement occurs, rather as in other parts of the economy, but the very heterogeneous nature of oil fields makes it very hard to produce a reliable estimate of the long-term trend.

Which is where prices provide valuable insight. As far back as the 1963 publication of Barnett and Morse,[6] it was noted that mineral prices did not show any trend toward increase, which led many such as Adelman[7] to conclude that science and technology were roughly offsetting depletion. Although not very precise, it is nonetheless a valuable insight, as Adelman noted in the quote above.

Off the Radar

Long-term supply curves, showing resources and costs, are helpful but almost always misleading because they usually exclude significant, even enormous, resources. Few such curves include methane hydrates, for example, although the resource is thought to dwarf all other fossil fuels. The technology to produce them at competitive prices is not currently obvious, so they are typically ignored. Table 15.4 lists what could be considered "total resources" although even it is undoubtedly incomplete. There are still many blank spots on the map, and the sudden "appearance" of shale oil and gas resources as recoverable when most assessments had completely ignored them should be a lesson in humility.

This approach can backfire because resources move into the "available" category all the time. Traditionally, available resources increase gradually as, for example, the industry becomes slowly capable of drilling in deeper and deeper water, or raises recovery factors almost glacially. Similarly, infrastructure develops and brings costs lower or even allows entirely new regions to be opened up, while a better understanding of geology will sometimes send drillers to new areas, again, moving an existing but invisible resource onto the supply curve.

Similarly, any new oil resource supply curve should include oil from shale, not just kerogen or oil shale. Before the advent of hydraulic fracturing of shales, costs would have been prohibitively high, and the shale oil would not appear on the supply curve—although it was in fact present as a resource. With fracking, the recoverable portion of the shale oil resource shifts down and to the left, extending the supply curve substantially, shown in Figure 21.7.

THE MODEL

Thinking in terms of an equation for different regions, basins, or resources (to be defined), it would be

$$\text{Cn} = a + b(\text{geology}) + c(\text{decline rate}) + d(\text{technological advance}) + e$$
$$(\text{infrastructure change}) + f(\text{regulation}) + g(\text{error term}).$$

Note that terms d and e should be negative.

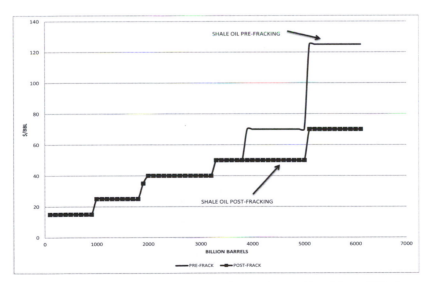

Figure 21.6   Movement of Oil Supply Curve Due to Fracking of Shale

Table 21.1   Subjective Judgment of Model Parameters (Author's Estimates)

		United States	Saudi Arabia	E. Africa	Shale Oil in United States
Geology	b	High	Low	Medium	Medium
Decline Rate	c	High	Low	Medium	Low
Technology	d	Low	Low	Low	High
Infrastructure	e	Low	Low	High	Low
Regulation	f	High	Low	Low	High

Table 21.1 suggests roughly what the various parameters should be in a variety of areas, and these cases are relatively obvious. A new producing area, like East Africa, should have declining infrastructure costs, but conventional oil in the United States should have achieved its maximum economies of scale. On the other hand, while the eastern sections of the British North Sea have probably attained the maximum reduction in costs from density of infrastructure, West of Shetlands might have some way to go.

Of all of these, regulation is the hardest to estimate. In theory, advanced industrial nations like the United States and European Union will have the toughest, yet might progress more slowly than in other areas where regulation is much less developed.

Supply Curve Movement

For basic economics, supply is visualized as a curve, as in Figure 16.10. Different sources have different costs and amounts, and organizing these from cheapest to most expensive gives an idea of how much can be produced at a given price. In Figure 16.10, the implication would be that at a price of $50 a barrel, about 60 million barrels per day can be produced.

What has confused many is the apparent directionality of the curve, from left to right, which seems to imply it can be extracted further to the right to explain future costs. However, this is not correct; the curve represents a snapshot in time, not a dynamic process, and most especially, the most expensive fields do not represent the future costs, but rather the outliers at the present time. A few fields have proven to be very expensive, but, once discovered or partly developed, the decision to complete them was made; it does not mean that future costs will be higher than those examples.

More realistically, the curve should be portrayed as moving toward the left over time. That is, as wells, fields, and regions decline, their costs will rise. This is the general meaning of resource depletion for existing production.

And while it is difficult to estimate overall productivity improvements, there have been ample improvements to offset the depletion that has occurred to make it clear that this is a significant effect that needs to be better quantified.

It is crucial to realize that the parameters in the equation above are *not constants* but dependent variables. That is, the impact of depletion can grow or shrink according to how rapid production is; this won't change much for most countries, but others, like Saudi Arabia, whose production tends to be the most responsive to market conditions, will see it change significantly (although the value should also be low).

Then, too, infrastructure will change for many countries reflecting a wide variety of factors. The isolation of the Alberta oil sands operations means that infrastructure (or geographical) costs will be high, where the development of a pipeline system in the Marcellus will bring down costs. Thus, we would expect the value of e to be high for the Marcellus, but low for the Barnett. Empirical research will be needed to establish values for the curves.

And in new areas, such as the presalt resources offshore Brazil, technological advance (partly learning curve also) should mean that initial costs are high, but they decline rapidly as the first few fields are developed and operators come to understand ways to optimize production, given the geology.

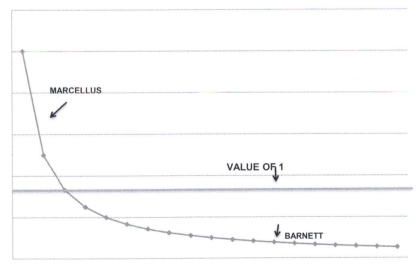

Figure 21.7   Subjective Representation of Infrastructure Cost Multiplier

Some proxy variables such as population density or well density could also be used to stand for some of the parameters.

FUTURE CHANGES IN RESOURCE AVAILABILITY

Some might think that the recent case of shale oil and gas becoming economically viable is an outlier, but arguably, this is the natural progression of mineral extraction over the long run. Many of the changes, as noted, are incremental and do not result in a noticeable impact on the supply curve, simply moving one segment up or down slowly. An obvious example is the gradual development of offshore drilling, which meant a slow increase in the resource that could be considered recoverable.

The development of taconite production is the classic example in mineral extraction. Because it is a low-grade iron ore and a hard rock, it was thought that iron extracted from it would be far more expensive than conventional ores, and the supply curve would presumably resemble the kind of step function seen in 1970s oil economics (see Figure 21.1). However, E. W. Davis discovered that the iron could be concentrated by crushing and applying magnetism, lowering costs to make it competitive with "conventional" ore.

Breakthroughs

There are a number of major hydrocarbon resources that might be shifted onto the "visible" curve should an advance in the production technology occur. These are listed in Table 21.2, although some potential sources, like coal-to-liquids or gas-to-liquids (with methane hydrates) are not shown.

Predicting such breakthroughs is extremely difficult, however, and new technologies are usually analyzed in retrospect. Shale gas was under development by Mitchell Energy for about a decade before most analysts even acknowledged it, and the Bakken oil shale nearly a decade ago, but most forecasts ignored it even two or three years ago. The National Petroleum Council's 2007 review of world oil supply and challenges, *Facing Hard Truths*, barely mentioned shale.

The impact can be seen by considering a U.S. gas supply curve before and after shale gas becomes accessible through hydraulic fracturing. Again, considering the resource supply curve rather than the capacity supply curve allows for the transformation to be seen more clearly.

Four potential sources of new hydrocarbon supplies can be seen at this point, again reiterating that surprises, such as shale, have occurred in the past. The first two, methane hydrates and geopressurized methane, are enormous to the point where uncertainty about the precise size is irrelevant. Instead, the operative factor is how fast they could be brought online if viable production methods were developed. The former, in particular, is the subject of significant research inasmuch as it is present in many otherwise research poor countries. Most notably, Japan has been carrying out experiments in its offshore region, and some oil companies have been working in Alaska's North Slope, both testing production methods. Prospect for exploitation remain unpredictable, but there is no doubt that no impact on markets is likely for at least a decade.

For liquid fuels, there are two potential resources of quite different nature. First, biofuels are the subject of much effort, in part reflecting U.S. government mandates for the use of cellulosic ethanol, but more probably, the potential for enormous profits if recombinant DNA techniques can be employed to convert otherwise useless waste material into liquid fuels like ethanol or even petroleum products.

The nature of biological research is such that discovery of the appropriate genes is not likely the result of gradual, evolutionary improvement but rather a revolutionary breakthrough, a eureka moment when scientists find the right gene to turn a bacterium into an oil refinery, for example, or a yeast that breaks down cellulose cheaply and efficiently, without the energy inputs that current ethanol plants require.

The final is our old friend kerogen, a form of shale that is thermally immature and must be extracted as rock and processed to become crude oil. It has been known for over a century that kerogen could be utilized to provide liquid hydrocarbons, but the expense of doing so has generally

Table 21.2   Petroleum Resources (Billion barrels)

	Reserves	Ultimately Recoverable Resource	In-Place Resource
Petroleum			
Conventional	1,200	2,700	6,750
Heavy	370		3,500
Shale oil	40	350	7,000
Kerogen	20	100	13,000
Biofuels	Not fixed but very large	Not fixed but very large	Not fixed but very large

exceeded that of producing conventional oil. Research into various new production methods, such as radiating the rock in situ with microwaves or using a process similar to SAGD, employed in the Athabasca oil sands, has proceeded since the first oil crisis, but without producing an economically viable method as of yet.

Breakthroughs are required in most instances to make these resources viable, and while scientific advance is unpredictable, the use of scenarios should allow some insight into what would happen should a given advance occur.

RESOURCES TO RESERVES

The second major step, after determining a resource-based supply curve, is to calculate investments that will create reserves. The role of politics in deciding whether or not to encourage investment and how much access to allow is the main obstacle at this stage.

$$\text{Reserves}_t = \text{Reserves}_{t-1} + (\text{Investment/exploration cost}) - (\text{Production}_{t-1})$$

Investment levels will be a function of prices relative to costs, with constraints including, first, past investment levels, second, access allowed by governments. Investment will require that resources are sufficiently cheap relative to price to allow significant returns to both investors and the government. Potentially, a sliding scale can be allowed relative to prices, reflecting the willingness of companies to invest, which arguably rises as prices rise.

$$\text{Investment}_t = (\text{Investment}_{t-1}) \times (\text{Government access multiplier}) \times (\text{Price/cost multiplier}) \times (\text{Price}_t/\text{price}_{t-1})$$

Presumably, the higher the price relative to cost, the more investment is likely to grow, whereas falling prices would mean lower investment due to reduced cash flow. The cost multiplier needs to include full cycle costs, plus fiscal terms, which as a policy variable can be difficult to model. Government access to resources is another policy variable that can, in theory, be modeled (see the discussion of Modeling Politics below), but imperfectly at best.

There should also be an adjustment to allow for higher multiples in countries with relative low levels of past investment; smaller producers can experience much more rapid change than a larger producer like the United States or Canada, for example.

MODELING POLITICS

> [T]here is no guarantee that the current free-market orientation of economic ideology will not swing back towards the left and greater governmental interference in markets or ownership of industry . . . . Also, though, continuing economic problems due to international
>
> debt, the recession, bank failures, etc., may cause . . . policymakers in many developing countries to abandon efforts at market reform.
>
> Lynch, 1991[8]

Predicting political trends is almost impossible, but there are some ways in which they might be modeled to provide sensitivity analysis or test scenarios. First, resource nationalism could be assumed to be directly proportional to oil prices: higher prices give more power to resource owners, who become more aggressive in their fiscal demands and slower to permit access.

Another approach is to assume that larger political cycles, roughly categorized as left to right, or more or less liberal or conservative, influence thinking about resource policies over the long term. Arguably, events in countries like Russia, Venezuela, and some others suggest we are in a general leftist trend, but growing problems in such countries imply that we might be getting ready to swing backward.

Third, at the national level, relative revenue needs can be assumed directly proportional to openness to foreign (or even domestic) investment. Countries like Kuwait or Norway would be less eager to increase production than nations like Colombia or Nigeria. But additionally, countries where oil revenue has risen significantly will become more complacent about releasing acreage for drilling. This can mean higher prices reduce investment or, more likely, leads to a period without new production.

Unfortunately, there is no method that can come close to predicting what decision-makers will do. This is the primary obstacle to predicting supply, especially at the national level (or below).

CONCLUSIONS

The primary point of this review is to show that supply curves can be easily misinterpreted to suggest long-term prices must be high and rising. Most curves are actually flat over the bulk of their coverage, the highest cost points appearing to suggest rising costs over time, when in fact they are more likely outliers. New supply is often added in the moderate cost segments of the curve, not the right-hand side, and breakthroughs, such as for shale drilling, can bring unseen resources into the viable/visible portion of the curve.

Further, much of the world's resource base is not represented on supply curves, especially where it is judged to be uneconomic or not technically viable. But aside from gradual movements of resources onto the curve through technological advances, breakthroughs like the application of hydraulic fracturing can seriously change the curves' shape. The main lesson is for abundant petroleum resources into the distant future, which has serious ramifications for climate change and other policies.

Modeling supply is extremely difficult because of the political control over both exploration and production levels, but building a supply curve with costs modeled according to the factors that drive them, rather than simply assuming geology explains everything, can improve understanding of what is possible.

TWENTY TWO

Lower Energy Prices: The Good, the Bad, and the Ugly

God grant me one more boom and I promise I won't screw it up.
                                    —Texas bumper sticker in the late 1980s.

"Oh, yeah?" My rejoinder

What would lower oil (and other energy) prices mean for the U.S. (and global) economy? The impacts are many and intertwined worse than a Gordian knot, but the first order of logic is to think of all the problems ascribed to higher oil prices and assume them reversed. In other words, the world would see:

- Lower inflation;
- Stronger economic growth;
- Lower borrowing costs;
- Lower manufacturing costs;
- Higher consumer income and spending;
- Cheaper vacations; and
- Lower movie ticket sales, as people hit the road.

It would be nice to provide estimates of the precise economic effects, and there are those who could do so with their computer models, simple or complex; however, there are many and complex feedback loops that introduce uncertainty at nearly every level. But it is safe to say, "High oil prices bad. Low oil prices good," unless you're an oil company, of course.

But we shouldn't take this to absurd levels. Jeff Rubin suggested food like salmon and avocadoes would cease to travel given his projection of higher oil prices, but it seems unreasonable to predict the reverse—that heavy, bulky commodities like cement will be flown overnight thousands of miles to consumers.[1] And it's not clear what would be the opposite (converse) of "the extinction of mankind" that some peak oil advocates have warned us about.

On the other hand, for those who remember the 1990s as a golden age of stable economic growth, low inflation and interest rates, and political optimism, that should be the closest analogue for what the world will face in the next decade (hopefully without the dot.com bubble, although bubbles bloom eternal in investors' breasts).

Peak oil advocates are correct in one instance, oil and gas are extremely important elements of the world economy, and price changes create ripples and waves through many industries. Some of the effects are direct, others indirect, and the precise impact will depend on how companies and consumers respond to them.

THE GOOD

Given the widespread reliance on petroleum, lower oil prices will have quite a number of positive effects, primarily for importers and consumers of oil, including companies and industries. These impacts can be split between economy-wide or macroeconomic, and industry-specific, or microeconomic. In the United States, which consumes nearly 20 million barrels a day of oil, the savings will be enormous, equal to about 2% of GDP as the data in Table 22.1 shows.

Macroeconomics

As Table 22.2 shows, oil expenses make up a significant portion of expenditures in the industrialized nations and are hardly trivial in developing countries like Brazil and India. There are differences in imports versus total consumption as some nations are notable oil producers (the United States, Britain, and others), which means that some of the savings will also count as losses to the oil-producing sectors of those economies.

Table 22.1   Savings from $50 Drop in Oil Price (EIA)

	Million barrels per day	Billion $	% U.S. GDP
Production	8.7	−$159	0.95%
Oil imports	9.2	$168	1.00%
Consumption	19.0	$347	2.06%
Government	0.3	$6	0.04%

The lower costs from oil and gas will have a beneficial macroeconomic effect, as lower inflation can create a "virtuous circle" (assuming policy-makers don't undermine it). Lower inflation results in a more stable investment climate and encourages expansion of output and employment. True, many other factors affect prices and economic growth, but energy remains a prince among knaves.

The 1990s saw robust economic growth in many parts of the world, and the United States clearly prospered (Figure 22.1), partly because of low and relatively stable oil and gas prices. From 1974 to 1985, when oil prices were elevated, U.S. GDP growth averaged 2.9% per year, while from 1986 to 2000, it grew to 3.4% per year.

Table 22.2   Impact of Lower Oil Prices in Various Countries ($billions)

	Oil Income at $100	Oil Income at $50	GDP	Loss
Canada	67	33	1827	1.8%
Mexico	33	16	1261	1.3%
Norway	60	30	513	5.9%
	Oil Imports			**Savings**
Australia	20	10	1560	0.6%
France	58	29	2806	1.0%
Germany	83	42	3915	1.1%
Italy	39	20	2149	0.9%
China	249	124	9240	1.3%
India	108	54	1877	2.9%
Japan	157	78	4920	1.6%
Korea	90	45	1305	3.4%
Argentina	1	1	610	0.1%
Brazil	32	16	2246	0.7%

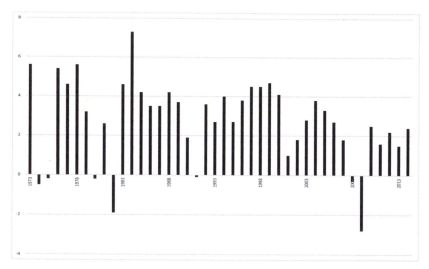

Figure 22.1   U.S. Economic Growth (Council of Economic Advisers [2015])

While it is normal for a post-recessionary period to see higher-than-usual growth (although that hasn't happened as of yet), lower oil prices could lengthen and strengthen this tendency. And while U.S. government royalties from oil and gas production would decline, so would its expenditures on fuel (especially for the military), which would improve the budget balance. Higher economic growth would also generate more tax revenue, which would reduce the deficit further and in a positive and beneficial feedback loop. The 1990s were a time of U.S. government budget surpluses, although many factors besides low oil prices were involved.

Many countries like India will benefit as lower oil and gas prices reduce their bill for fuel subsidies. The IEA has noted that global subsidies for fossil fuels are enormous, although much of that is in the form of cheap oil products in OPEC nations.[2] But in quite a few industrializing nations, governments have kept prices low for some products, usually as a form of social welfare. Table 22.3 compares subsidies by fuel for 2013, the latest data available, and shows the approximate savings from a $50 oil price. The biggest subsidies are for petroleum products in OPEC nations; other countries like Egypt and India spend billions of dollars, which they arguably cannot afford; the savings will be especially valuable to them.

Negative macroeconomic effects do include the layoffs in the oil fields and the drop in capital investment in that sector, which Goldman Sachs estimates will chop 0.5% from GDP in 2015.[3] Few go so far as economist A. Gary Shilling, who has suggested that low oil prices might actually trigger a recession, given the combination of deflationary effects and damage to oil-producing sectors and economies.[4]

Certainly in early 2015, it appeared that the anticipated (and hoped-for) consumer spending boom had not materialized, perhaps in part because of bad weather in much of the United States during the first quarter (when GDP growth was 0.7%). However, as one reporter noted, "Lower oil prices have proven to be more of a bane than a boon for the U.S. economy. But that is about to change."[5] Second quarter growth is now estimated to have been 3.7%, and overall, the U.S. economy is weathering the economic downturn afflicting much of the rest of the world.

Some regions have shown the effects of lower oil prices already. Calgary, for example, had the biggest recorded monthly drop in home prices in May 2015, and support for the Stampede, a cattle-and-cowboy-themed annual festival, was "subdued" in 2015.[6]

The *New York Times*'s Clifford Kraus described low oil prices as "lowering the boom" in Texas, where a combination of industry layoffs and lower royalties to leaseholders was depressing what had been a boom area.[7] Texas grew by 5.2% in 2014, but is expected to grow by only 1.3% in 2015, according to Mine Yucel of the Dallas Fed, which isn't too bad, and is the result of the diversification of the economy since the 1986 oil price collapse.[8]

Other areas that had experienced a shale boom, such as Pennsylvania and North Dakota, are seeing a serious deflationary pressure and some

Table 22.3 Petroleum Subsidies ($billions) (International Energy Agency, "Fossil Fuel Subsidies Database." In: http://www.worldenergyoutlook.org/resources/energysubsidies/fossilfuelsubsidydatabase/)

	Subsidies	Savings at $50
Argentina	1.3	0.65
Bolivia	1.6	0.8
China	11.8	5.9
Ecuador	5.2	2.6
Egypt	20.9	10.45
India	36.6	18.3
Indonesia	21.3	10.65
Malaysia	4.9	2.45
Mexico	9.5	4.75
Nigeria	5.8	2.9
Thailand	2.8	1.4

unemployment, but Pennsylvania's economy is large enough that the reduced drilling has not had a major impact.[9]

Microeconomic Effects

The impact of low oil prices varies from industry to industry, depending in part on how much energy they use, although indirect factors should not be ignored. Obviously, industries and companies that are most heavily dependent on oil consumption (their own or their customers') will benefit the most.

Energy-Intensive Industry

The most obvious winners will be those industries that rely heavily on oil and energy to produce their products and services, such as those that involve transport, most notably shipping, but also airlines. Trucking companies will benefit the most, railroads less so, as the diesel fuel they use is a smaller component of costs.

The airlines are a primary case study, since the past decade has seen them work exhaustively to offset the high costs of jet fuel. In this instance, however, the extremely competitive business environment has been a major source of losses, but most of the U.S. carriers have worked to improve their seat occupancy rates (i.e., the seat next to you is more often occupied). The question remains as to how much lower fuel costs will translate into higher profits, and how much lower ticket prices, as the airlines often compete away any cost savings.

Both large trucking firms and package delivery companies like UPS should see much lower costs, and again, a combination of lower charges and higher profits should ensue. In the 1990s, this sector definitely benefited from low diesel fuel costs, and their stock prices reflected that.

Retailers would obviously benefit from lower shipping costs as well as the extra money in the pockets of consumers. Not to suggest that Cartier and Mercedes Benz will see a huge boost in sales, but Target, Walmart, and similar stores, as well as the fast food industry, will benefit from the extra $25–50 a week that low gasoline prices will translate into for U.S. consumers.

To date (September 2015), the benefit does not seem to have had an impact on retail spending, with the anticipated boom delayed, at the least.

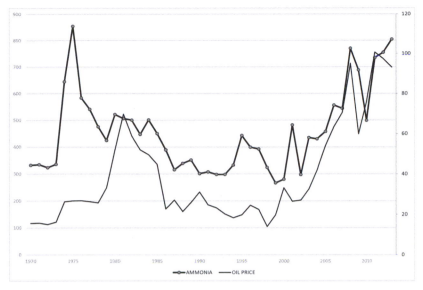

Figure 22.2   Price of Anhydrous Ammonia ($/ton left hand scale) (Agricultural Prices, National Agricultural Statistics Service, USDA. In: http://usda.mannlib .cornell.edu/MannUsda/viewDocumentInfo.do?documentID=1002)

Agriculture and Fertilizers

Oil and gas are major components of fertilizer, and with lower prices, farmers should receive a major break from at least some of the problems of the past decade. Figure 22.2 shows how fertilizer prices grew during recent years—and how low they were when energy prices were low in the 1990s. Specialty foods that are delivered by air, from flowers to lobsters, should benefit as shipping costs drop.

This is not to disregard the role of bad weather in raising agricultural prices, but, as peak oil advocate Richard Heinberg notes, every calorie of food produced requires 10 calories of fuel.[10] In other words, each Big Mac contains about 35 cents worth of energy with oil at $100, and half that at current prices. Thus, a drop in oil prices would mean cheaper transportation costs for food (not good for local farmers in high-cost areas like New England, granted) and conceivably lower fertilizer costs, which would help farmers and ultimately consumers. Fertilizer prices, as shown in Figure 22.2, are strongly correlated with global oil prices.

Travel and Tourism

The past decade has seen the birth of the Staycation (although the term may be more prevalent than the behavior), as people find themselves unable or unwilling to pay high airline fees or the cost of gasoline for long-distance vacations. Instead, backyard barbeques and trips to the nearest state park have taken the place of flights to Orlando and San Francisco, unless you have kids (of whatever age) who can't survive without Mickey Mouse or Ghirardelli sundaes.

But the better economy and lower oil prices should mean much more travel to the larger parks and more distant locations, so that sales of barbeque equipment and backyard furniture will decline in favor of luggage, airplane, and amusement park tickets. Ski resorts in the Rockies should win out compared to those on the East Coast, always depending on snowfall, naturally.

Automobiles

The U.S. automobile industry has been whipsawed over the past few decades by changes in oil prices and the shift in consumer demand for fuel efficiency. In the early 1980s, when U.S. carmakers were unable to supply enough small cars to meet soaring demand, Japanese automakers made huge inroads into the U.S. market. However, when oil prices fell, U.S. firms responded with, first, minivans and then SUVs in response to consumer desire for bigger, more luxurious cars.

Environmentalists often attack the industry for its reliance on large vehicles and when, in the 2000s, high oil prices saw a decline in demand for them, they mistakenly argued that this reflected the industry's failure to anticipate a long-term shift to more efficient vehicles. The problem, rather, was that the auto companies could not adapt swiftly to changing oil prices.

The question will be what will happen to the regulations that are now set to require vehicle efficiency to double by 2030. In the 1980s, the Reagan administration decided to overlook the failures of Ford and GM to meet the targets, which became relatively meaningless as car makers turned to pickup trucks, minivans, and SUVs, which were not classified as "cars" and thus didn't count toward the efficiency targets.

The past few years have seen a notable shift from the largest vehicles, primarily SUVs, to not just small vehicles but the crossover class, which is larger than a sedan but smaller and more fuel efficient than the SUVs that were popular in the 1990s. It is quite possible that these will grow larger and more power-hungry when gasoline prices return to pre-crisis levels, but the shift in tastes for smaller cars could well persist.

Energy Winners

There are two points where the energy sector could prosper from low petroleum prices. The first is the refining sector, which relies on crude oil as its primary feedstock and requires significant energy inputs for its operations, and should thus theoretically prosper. Recently, the European refining sector has been making so much money that they postponed much of the normal fall maintenance in 2015 because they didn't want to miss out on the high margins.[11] Of course, like the airlines, the refining sector has a long record of competing away profits, in part due to governments seeking autonomy in petroleum products or trying to capture the value added from processing.

Another major winner is the production of electric power from natural gas, especially with the use of gas turbines. Cheap natural gas in the United States has meant a two-thirds increase in power produced from natural gas over the last decade, about 500 billion kilowatt hours, or three times the growth in wind power. With the spread of cheaper natural gas internationally, the sales of gas turbines should rise sharply, given the many areas still relying on coal as well as the billion or so without access to electricity.

THE BAD

The oil industry and oil-producing nations will certainly suffer from lower prices, although the effects will vary widely and the apocalyptic visions of some will probably not be fulfilled. After all, such dire consequences did not occur during the last period of prolonged "low" oil prices, from 1986 to 2000, when the price roughly averaged $30/barrel.

Needless to say, the oil-producing nations in the Middle East would rather have more money than less money, but it is important not to exaggerate the impact on political stability. Libya, after all, entered the current period of turmoil when the price was still $95/barrel. And the suffering of that country might act as a deterrent to at least some who might consider opposing their own governments in the Arabian Peninsula.

Not that there won't be negative consequences with cutbacks and lay-offs throughout the area, although few have become public. Probably the most prominent has been the staff reduction of al-Jazeera, the Qatari-supported news organization, said to be cutting hundreds of jobs.[12]

Rosy as the low-price scenario is, there will be losers as well as winners, although perhaps not as bad as during the 1986 oil price collapse when the real estate market in Houston went into a tail spin and whole neighbor-hoods were abandoned. Smaller oil field service companies went bankrupt or were bought out by larger competitors, and unemployment in the indus-try soared. Companies that had a lot of debt, including Phillips and Uno-cal, which had been greenmailed by T. Boone Pickens into raising their debt levels in 1985, struggled for years to stabilize their finances when prices fell in 1986.

Clearly, there will be troubles again, especially where some companies have accumulated a lot of debt. Chesapeake Energy, which adopted a strategy geared for a high natural gas price and is now paying the price, stands as the example of likely prospects for companies like these. Most of the major oil companies have large cash hoards, thanks to the high oil prices of recent years and a fair degree of capital discipline, but some have been more profligate.

At present, the industry is clearly hurting, with $500 billion in debt out-standing and a much reduced cash flow. Several smaller companies, like Samson, are thought close to bankruptcy, and Chris Helman at *Forbes* identified nearly $400 million in distressed oil company debt held by four major banks.[13] Bloomberg estimates that 30 producers have debt levels equal to more than 40% of their value.[14]

As banks perform regular evaluations of their borrowers' financial sta-tus, the drop in reserve values could leave many of them in trouble. Reg-ulators have been described as concerned about banks' exposure, although banks have complained that they are overstating the risk of reserve-based loans.[15] Reappraisals of their loan portfolios will probably leave most banks reluctant to make further lending to the smaller compa-nies, but whether they will restructure loans or insist on bankruptcies for troubled institutions isn't yet clear. Still, these are mostly the companies that sprang up recently during the shale boom, and even dozens of bank-ruptcies won't have a major impact on the industry.[16]

The result has been a strong cutback in employment throughout the indus-try, including support companies. Swift Worldwide Resources in Houston estimated in 2015 that worldwide oil field layoffs totaled 176,000.[17]

Service companies will be hit as well, since they have been benefiting from high rig rental rates as high oil prices encourage companies to pay premiums for drilling and related services. It has long been observed that even a small cooling in the drilling rates leads to significant drops in prices for rentals, particularly if the sector has been overheated, operating at high-capacity levels. In response, the industry has begun layoffs, including 6,400 by Halliburton, and mergers, led by Schlumberger's proposal to acquire Cameron for $15 billion.[18] The sand industry, something few think about, is also being hit hard by the drop in fracking, with the price down by about one-third and a projected drop in the market of over half.[19]

The coal industry will be hurt in many places, especially if international gas prices fall as predicted. This has already occurred in the United States, where gas has begun to steal market share in the power-generating sector from coal. Environmentalists have failed to reduce coal consumption for decades, but now the oil industry has reduced it by the simple expedient of having a better cheaper fuel. As one news report put it, "Bargain-basement coal deals are proliferating as producers bail out of an industry stuck in its worst downturn in decades."[20]

Sadly, the price of gas in many parts of the world remains elevated as the result of a misguided pricing policy adopted in the 1970s, wherein delivered gas prices tied explicitly to oil prices, so that when oil supply disruptions sent oil prices up, instead of gas users getting relief, they saw their costs rise too.

Crash or Pause?

Of course, many in the petroleum industry are convinced that the oil price will "recover" to levels that will at least cover their capital budgets and conceivably make them nice profits. To date, however, there have been a number of very specific moves that suggest where the industry is going in the next few years.

Anticipated spending in the Canadian oil sands fields has been reduced, and the industry expects to see 1 million barrels per day less production in 2020 than it did last year (Chapter 20). Shell has announced the cancelation of one oil sand project and is ending its offshore Alaska exploration, although that might be due as much to poor drilling results this summer.[21]

Indirect effects will include some of those who have benefited from high energy prices, such as competing types of energy, but also, as mentioned above, companies that provide services to the oil sector. Flatbed trailers that carry equipment to the oil fields are having a hard time, and a wave of consolidation is possible.[22] Manufacturers like Caterpillar, who sell equipment to oil service companies, have been hit with lower sales, and steel companies are experiencing a downturn in pipe and other sales.[23]

Green Power Bubble Collapse?

[R]enewables producers might seek to invest in products which are not economically viable expecting [looming resource scarcity] to improve the market environment.

Lynch, 1999[24]

The past decade has seen a great expansion in both the use of and the interest in various types of renewable or environmentally friendly technologies and fuels, although not all deliver the environmental benefits promised. Governments have pursued a variety of policies from outright subsidies to guaranteed high prices for the power produced, as well as loans for both producers and consumers of so-called Cleantech products and mandates for their sales or purchases.

These mainly fall into two categories: power generation and automotive. The former tends to be low or nonemitting, while the latter often involves merely moving the emissions from the vehicle to the creation of the energy, such as the production of hydrogen or electricity. All, however, are motivated by both expectations of continuing high and rising petroleum prices and concerns about climate change.

The latter seems unlikely to change significantly with lower petroleum prices, although cost/benefit analyses of greenhouse gas emission reduction policies will change somewhat. But lower oil and gas prices will undoubtedly have an impact on programs to encourage alternative fueled vehicles, such as those using batteries or hydrogen fuel cells.

Already, there have been signs that consumer interest in vehicle mileage and especially advanced systems is declining. Honda has abandoned its compressed natural gas vehicle, and hybrid electric sales are said to be weakening, even as automobile sales generally remain strong.[25] Toyota's Prius line-up, the dominant make of hybrid vehicles, saw sales decline by 17% during the first eight months of 2015.[26]

This is hardly new: promoters of electric vehicles have been repeatedly shocked to find that consumers have minimal interest in a product that performs poorly and is extremely expensive. Watching the movie *Who Killed the Electric Car?* with a critical eye should leave one feeling it is less a documentary than a comedy. The film shows a young saleswoman saying the GM Impact should have been popular and a handful of owners waxing eloquently about it. But it also mentions the problems of cost, limited range, and lengthy recharge time, though only in passing.

And although Toyota has great hopes for its newly released hydrogen fuel cell vehicle, its success would probably require both strong government support and very high petroleum prices, in conjunction with cheap electricity to generate the hydrogen. Because neither hydrogen nor battery electric vehicles reduce greenhouse gas emissions very much and do so at great expense, government support for them could very well be reduced as oil prices remain low for an extended period. (In keeping with my philosophy of attempted objectivity, it is only fair to mention that Joe Romm, who has slammed me for my disbelief in peak oil, has also argued that a hydrogen economy is not feasible, demonstrating that he's not a mindless ideologue.[27])

Solar power is already finding itself in difficulty in some countries such as Japan, Spain, and the United Kingdom. These governments offered prices for solar power that were so high that they led to a boom and cost the governments and/or consumers much more than anticipated. As a result, all three are reducing the prices paid, with one solar company referring to the UK government's shift in policy as a "damaging set of proposals."[28] Japan, as well, has reduced the prices offered for solar-generated electricity by about 20%.[29] No government has suggested that low oil prices are the reason for reducing support for renewables, but it is probably a factor.

The biofuels sector can expect to struggle. The United States might abandon efforts at mandating the use of cellulosic ethanol, and the emphasis will return to the laboratory, trying to make it commercially competitive. In other parts of the world, biodiesel production should decline, but in nations like France, Germany, and the United States, where biofuels is a form of support for farmers, programs should continue.

Finally, although coal is not directly competitive with oil in most places, the shale gas revolution has already done severe damage to the industry in the United States, where the industry is said to be "a shambles" cutting 10,000 jobs a year since 2011.[30] Cheaper petroleum will also mean that natural gas prices should decline worldwide and gas turbines for electricity generation should displace coal plants in other parts of the world.

THE UGLY

Wildcatters and roughnecks will no doubt have their share of problems, along with any pensioners foolish enough to concentrate on energy stocks,

Table 22.4   Financial Impact for OPEC Nations of Low Oil Prices (Calculated from BPSR and World Bank data)

	Oil Exports thousand barrels per day	Oil Revenue At		Lost GDP Billion $	Percentage
		$100	$50		
Algeria	1,330	49	24	210	11.6
Angola	1,644	60	30	124	24.2
Ecuador	273	10	5	95	5.2
Gabon	251	9	5	19	24.1
Iran	1,489	54	27	370	7.3
Iraq	2,621	96	48	229	20.9
Kuwait	2,313	84	42	176	24.0
Libya	274	10	5	74	6.8
Nigeria	2,140	78	39	522	7.5
Qatar	1,824	67	33	203	16.4
Saudi Arabia	8,663	316	158	748	21.1
UAE	2,776	101	51	402	12.6
Venezuela	1,154	42	21	438	4.8

but in all likelihood, the worst affected will be those in countries where oil income is so important that a decline in prices will lead to recession and possibly political unrest. Iran and Venezuela stand out, partly because economic mismanagement has left them in poor shape even with high oil prices, but also because their political systems are unstable enough that a decline in oil prices could lead to violent upheaval.

Although many analysts have argued that the fiscal breakeven for oil-exporting countries ranges from $50 to $150, the reality is that many are quite capable of coping with lower prices. Table 22.4 shows the oil revenue for OPEC member countries and the amount lost from a drop in oil prices compared to their GDP. (The numbers are approximate, calculated from oil exports and not adjusting for quality differentials.)

While it is true that countries like Kuwait, Qatar, and the United Arab Emirates would lose double-digit amounts of their GDP from a $50 drop in oil prices, they also have the highest per capita oil income and can more easily afford to lose the money. Additionally, most of these countries (including Saudi Arabia) have significant expenditures on capital goods and infrastructure that can be cut back without too much damage. After all, they survived the years 1986–2000 on lower prices and lower production levels.

The Weakest Links

We are terrorized by the drop in oil prices.

Maryclen Stelling, 2015[31]

Venezuela is an amazing case study of ignorance triumphing over reason, with its recent leader, Hugo Chavez, repeating almost precisely the mistakes of a predecessor, Carlos Andres Perez.[32] Perez came to power in the early 1970s with ambitious plans to develop the Venezuelan economy, just as oil prices soared and gave him the means to attack his goals.

Unfortunately, he not only had little practical experience in building a new economy, but also had enough ego to believe that those who warned him not to be too ambitious were foolish, and sidelined any such advisors, moving increasing amounts of power into the presidential office. He then proceeded to spend the country's oil bonanza on ill-considered investments such as large steel plants that proved to be white elephants. A modern economy cannot be dictated by the government.

Hugo Chavez repeated these mistakes with a vengeance. His distaste for the power of the country's elites has made him dismissive of advice from professional economists or bureaucrats, and reinforced his belief in socialist ideology. He admired Castro's Cuba, but more for the control that his government exercises over its citizens than for the economic progress— and mostly lack of it—which that country has experienced.

As a result, the greatest oil bonanza in history has perversely left the country all but destitute. Gold reserves are shrinking, inflation soaring, and the nation's infrastructure crumbling. Except for some parts of the social system, especially where Chavez has traded oil to Cuba for medical personnel, the country appears to be increasingly dysfunctional, with more and more of the economy under the control of the president, often managed by the national oil company or the military.

The question is: can the country recover from lower oil prices? The fact that much of the expenditure is controlled directly by the president with little or no accountability means that it is difficult to guess how much money is being spent on big capital projects that might be canceled. Promises to build housing for the poor are not being met and the crumbling infrastructure doesn't seem to be receiving the funds it needs.[33]

Additionally, Chavez spent large sums on foreign aid and these are already being ramped down or canceled, freeing up some money, but they do not appear to amount to more than a small fraction of the overall budget. In the longer term, Chavez's successor Nicolas Maduro might quietly swallow his pride and encourage new investment by foreign companies in the upstream area to generate more oil exports and revenue. However, this would take years to yield results and he would have to overcome some resistance to investing in a country whose government has been less than stringent in honoring contracts.

Another possibility is regime change, either if Maduro steps down in the face of widespread chaos (i.e., greater chaos than is present now, impossible as that might seem), or resigns due to pressure from his supporters and/or the military. In that case, there might be unrest from his supporters, but expect an effort to reform the economy and probably increase oil production and exports, even if one of his close supporters replaces him.

Russian Bear on a Diet

The end of Communism and the dissolution of the Soviet Union has brought many changes to Russia, good and bad. High oil prices have, as in the case of Venezuela, helped to gloss over many of its persistent problems, including poor infrastructure and political institutions that remain corrupt and retard economic progress in most sectors of the economy.

At the same time, the non-oil economy is stronger than in Venezuela, where it has been backsliding. Higher incomes have meant more demand for products, and at least some of that was produced domestically, but for every success story there is one of an entrepreneur frustrated by bureaucracy and/or corruption. One response to lower oil prices could be a crackdown on corruption and a serious effort to encourage indigenous enterprises, most of which are now largely neglected.

Also, although Putin remains the face of the government, there are much stronger institutions in Russia than in Venezuela, and the likelihood of major political unrest—riots, attacks on party offices—is lower than in Venezuela. An increase in protests is certainly to be expected, but there is a much better chance that opposition will be expressed through normal political channels. Parties that are relatively moribund now should see a revival in support and pose a serious challenge during elections.

This could lead to increased efforts at repression, should Putin's clique decide that they do not wish to give up power or even fairly contest it. Already, the government has moved to suppress dissent and curb the activities of foreign-based nongovernment organizations, while strengthening laws.

Pressure

A number of other countries will suffer from lower oil prices and could face varying degrees of instability. Again, the situation should be more like 1986, when prices fell from elevated levels, and saw no major political turmoil, than in 1998, when they dropped from normal levels to extremely low ones.

The worst off in this group will undoubtedly be Nigeria, where the government faces a number of internal political challenges even at high oil prices and has been using oil revenue to try to maintain stability. However, the siphoning-off of large amounts of that revenue through corruption reduced the benefit of higher oil prices, although it's not clear if lower oil prices will translate into lower—relative—corruption.

But other countries, such as Algeria and Oman, possibly Kazakhstan and Azerbaijan, would find that lower oil revenues could worsen existing political opposition to the governments. Though none of these is likely to see massive turmoil, the recessionary effect of a drop in oil revenues by 30–50% could be enough to stimulate dissent and, ultimately, lead to anything from labor actions to civil unrest. Remember that starving people don't revolt, but hungry people do.

And there is always the possibility that a number of smaller countries experience instability that disrupts enough production, combined, to put pressure on prices. This is the case at present when Yemen, Syria, and South Sudan have seen a combined drop of nearly 1 million barrels per day in production, which would be enough to put pressure on prices, if it were flowing to the market.

Still, it is worth remembering that there have been numerous disruptions of oil production in the past few years in a variety of countries despite high oil prices, proving that oil revenue is far from the only factor affecting political stability. To say that high prices actually encourage instability is perhaps too much, although the imprecision of "encourage" provides a lot of leeway.

The Ugliest

A great fear is that lower oil prices will have a contrary effect, as in 1998, when the oil price collapse led (in part) to the rise of the hawkish Hugo Chavez in Venezuela and the weakness in non-OPEC production after 2000. The current price drop could, in theory, lead to renewed oil market tightness and high-security premiums, returning prices to the elevated levels of recent years: $100 and more.

There are two primary reasons to question such a scenario: First, oil prices in 1998 did not drop from elevated levels, but rather from prices that appeared to be sustainable over the long term. A drop in prices now would be more like the drop that occurred in the early 1980s, and the industry almost certainly will see lower costs, which should leave investment levels lower but activity levels—actual drilling and development—more or less unchanged.

The potential for the rise of new hawkish oil ministers seems less threatening also. Until recently, two of the more important members—Iran and Venezuela—had ministers who pushed for higher prices whenever possible, although the current Iranian minister is more hawkish about raising production. Changes in policy in countries like Angola, Libya, and Nigeria are unlikely to be very important except on the margin—assisting other hawks, but not driving price movements.

The big threat is that a major producer like Iraq or Saudi Arabia could change their stance, and carry along Kuwait and/or Abu Dhabi. Their historic dovishness on prices has reflected a combination of their relative pro-Western politics (excluding Iraq), and large resource endowments, which translates into a desire to avoid price levels so high that they would reduce long-term demand.

One expert, Frank Parra (a former OPEC secretary general) once suggested that OPEC should take advantage of high prices to sock away savings whose income would replace future petroleum revenue.[34] Unfortunately, few of the current members have done so, especially outside the Arabian/Persian Gulf region.

LOOKING AHEAD

After a transitional period, the global economy should settle into a new equilibrium, as industries and consumers adjust to lower energy prices. The macroeconomic effects will be broadly positive, with the exception of the largest oil exporters, and there, the pain could result in political unrest. Perversely, that could see oil prices rebound as they did after 1998, although only temporarily.

Whether or not demand booms remains the biggest question, as it would imply the need for policies to address security and environmental challenges, or at least perceived challenges. Governments are prone to either not acting or overreacting and there are few signs of organizational learning. Indeed, the next boom will probably see new declarations of a new paradigm, resource scarcity and ever-higher prices.

TWENTY THREE

Assessing the Viability of New Energy Technologies

The past few years have seen great enthusiasm for various energy technologies and fuels that are thought to be the answer to the energy/climate crisis we're facing. "Cleantech" was coined to describe many of these, such as renewables and electric vehicles, and they attract money the way the Internet did in the 1990s. The difference is, of course, that many of these technologies are now being sold and producing power or providing services, with actual revenue and, in some cases, profits.

But the field remains fraught with risk, which few seem to be able to judge, as the oft-touted case of Solyndra showed. That company planned to market a particularly expensive form of solar panel, thinking it would be attractive because market tightness for conventional photovoltaic panels temporarily raised their prices. Once those prices fell, the company's business plan was untenable.

Comparing the past decade with the 1970s, when fast breeder reactors, coal gasification, and photovoltaics were all enthusiastically promoted but then dropped by the wayside with lower oil and gas prices, would seem appropriate—but only up to a point. The cost of solar power has dropped sharply and battery advances mean that electric vehicle capabilities are much improved. But is that enough to weather the current round of lower fossil fuel prices?

HISTORY LESSONS LEARNED AND UNLEARNED

Experts predicted in 2000 that wind generated power worldwide would reach 30 gigawatts; by 2010, it was 200 gigawatts, and by last year it reached nearly 370, or more than 12 times higher. Installations of solar power would add one new gigawatt per year by 2010, predictions in 2002 stated. It turned out to be 17 times that by 2010 and 48 times that amount last year.

The New Optimism of Al Gore[1]

Over a century ago, it was estimated that crude oil could be extracted from kerogen by retorting at a cost slightly above the price of conventional oil. In the 1970s, when attention returned to this large, but problematic resource, it was again estimated that the cost would be higher than that of conventional oil—even though oil prices had trebled. When they trebled again during the Iranian oil crisis, it seemed that oil from kerogen must be then economically viable, but, no, costs were said to have risen again, once more leaving them above oil prices. This created the aphorism that shale oil from kerogen would *always* cost more than conventional oil.

This is rather simplistic, but widely believed. Interestingly, advocates of other technologies, like photovoltaics, do not adopt the same principle for their favored approach, but instead continuously predict that this expensive technology will become viable and soon.

This chapter will attempt to show how potential technologies and fuels can be judged, given limited information availability. The first lesson is to realize that enthusiasm about different technologies is not new and must be treated with caution. Next is to recognize that interested and biased parties abound on all sides of the issues, but especially environmentalists who exercise poor judgment about so-called Cleantech. Then, the question of how much progress can be expected in different areas will be assessed, given certain objective factors. Finally, these rules will be applied to potential new technologies.

The Dark Side of Desire

Before considering the difference between viable and unviable technologies, it must be admitted that sometimes products go beyond not being commercial to actually fraudulent.

The enormous attention that energy has received in the past decade (and earlier in the 1970s) has resulted in any number of alleged new technologies or fuels appearing that go beyond impractical to ethically dubious.

And clean energy has sometimes been a target of frauds, perhaps because it is popular, possibly because so many technically unsophisticated—even gullible—people crave it.

Whenever oil prices soar, gadgets are touted that claim to provide enhanced mileage for cars, accompanied by persistent rumors that the oil industry has suppressed an invention, such as a carburetor, that gets phenomenal gas mileage. The pseudo-documentary *Gashole* describes some of these claims while setting a new standard for gullibility. Some of the inventions presumably are at least minimally effective, but listening to their promoters leaves the sense that, at best, their apparent utility is more due to operator bias and/or measurement error.

Most amazing are the perpetual motion machines that pop up now and again, in a variety of forms, but all supposedly putting out more energy than is put into them. The claims for "cold fusion" by Fleischman and Pons received worldwide attention, but many others are less famous or more accurately infamous. In recent years, claims of cars that run on water or air have appeared repeatedly, except that they prove to be running on compressed air ("Gasoline is already the fuel of the past") and liquid nitrogen or using electrolysis to break down water into hydrogen and oxygen, running off the energy in the latter but, in every case, requiring much more energy than is actually consumed by the engine.[2]

As with financial investments, always question closely anything that seems too good to be true. A tablet converting water to gasoline is essentially impossible, as are perpetual motion machines, and experience suggests that they are virtually all frauds.

ENTHUSIASM AND $4.30 WILL BUY YOU A GRANDE LATTE[3]

Certainly, the briefest survey of energy R&D will find many writers waxing eloquently about a given effort, especially in the field of Cleantech, but also other new technologies such as modular nuclear power plants. Writers and pundits often quote these, as do policy-makers, without seeming to understand that they are highly unreliable. Few are analytical in nature, and myths and factoids are endlessly cited.

Trust Me, I'm from Academia/Silicon Valley

Once at MIT, some industry executives complained that the researchers speaking to them were all just promoting their own work, rather than providing an overview of the field. One professor responded that, as the head of a multidisciplinary research lab, he was by definition unbiased, and therefore his judgment that his work was of prime importance could be trusted. The audience was unimpressed, and his work remains in the laboratory two decades later.

The point is that no inventor or researcher is ever going to admit that his or her work has not yielded usable results—well, almost none. Nor is a bureaucracy that has committed to a given technology or fuel (or weapons system) going to announce that it has wasted its time and (possibly your) money, again except in rare instances. The Sergeant York M247 artillery system stands out in the history of military procurement because it was canceled. Even weapons systems that perform poorly usually get funded. As someone remarked at the time, you don't get promoted in the army for canceling a project.[4]

The interesting thing is that many will be skeptical about pronouncement from oil industry executives, but think that someone from a renewable energy company will give an unbiased view of energy policy and technology choices. And despite the fact that one of Silicon Valley's greatest strengths is the ability to learn from failure, all too often efforts from Silicon Valley are treated as sure things. Elon Musk is a genius, but that doesn't mean he can solve the challenge of low energy density in batteries.

How Green Was My Ideology?

> Friends of the Earth Europe, Greenpeace, and Transport & Environment have therefore jointly commissioned this study to look into how the full potential of electric cars can be realized.
>
> From "Green Power for Electric Cars:
> Development of Policy Recommendations to
> Harvest the Potential of Electric Vehicles," 2009[5]

The "green" industry, particularly renewable energies, is now large enough to have significant political clout, although still less than many other industries. Oddly, however, it has an entire cadre of proponents who are usually tangentially involved in the industry, namely, the environmental community. Few other industries have outside activists who promote their interests, one exception arguably being health care. The Red Cross doesn't lobby for tax breaks for sofas, the League of Women Voters is not known to promote lawn care equipment, but Greenpeace feels no compunction about touting electric vehicles.

Enthusiasm is not always a good indicator of technological viability, although the breadth of enthusiasm can be indicative of potential sales.

But it should always be assumed that the creators or analysts of a given technology will enthusiastically endorse it (waving off all shortcomings as immaterial) and insist that consumers will naturally agree with their judgment of the value of the product or service they are marketing.

Similarly, there are several pundits who are willing to endorse a given fuel or technology simply because they find it appealing in a general way. Jeremy Rifkin, with little or no expertise on energy issues, argued that the country could switch to hydrogen as its primary fuel just as it created the Internet, an absurd comparison.[6] The economic, technical, and environmental challenges did not impress him, but the idea of a fuel that, when burned, yielded only water turned him into a fervent advocate.

Note that my own arguments in favor of increased petroleum use (natural gas and propane especially) can be called into question on this basis, and certainly I would not disagree. No matter how reliable or trustworthy you might judge my opinions, they should be held up to the same scrutiny as any others', the primary caveat being that I do have long experience in energy economics and policy. The difference between skepticism and blind rejection is the difference between good policy making and ideology.

Home and Garden

I always love seeing articles in newspaper sections about wonderful solar installations written by journalists as puff pieces, usually in the "Living" or "Home and Garden" sections of newspapers. These talk about the beauty of the piece or how it fits in its environment, with glowing words from the owner, usually followed by an exhortation for more tax breaks, which is an implicit admission that the product is not really commercial, but more of a status symbol.

Similarly, some magazines such as *Popular Science* are intended for enthusiasts of new products and technologies, rather than providing critical assessments of their merits. *Technology Review*, the official magazine of MIT, sometimes has realistic articles, but long-time readers must be amazed that photovoltaics are not yet the dominant source of electricity, given the hundreds of advances described in its pages over the years.[7]

Survey Says

[T]he potential market for EVs among hybrid households will be no less than 7 percent of the new light-duty vehicle market. We believe therefore, there is sufficient household consumer interest in EVs to satisfy the mandated 2 percent level of sales of zero emission vehicles (ZEVs) in the year 1998 as well as the 5 percent level in 2001 given current EV technologies.

Turrentine and Kurani, 1995[8]

Surveys are notoriously malleable by those with a vested interest in the results. My favorite came during the Cold War when I received a survey from a conservative group that asked my opinion on a proposed arms control agreement with the Soviet Union. If I opposed it, I was to return the survey with a small U.S. flag tucked into the envelope; in favor, they asked that I return a Soviet flag. No doubt they raised some money from a few diehards, but the transparency was pretty laughable.

Similarly, electric car advocates often point to surveys in which consumers "express interest" in electric vehicles. Notice the use of the word "interest" and consider what it really means. I'm interested in the Boeing Dreamliner but have no rational expectation of buying one.

In "Who Killed the Electric Car?" this point comes through clearly, as proponents insist that consumer interest was very high, yet automakers such as General Motors claimed that this did not translate into intended purchases. The research cited above claimed there was sufficient interest to provide a 5% market share in California by 2001. Now, a decade and a half after that, the actual market share is about 5 out of every 1,000 registered vehicles, meaning it took the enormous technical advances of the last two decades and heavy government subsidies, and the 5% market share (of sales) has roughly been reached.

Can't Always Get What You Want (or Shouldn't)

Historically, there have been many efforts to explain why consumers behave the way they do, and a subset of this is the attitude "why don't people buy what I think they should buy?" Many argue against the consumption of material goods as having a hint of immorality, based (consciously or otherwise) on various religious ideologies that urge a concentration of spiritual improvement. A typical comment is "you don't need x," which assumes, first, that all consumption is driven by needs; second, that the author can determine the needs of others; and third, that average needs are identical to total needs.

In electric vehicles, this last assumption takes the form of noting that most car trips are short and most commuters don't travel very far in a given day. These are both true, but seriously understate the importance to consumers of having the freedom to travel when and where they want. One BMW executive recently admitted, "What we learned is people want to use their cars beyond a single charge range; they want to take longer trips, to drive like they were always used to driving."[9]

It also represents a misunderstanding of the cost/benefit analysis that consumers make. Consumers in huge cities like New York and Tokyo, where commuting is difficult and parking horribly expensive, choose to commute by subway or train in huge numbers. In other cities like Boston, there is still a much larger proportion who prefer to use cars. They don't *need* to, they *want* to, given what they perceive to be the value and the costs (in time especially) of driving.

The word "perceived" is also an operative one. An electric car advocate is likely to have a different perception than the average consumer of the value of the flexibility that a gasoline car with its greatly superior range offers. Given the intangible nature of flexibility, people can make nearly any claims they want. There is a test, however: go to the nearest main street and see how many large vehicles (SUVs, pickups, etc.) there are. In most places in the United States, compact cars are still a fraction of the market and you almost certainly can point to acquaintances with more power or size than they "need" in a vehicle.

Proponents of technologies that fail often blame advertising for the behavior of consumers and the desire for material consumption. Again, the presumption that people should not want things or that one person's tastes and desires should be universal is used to explain consumers' "bad" tastes. In "Who Killed the Electric Car?" the argument was that GM didn't use sexy models in its ads for the EV1; otherwise they would have been more successful.

This is more or less nonsense and is just rationalization for the failure to understand consumer desires and a new product's ability to meet them. The development of the compact fluorescent light bulb (CFL) is illustrative. They are much more efficient than incandescent bulbs but were slow to catch on with consumers. The reality is that they were initially quite expensive, even when subsidized; had poor-quality light; and, despite promises of long lives, tended to fail frequently. Extensive promotion by utilities and large rebates were needed to build a market, and even though they were claimed to deliver a huge economic benefit for consumers, they struggled to find a market.

The idea that advertising has more than a passing influence on consumers also contradicts the many who would never accept the concept that heavy metal music or video games cause violence. It almost descends to the level of predestination, the religious philosophy that we have no free will, or the Cold War concept of brainwashing, which proved to be unsupported empirically.

Lack of information should also raise one's antennae. Companies that tout innovations but are disinclined to provide detailed information usually have a good reason for doing so. Arguably, Toyota is the exception to this rule, as they were largely silent on the cost of hybrid electric vehicles, yet these have proved a great success.

A Laboratory Is Not the Real World

Most of the misplaced enthusiasm for new technologies reflects a failure to recognize that laboratory success is far from commercial viability. A process that works beautifully once might succumb quickly on multiple uses or when exposed to real-world conditions. Steve Levine, in his excellent book *The Powerhouse*, describes how advances in battery technology that appeared revolutionary proved inadequate upon extensive testing.[10] Any number of exotic technologies, such as spray-on paint with photovoltaic cells embedded in them, have been trumpeted based on laboratory results, but commercialization continues to elude them. In many such cases, it is the media, not the researchers, who downplay the challenges.

Crossing an Energy Rubicon: Cost Competitiveness

A major element of the misguided support for Cleantech is the insistence that, as renewable energy costs drop, fossil fuel costs will rise, meaning the crossover point, where renewables become competitive, will occur much earlier than if they were flat or declined.

That prices for conventional energy must rise is a truism that is repeated consistently, but is based on the presumption that depletion will raise costs more than advances lower them. This has been demonstrated repeatedly to be untrue on an empirical basis, starting with Barnett and Morse's 1965 examination of historical mineral price data and continuing to the present day.[11]

Making the simple change of assuming that prices for conventional electricity remain flat (line B in Figure 23.1) results in the crossover point of competitiveness for solar moving out nearly two decades, to 2044, far enough out to suggest the forecast is perhaps unreliable.

But it would be a mistake to think that the lesson of Chapter 3 and the "irrational exuberance" for new technologies in the 1970s mean that they are all worthless or will not become competitive any time in the near future.

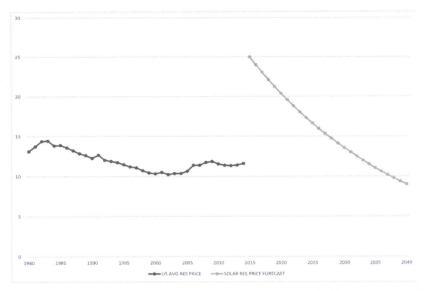

Figure 23.1   Photovoltaics Cost and Electricity Price (2010 cents per kwh) (EIA)

Again, the hybrid gas-electric vehicle, pioneered by Toyota, has sold millions and has now become widely adopted. When first announced, I was skeptical of the economics, in part because cost information was kept secret, but in many parts of the world where gasoline is very expensive, such as Europe, the vehicle's viability is clear. Conceivably, all new vehicles will have this type of power system at some point in the future. Similarly, photovoltaic costs have come down enormously, even if they are still primarily attractive in niche applications.

Thinking that environmental scientists are incapable of serious consideration of new technologies would also be a mistake. Indeed, some of the more experienced demonstrate skepticism, arguing that hydrogen energy is far from being economic or that electric cars are not yet viable (at least in 1979), and argue for more research to make them competitive.[12]

It is important to realize that, as with oil shale, renewables are not condemned to eternal uncompetitiveness just because they are now uncompetitive. Nor do the many failed claims prove that all such declarations are or will be false. As is the case with peak oil, the proof is in the pudding.

Specifically, wind power appears to be commercially viable in many (but not all) instances, although it still receives significant government support. Table 23.1 compares the cost of various types of generating

Table 23.1 Cost of Electricity Production from Different Sources (cents/kwh) (Energy Information Administration, Annual Energy Outlook. Washington, D.C.: Department of Energy, 2015. In: http://www.eia.gov/forecasts/aeo/electricity_generation.cfm)

Plant Type	Capacity Factor (%)	Levelized Capital Cost	Fixed O&M	Variable O&M (Including Fuel)	Transmission Investment	Total System LCOE
Conventional coal	85	6.04	0.42	2.94	0.12	9.51
Advanced coal	85	7.69	0.69	3.07	0.12	11.57
Advanced coal with CCS	85	9.73	0.98	3.61	0.12	14.44
Natural gas–fired						
Conventional combined cycle	87	1.44	0.17	5.78	0.12	7.52
Advanced combined cycle	87	1.59	2	5.36	0.12	7.26
Advanced CC with CCS	87	3.01	4.2	6.47	0.12	10.02
Conventional combustion turbine	30	4.07	0.28	9.46	0.35	14.15
Advanced combustion turbine	30	2.78	0.27	7.96	0.35	11.35
Advanced nuclear	90	7.01	1.18	1.22	0.11	9.52
Geothermal	92	3.41	1.23	0	0.14	4.78
Biomass	83	4.71	1.45	3.76	0.12	10.05
Nondispatchable technologies						
Wind	36	5.77	1.28	0	0.31	7.36
Wind—Offshore	38	16.86	2.25	0	0.58	19.69
Solar PV[3]	25	10.98	1.14	0	0.41	12.53
Solar thermal	20	19.16	4.21	0	0.6	23.97
Hydroelectric[4]	54	7.07	3.9	7	2	8.35

electricity, and clearly the primary success story is wind. Solar power is usually far too expensive, and only heavy subsidies or guarantees of prices far above market levels can entice investment.

EPIC FAILS AND QUIET SUCCESSES

A quick review of some of the fuels and technologies that have been promoted over the years allows generation of some basic rules. In Chapter 3, several major initiatives, public and private, were discussed including the fast breeder reactor, kerogen retorting (oil shales), and photovoltaics.

Battery Electric Vehicles

> The EV1 was a marvel of engineering, absolutely the best electric vehicle anyone had ever seen. Built by GM to comply with California's zero-emissions-vehicle mandate, the EV1 was quick, fun, and reliable . . . . In fact, battery technology at the time was nowhere near ready to replace the piston-powered engine.
>
> *Time*, "The 50 Worst Cars of All Time"[13]

Automobiles, perhaps because of their centrality to modern life, perhaps because of the petroleum they consume, are the constant target of inventors and innovators, but with minimal success. Electric vehicles, which technically existed a century and a half ago, were displaced by the internal combustion engine in the early part of the 20th century, yet continue to have an almost visceral appeal. In the past several decades, companies from Fisker to Tesla have thought they had solved the barriers to a mass market, but progress remains insufficient to make them competitive with conventional vehicles.

Electric engines are much simpler and cleaner than internal combustion engines and therein lies the attraction. The problem lies in the low energy density of batteries, their cost, and the difficulty of recharging them. Proponents have tended to play these down, arguing that conservatism in the automobile industry is the biggest obstacle.

The fact that the electric vehicle has been touted for a century as a superior technology and "just around the corner" for decades has not made proponents any more skeptical.[14] Few question claims that problems will be solved soon, such as the 1997 observation that "In the near future, however, a new type of battery called a zinc-air battery should be available which is reputed to double the range of an electric vehicle for the same weight."[15]

Also, consumer preference is brushed aside with assertions such as most drivers don't go more than 40 miles a day. That's like selling an oven that won't cook a turkey or roast because most people don't make them very often.

Similarly, instead of talking about the difficulty of recharging the vehicles, proponents hail the expansion of the network of recharging stations. However, not only is the number of stations still a fraction of gasoline stations, but even with a fast charging station, it can take five times as long to "fill up" as for a conventional vehicle.[16]

Finally, the fact that the batteries are not a *source* of energy, but a storage mechanism, greatly reduces the environmental benefits from a "zero emission vehicle." Indeed, many critics call it a "remote emissions vehicle," because most rely on the local power system to recharge, and the United States still gets about 60% of its power from fossil fuels (coal and natural gas).

This means that the subsidies for electric vehicles are an extremely inefficient way to reduce greenhouse gas emissions. Some argue that installing residential solar power allows them to charge with zero-emission electricity, but it would still be better to use the renewable power to displace coal-based generation.

It is true that the technology has advanced substantially in the two decades since the California mandate was rescinded, killing the electric car for a time, but the industry remains seriously dependent on subsidies, including, in the United States, a $7,500 per-car tax rebate. However, that applies only to the first 200,000 vehicles a manufacturer sells in the United States, so the next couple of years should see the primary sellers begin to lose their subsidy, even if politicians don't remove it before then (i.e., Republicans control the White House and/or Congress after 2016). For its part, China has already announced a faster phase-out of its subsidies; since it was considered a major success story for the industry, this is suggestive of trouble ahead.[17]

The repeated failures of electric vehicle technology to either succeed or live up to expectations provide one lesson for those attempting to understand the factors that predict success of a new technology, but primarily concerning the way enthusiasts have glossed over the challenges and shortcomings of each. In particular, as the discussion of "chemistry not physics" below describes, the challenge of improving battery technology remains daunting.

At present, Nissan and Tesla are the primary electric vehicle manufacturers and Elon Musk's Tesla has many ardent disciples. However, Tesla's announced goal of 500,000 electric vehicle sales by 2020 appears unrealistic and seems to rely heavily on a belief that a larger factory will solve the problem of battery costs. The skeptics include this author.

Hypercar

> Andrews: Given the above, I have a serious question for the RMI Hypercar Brain Trust: How many hybrids and fuel-cell vehicles do you anticipate in 2010? (Round numbers of millions . . .)
> Lovins: Collectively these could well have a market share around half—more like two thirds if one is optimistic. . . .[18]

The case of hypercar is also illustrative. Amory Lovins, idolized for his promotion of energy efficiency in the 1970s, has long been promoting a redesign of the automobile that would provide extremely high mileage, as much as 200 mpg in the long run, although an initial target is 90 mpg. Still, that would roughly double the efficiency of even the most efficient cars now available. By the use of new materials, such as carbon-fiber composites, as well as better aerodynamics, a hybrid electric engine and a variety of more modest improvements, Lovins argued that the hypercar would be so efficient that it would dominate auto sales in a short period. None has been produced to date.

Several mistakes are involved here, aside from blind enthusiasm. Lovins originally argued that the hypercar would cost roughly the same as a conventional vehicle, but this was apparently not based on extensive analysis, and the current version would seem to have a six-figure price tag.[19] For example, Lovins considers the BMW i3 a hypercar based on the fact that it's electric and has a carbon fiber body.[20] The base price for that is about $45,000 for a very small car.[21] Volkswagon's XL1, which has also been described as a hypercar, costs $150,000—but it gets 100 mpg (diesel).[22] Costs would be lower for a production line model, but it's still not clear how saving less than $1,000 a year on fuel can justify spending $100,000. Or anything over $10,000.

In this instance, and his work more generally, Lovins has almost completely ignored economics, treating price and cost as nearly irrelevant compared to energy efficiency. His technological optimism is to be lauded, but his disinterest in realism has made much of his work of little value.

Hydrogen Vehicles

One day you may be able to drive your fuel cell car during the day, then connect the car's engine to your house to provide heat and electricity at night. Alternatively, electricity generated by the engine could be fed to the grid in return for credit. Thanks to the fuel cell engine's efficiency and reliability, an asset that usually sits idle for all but an hour or two a day could become a steady income earner.[23]

Naturally, most industries wish to create buzz around their operations and/or products, and quite a few are successful, but because clean technologies are perceived as being beneficial to society, they seem to get an extra fillip of publicity and receive less objective scrutiny than many others.

A case in point is the hydrogen fuel cell. Although fuel cells are not a new technology, having been invented in the 1830s, the expense remains exorbitant for most applications. Logically, there are those who continue research in an effort to bring costs down enough to make them mass market products, although without success as of yet.

In an extraordinary example of technological over-enthusiasm, two decades ago, Ballard Power announced a great advance in fuel cell technology, which was widely heralded:

[O]fficials from Daimler-Benz and Ford agreed in Stuttgart to form a new partnership with Ballard Power Systems, a tiny Canadian firm considered one of the leaders in the development of fuel-cell technology. With a combined investment of roughly $1 billion, the new consortium hopes to produce an initial 10,000–50,000 cars a year powered by fuel cells, starting commercially in 2004.[24]

Daimler-Benz increased their sales forecast to 100,000 cars annually by 2005.[25] Needless to say, this target proved completely unrealistic, and Ballard Power's stock dropped by 95% to its pre-excitement level after soaring earlier.

The reality is that hydrogen production, transportation, and storage remain much more expensive than for conventional fuels roughly two decades after the great excitement over Ballard Power's advances. While Toyota is planning to sell an HFCV in 2015, the cost will still be exorbitant and only a handful are expected to be sold this year. What is astonishing is that instead of announcing more intensive research, automobile companies not only invested huge amounts of money 20 years ago but had expectations of production line vehicles that were beyond optimistic, indeed, a poster child for irrational exuberance.

Cellulosic Ethanol

In the mid-1990s, James Woolsey, the former head of the Central Intelligence Agency, spoke at a Harvard conference about the benefits of cellulosic ethanol and the potential for recombinant DNA technology to develop a commercial process to produce it.[26] He is supported by enough other experts that Congress passed the Renewable Fuels Standards was included in the Energy Policy Act of 2005 and mandated the use of specific amounts of ethanol by the refining industry, as well as a small but increasing amount of cellulosic ethanol. The intent was to jumpstart the technology by providing a guaranteed market, however, this has been a failure.

The 2007 Energy Independence and Security Act set specific targets for cellulosic ethanol that would have meant consumption of 3 billion gallons in 2015, or about 1.5% of total biofuel consumption, but they lowered this to 100 million when it was realized the goal was completely unattainable.[27] At this point, it appears that the industry is nowhere near commercialization despite years of research and investment.

Methanol

Methanol is the latest in a line of investigated fuels, and the results of the research on the fuel have been so positive that the State of California is funding a $5 million pilot program to supply municipalities with methanol-fueled vehicles and study the results.

Automotive Fleet, 1982[28]

In the 1970s energy crisis, numerous books and articles proposed that the United States switch to methanol as an engine fuel (among other things). The basic case was that methanol could be produced from biomass like wood, coal, or natural gas, and used as a blending agent in transportation fuels, as a fuel on its own, or as a petrochemical feedstock. Indeed, there was already an existing major market in methanol, primarily in the petrochemical industry.

The closest methanol-as-vehicle fuel came to existence was the methane-to-gasoline project that Mobil launched in New Zealand after the 1979 Iranian oil crisis. The major impetus was the discovery of a huge natural gas field offshore New Zealand which encouraged the government to push for energy independence. Given that oil prices were assumed to rise with time, the project looked extremely attractive. So much so that the government bought the completed plant from Mobil rather than let it have the expected profits. Unfortunately, shortly thereafter, oil prices peaked and began their lengthy decline and the government, after absorbing major financial losses, sold the plant to Methanex, which used it primarily as a methanol producer.[29]

The problem: methanol is generally more expensive than conventional petroleum products, and the cost of setting up an entire distribution network and modifying engines to use it has been prohibitive. Since the environmental benefits are minimal and the only benefit is to lower oil imports, there has been almost no interest outside of the chemical engineering community to exploit it.

Fusion

Hot fusion is in a category of its own, though, becoming a joke to some people because for four decades, it has been described as being thirty years away. Many think that this means it simply can't be done, but "thirty years away" is really the same as saying "too far away to predict." That doesn't mean that current predictions of becoming viable in thirty years are reliable, but suggest skepticism should be moderated by its own skepticism. Quite possibly, fusion will not be commercialized in my lifetime and sometime beyond, but then, they are making marvelous advances in medicine these days.

Now several groups have announced progress with a new approach besides the traditional massive magnetic containment methods, claiming to have a compact design that can fit in a truck because its magnetic bottle requires one-tenth the power of traditional designs. Lockheed-Martin is supposed to be planning a test this year, and have a prototype within another five years, which implies great confidence in their design.[30]

Also, MIT scientists have said that the use of superconductors can create a much smaller, more powerful magnetic field that will allow for a 10-fold increase in power.[31]

Receiving less attention was the University of Washington's "dynomak" design, which the researchers say could be competitive with coal power plants. It also appears to rely on a redesigned magnetic containment field and an alternative proton-boron fusion.[32]

This all sounds interesting, but does leave me with a feeling of déjà vu. In the early 1980s, Phillips Petroleum invested in an unconventional design for small-scale fusion, without success. And in the late 1990s, at Rice University, a physicist argued that it was possible to build a room-sized fusion reactor using Boron atoms. (The room was full of oil industry executives and none of us had a clue as to whether he was a crank or a genius.) The fact that neither of them panned out could be due to promoters' excessive optimism, or to the challenges involved—which require a long-term development effort that few can manage. The failure of these to produce working models doesn't tell us much about Lockheed Martin's, MIT's, or UW's efforts, but definitely is a reminder to take announced breakthroughs with a grain of salt. Still, given advances in materials and software over the past four decades, this is well worth monitoring.

Cold fusion, or low energy nuclear reaction (LENR), as it is more formally known has its origins with work by Pons and Fleischmann in 1989, which was subsequently discredited. Numerous other groups have attempted to replicate their results, but without success. That is, some achieve excess heat, but not in a reproducible way, and the mechanism by which it occurs has not been explained. The mainstream scientific community generally considers this to be a case of "pathological" science and journals routinely refuse to accept papers on the subject.

The likelihood that anything will come of LENR seems extremely small, at least until someone can explain the process by which the excess heat seems to be generated. At present, it appears that bad experiment design, misreading of instruments, and other problems dominate the reported results. Those who claim success, like Rossi with his E-CAT, have been unable to produce results that satisfy unbiased observers. (Randall Mills, at Blacklight Power, has the same problem, but claims that, despite the many similarities with LENR devices, his device does not rely on fusion.) Unfortunately, because LENR confounds "conventional science" this field has attracted many who have more enthusiasm than judgment, which both hurts the reputation of reliable researchers, but also opens the way for the shadier element to exploit them.

So, the Mr. Fusion used by Christopher Lloyd's Dr. Emmett Brown in "Back to the Future" should be remembered as a funny movie bit. Small-scale fusion reactors might be possible (the Lockheed Martin version), and would certainly have an enormous impact, possibly being the biggest contribution to reducing climate change.

PROGRESS, BUT NOT COMPLETE SUCCESS

A technology's previous failure to become viable, and especially to meet the lofty goals which its passionate proponents described, should not blind anyone to the progress that has been made.

Photovoltaics, especially residential solar, fits into this category. Environmentalists often laud the success of solar energy, especially in countries like Germany and Spain, while glossing over the huge expense involved in achieving that so-called success. Costs for PV panels have dropped sharply, and modular installation has lowered the costs of attaching them to household rooftops, but without massive subsidies, covering one-third to one-half the cost of the units, the market would be relatively minor. When Spain stopped paying exorbitant prices for solar electricity, that market disappeared overnight.

Electric vehicles have also seen progress, though they are much further from commercialization in most uses than photovoltaics. Sales still represent a tiny fraction of vehicle purchases, even with tax credits and subsidies of $7,500–10,000 in various parts of the United States. In nations like Norway, with very cheap hydroelectric power, they have achieved more, but global electric vehicle sales were still only 320,000 in 2014 compared to total of 88.5 million.[33]

WINNERS

My favorite "green" success story is low flow toilets. This was a case where the government mandated toilets that required less water per flush, which outraged conservatives and led to many complaints of inadequate clearance. The animated series "King of the Hill" attributed the local regulations to corruption, but this was a clear case of environmental benefit. In fact, I have a late generation low flow toilet which was simply redesigned to increase the water pressure. It functions extremely well.

Windpower is another example of a technology that has come of age. Despite the fact that windmills are millennia old, the technology was largely ignored in modern times until the energy crises of the 1970s. Unfortunately, too many installed windmills (or turbines) prematurely, but now, high-tech materials and advanced designs has made wind competitive in many places around the world. This reflects, in part, the fact that wind is really concentrated solar power, especially in locations like mountain passes or the North Sea, although offshore U.S. windpower appears to be very expensive.

LED lights appear set to be the next big winner, delivering high-quality light (compared to compact fluorescents) and at very high efficiency. Costs have come down to the point where they are attractive in a large number of uses, from public lighting to reading lamps. The invention of the high-brightness LED in 1994 (for which three Japanese researchers won the Nobel) is a good lesson about technology choice: governments had long promoted CFLs as the next generation of lighting, but this now appears to be a technological dead end.

Probably the best illustration of how to develop and market a new technology is the hybrid-electric vehicle. Because internal combustion engines operate best at a constant rate, providing an auxiliary engine to smooth out the peaks and valleys of power demand improves engine efficiency. After decades of work, companies like Honda and Toyota improved both the electronics controlling the system, the batteries, and the power trains so that a hybrid-electric vehicle delivers improved efficiency at a small price. In many parts of the world, such as Europe, where fuel prices are high, the extra cost is easily returned by the fuel savings. And even in markets like the United States, where gasoline is often cheap, the extra cost is not significant.

SOME GUIDELINES FOR EVALUATING TECHNOLOGIES

The first lesson is always, "Trust no one." This is especially true of those promoting the technology, whether they are the developer, an NGO with a vested interest, or even an oil company (presumably criticizing same). Science and technology magazines tend to be too gung-ho about anything new, and first adopters are not always an accurate sign of future market success.

An aversion to data is usually a bad sign. Companies that don't want to discuss the costs of a proposed technology should be avoided. And those that won't permit independent evaluation of a miraculous product, like Blacklight Power, should not be taken seriously.

Rationalizations usually mean that shortcomings cannot be overcome. Arguing that people don't usually drive very far is an implicit admission that a car's range is limited (electric vehicles mostly).

Economics are very important, although not everything. Promoters who tend to downplay costs or include subjective elements like externalities are all but admitting that their product is not economically attractive.

Promises are nice, but reality is what counts. From predictions that hydrogen fuel cell vehicle sales would reach six figures by 2005, to statements two decades ago that lithium-air batteries would be available soon, many fail to realize that from the laboratory to the showroom is a difficult and unpredictable journey that many fail to make.

There are some ways to judge the potential for new technologies to become viable within the near term of 5–10 years, which primarily involves what the cost is and the likely trend in costs. Far too many proponents will point to rapidly falling solar costs while not mentioning that they are falling from a high level, for example. And as mentioned, it is important to remember that all energy costs should decline over the long term, so that gradually falling costs for a particular technology should not be treated as revolutionary.

Technological Maturity and Breakthroughs

> There is no reason to expect any material improvement in the breed of horses in the future, while in my judgment, the man is not living who knows what the breed of locomotive is to command.
>
> Horatio Allen, early 19th century[34]

Costs usually drop much more in the early phases of development of a technology, as Figure 23.2, which shows photovoltaic panel costs, demonstrates. Granted the initial exorbitant costs reflected the use of the panels in satellites, but even so, from the 1970s when interest in use for commercial power production began, they have come down sharply, although much less so in recent years.

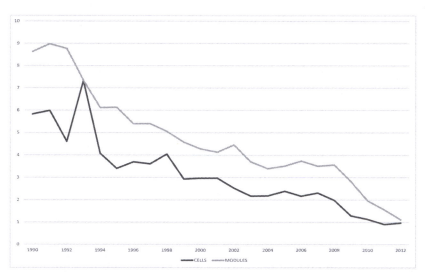

Figure 23.2   Price of Photovoltaic Panels ($/watt) (EIA)

Technically, of course, the photovoltaic cell is not "new" having been first developed by Alexandre Becquerel in 1839, but the amount of research aimed at developing them as a commercial product was minimal for long after that, especially as electricity prices were dropping sharply in the first three-quarters of the 20th century.

And the same can be said of wind: although the Romans had early windmills and they were quite common in parts of Europe centuries ago, no major effort to develop them into a mass market product began until the oil crises of the 1970s. The windmills or turbines deployed in the 1970s were, thus, relatively primitive compared to those made today, reflecting a combination of better design, improved materials, and control systems which allow the power to be integrated much easier.

Conversely, hydraulic fracturing of shales involves technology that is mostly decades old, but the application is being done in a new setting (shale), so that rapid advances in cost and productivity are still occurring (Chapter 19), although this should slow with time as the technology is optimized.

A subset of this is not the absolute age of the technology, but the maturity of the research effort. Rapid advances and breakthroughs are much more likely where little research has been done, as in the case of photovoltaics in the 1970s/1980s. Conversely, fusion power has been studied for several decades now, and while progress can be expected, it is unlikely that a sudden breakthrough will appear, where a researcher finds something long overlooked.

This justifies skepticism about the prospects for electric vehicles. Batteries have been a major commercial product for over a century, and used in automobiles for roughly the same length of time. Numerous companies have been trying to make them better—cheaper, more capacity, more recharges—since before any of us were born, and so there is unlikely to be some element of the technology that was not already studied.

The rule then is that rapid progress occurs only in the early years of a technology, although starting from the point at which a serious effort is made. Technologies that have been around for a long time and subject to much research are unlikely to experience sudden rapid progress, short of a breakthrough, that is, a new technology such as a different type of battery.

Physics

An inability to distinguish between engineering and scientific challenges, or the degree of a scientific challenge, helps to explain why some expectations have been met and others not. The best example would be the flying car, which numerous pundits have either predicted or complained about its failure to be realized. It could be argued that the flying car does exist, but it's called an airplane.

Of course, the 30+ foot wingspan means the typical small plane will not function on the roads, so what most people consider a flying car is one that has the power to lift from the ground without the assistance of wings. Alternatively, a helicopter that weighs 2,700 pounds requires an engine that produces 420–650 horsepower.[35] A Toyota Camry, by comparison, weighs 3,200 pounds and has a 174 horsepower engine.[36] Obviously, the two have physical differences, especially the helicopter's speed, but there is no doubt that the energy needed to lift a car from the ground is a fraction of that which is necessary to make it roll on the ground. You can easily push a car, but lifting it is far more difficult.

Chemistry Versus Electronics

The so-called Apollo Guidance Computer (AGC) used a real time operating system, which enabled astronauts to enter simple commands by typing in pairs of nouns and verbs, to control the spacecraft. It was more basic than the electronics in modern toasters that have computer controlled stop/start/defrost buttons. It had approximately 64Kbyte of memory and operated at 0.043MHz.[37]

A major fallacy often seen in energy technology policy is that the potential improvement in any given product or method can be as great as what is seen in the electronics industry. As an auto executive once remarked to me that the first computer put in a car several decades ago occupied the trunk and back seat, but would now be duplicated by a single chip. The implication was that the hydrogen fuel cell would advance at a similar rate.

This has played out quite well in the electric power industry, where electrical control systems have advanced sharply (the "smart grid") while generation has changed very slowly. Similarly, one of the factors behind the development of hybrid-electric vehicles is the better control of the two motors, electric and gasoline.

The many pundits, like Al Gore, who think that progress in computers and cell phones is a good analogy for future progress in photovoltaics, electric vehicle batteries and fuel cells are demonstrating a poor comprehension of science.

Mandate Engineering, Not Science

The use of mandates to spur innovation has been tried a number of times, with mixed success, the primary explanation being that improvements will result from mandates, but scientific advances will not. Perhaps it would be better expressed as incremental change can be forced, but exponential leaps are less amenable to government mandate. Thus, requiring carmakers to improve efficiency by 3 or 5% per year for a period simply directs engineers to focus on making a series of changes, which they themselves choose.

On the other hand, asking that an entirely new battery type be developed that would vastly improve electric vehicle performance has clearly failed. Much progress has been made, but the need for a revolutionary advance in energy density, cost, and recharging ability is something that requires entirely new technologies, which cannot be dictated.

This explains why programs like the 1990s mandates for electric vehicle sales were such a disappointment. Despite the various conspiracy theories, the fact that even now the technology requires extensive government assistance proves that the electric car died in the 1990s because the technology was not commercially viable. And while proponents of these mandates, including the current zero emission vehicle targets in California, argue that they produce some progress, they ignore the costs to both manufacturers and their customers, as well as the taxpayers.

Unfortunately, many Cleantech advocates think that mandates represent success for the development of their technologies or fuels, when they should treat them as success for their lobbyists. The CEO of one biofuels company described the Renewable Fuels Standard as a success, at the same time opining that without stronger government demands on the oil and auto industry, "it has chilled the outlook for us."[38]

LESSONS FROM A HISTORY OF SUCCESS AND FAILURE

We learn from failure, not from success!

Van Helsing, *Dracula*[39]

There are very clear factors that influence whether or not a technology succeeds and that is primarily the cost-benefit calculation. Costs are highly uncertain for technologies still in the laboratory, and benefits are easy to exaggerate. What often fools lay people is the failure to realize that many things are technically feasible but not economically practical. Worse, promoters often overlook or rationalize away this fact, and this includes not just the inventors and manufacturers but, in the case of energy, a large environmental community.

And the difference between, loosely defined, the scientific and engineering challenges is enormous, but waved aside by many, as in the constant insistence that car batteries can improve the way cell phones have. A low-flow toilet was simple to design because it's not overcoming gravity, but actually utilizing it, while a flying car must do the former.

Too many technologies failed because of the presumption that competition will lessen, not just that other, conventional technologies won't also improve, but due to misinterpreting temporary market cycles as new market realities. This was the case with Solyndra, whose product only looked appealing because the boom in photovoltaic installations had briefly raised the prices of photovoltaic panels.

The three biggest technological successes in energy in the past several decades are gas turbines for power generation, hydraulic fracturing of shales, and the LED light "bulb." Mandates for renewable use (including ethanol) have not improved the technology, only hidden the costs from the public, and government programs have produced relatively little progress outside of basic research. The current period of falling conventional energy prices is all but certain to see much of the recent efforts towards a new energy system abandoned, just as in the late 1980s.

THE FUTURE

A host of writers have opined on the fact that the predicted flying cars have not appeared, and yet, there have been any number of successful visions of future technology, from Isaac Asimov's vision of hand calculators to Arthur C. Clarke's description of geostationary satellites. What unites the successful predictions is a combination of understanding technological progress and recognizing what is physically achievable. Time machines and anti-gravity might someday be available, but the path by which they might be developed is not visible. Submarines, on the other hand, presented no major scientific challenges when Jules Verne was writing about them.

In the distant future, it seems highly likely that the world will rely on some variant of emission-free electricity generation, probably some mix of photovoltaic and nuclear (fission and/or fusion). But for the next decade or two, natural gas turbines and nuclear will be the major source of new power, partly replacing coal plants, but also extending electricity supply to gas-rich and power-poor areas in Asia, Africa and South America. Improved controls and software (the "smart grid") should also lower power consumption, and electrostatic batteries might help balance fluctuations in power demand and supply.

For automobiles, the next decade or two will see a growing dominance of hybrid-electric vehicles for personal transportation, but it seems likely that full battery electric vehicles will be mostly confined to fleet use. That said, urban centers of any size will probably have automated electric vehicles available on demand, which will reduce the usage of private cars in areas like New York and Tokyo (and greatly improve the quality of life except for cab drivers). A phone app will allow customers to call a vehicle from a nearby charging station, with nearly the convenience of a private car and *no parking worries.*

During that same time period, natural gas should increase its presence in both ocean and ground transport, the former quickly and in the form of liquefied natural gas. Both the economics and the environmental imperative favor it already. Fleet vehicles will see some switching to natural gas (a mix of compressed and liquefied), but cheap oil will continue to be play a major role, as in air transport.

In the very long term, a transition to either a fully electrical transportation system or fuel cells should be expected, depending in large part on the relative progress in battery and fuel cell technology. Fusion and solar could prove to be the primary sources of energy for the globe, while petroleum becomes relegated to feedstock for the chemical industry. The enormous methane hydrate resource could also become economically viable, which would rapidly displace coal use in Asia.

Climate change remains a concern, but given that carbon capture and sequestration are still in early stages of research, the potential for improvement in what is now a very expensive solution should not be ignored. Granted, the extraction of carbon from the atmosphere, so easy for plants, appears to be a significant challenge, and a breakthrough is hard to predict. Geo-engineering, such as spraying sulfur dioxide into the upper atmosphere, will almost certainly provide some assistance, despite the moral objections that dominate the current debate.

But ultimately, the obsession with the silver bullet, lone scientist defeats aliens, approach to global challenges needs to be discarded. A wide variety of policies and technologies will be needed, and will probably deliver success.

TWENTY FOUR

Conclusions

Everything suggests to me that the world is very likely to experience a new economic boom, with low energy prices being a primary factor. Except for increased emissions, a problem that can be mitigated, this should be beneficial for the world as a whole. Higher economic growth will hopefully mean stronger employment, falling poverty levels, and less income inequality. Cheaper conventional energy, while bad for some Cleantech industries, will reduce deforestation and indoor pollution and improve the health and lives of the energy impoverished.

But this happens only if policy-makers behave responsibly, which they often fail to do. Understandably, self-interest guides some decisions, such as the subsidization of petroleum product prices, but bad research can send governments (and other organizations) down the wrong path, whether motivating or merely providing a rationalization.

Despite all the progress in economics over the past century, the public—and the expert community—is still prone to believing things that are simply wrong. Constant references to the inevitability of ever-rising fossil fuel prices and the depletion of scarce resources have obscured real concerns, such as environmental degradation, including from some types of renewable resources. Huge amounts of money have been diverted to projects and technologies that were unwise, usually by preselecting a specific approach and then pushing it against market forces and consumer sentiments.

Many have argued we should take note of arguments such as those of the peak oil advocates, since conservatism in policy making has many benefits, or that the precautionary principle suggests taking action in the face of uncertainty. Some, like Matthew Simmons, talk about the need for a "Plan B" to deal with a potential threat.

But this ignores the very real scarcity of financial resources. There is no government in the world that thinks it has enough money (Norway a probable exception), and few organizations that wouldn't like bigger budgets. Spending money foolishly means ignoring real priorities, of which there are many, locally and globally.

Aside from the many billions that have been wasted because of a belief in petroleum scarcity, the global environment has also suffered. The United States was in the forefront of encouraging substitution of coal for oil in power generation under President Carter in the late 1970s, primarily because of the mistaken belief in natural gas scarcity, which was very obviously due to price controls. Similarly, denying the energy-impoverished access to conventional energy and encouraging them to wait to develop expensive alternatives when they have the finances is rather like telling a starving person they can't have a hamburger because the beef isn't organic.

A big part of the problem stems from the fact that so many writing about peak oil and resource scarcity more generally are not experts. It's almost as if those in a sports bar were able to influence the decisions of baseball team managers. And the mere publication of a book or sometimes just an article is often taken to demonstrate expertise, at least by those unfamiliar with a subject, that is the case for most people on most subjects. Informed discussion on policy issues has become replaced with "He said, she said" types of debates.

Most peak oil advocates remain little known to the general public, although some neo-Malthusians are. Those like Paul Ehrlich and Lester Brown, whose work has been shown to be riddled with errors, maintain many fans and win numerous accolades. Ehrlich especially practices what some would call "pathological science," and yet many who would never believe in Bigfoot treat his work as authoritative.

That is why the crucial point of this book is not about petroleum or energy, but research or, more specifically, evaluating research. Public debate is too often dominated by beliefs that have minimal support from serious research. Genetically modified organisms, vaccines, needle exchanges, the teaching of abstinence-only sex education, and many other policies have clear, well-established principles and results that are often ignored by many, usually with the support of some specious, supposedly scientific research.

Granted, in most cases, the research is seized upon as a rationalization for action that was intended regardless of the science. People who oppose needle exchange usually do so on moral or religious grounds apparently, but want to appear nonideological, so they cherry-pick the Internet for something that can be used to justify their actions. But while it might not be feasible to teach the general public resource economics (or medicine, statistics, etc.), those in policy-making positions should be able to rely on valid research.

REAL RESEARCH

Given the enormous uncertainties about oil price, supply, and demand, many might be inclined to throw their hands up and abandon any effort at prediction. Indeed, some have interpreted my earlier reviews of oil market forecasting as suggesting it couldn't be done, when the real lesson was that bad theories and models will lead to bad results.

This is not to say that anyone can predict the long-term oil market with any precision. To do that, both economic growth and resource politics would need to be forecast with some reliability, and that has so far been impossible. But like the sculptor who carved a statue of an elephant by cutting away all of the stone that didn't look like an elephant, uncertainties in the forecasts can be reduced by minimizing errors. A number of lessons about research and policy studies, both general and specific to energy, can be drawn from this review.

Methodological Errors

Probably the biggest challenge facing any policy analyst or researcher is overcoming their bias, which includes automatically assuming others are guilty of bias when their views counter your own. In any complicated issue like long-term economic behavior, the uncertainty about the many factors involved allows cherry-picking of data and research with ease. Indeed, ignoring contradictory opinions and research is a good indicator of bias and, thus, inadequate research.

Consensus can be misleading, especially where there is clearly a lot of uncertainty about a complex area like economics and long-term oil market behavior. The fact that, again and again, there has been a consensus about long-term oil prices, which then completely changes when the price shifts dramatically, demonstrates the psychological need for herding among forecasters.

Related to this is the tendency to become insular. Just as Bigfoot specialists tend not to participate in zoology conferences, so peak oil experts are primarily seen at specifically peak oil conferences, meaning there is a presumption of the conclusion rather than a critical examination of the facts. Surrounding oneself with yes-men is considered a major leadership failing, yet one that is difficult to overcome, given human nature.

The confusion of statistical analyses with science is also a puzzling mistake. The curve-fitting done by the disciples of Hubbert is essentially the same as economists' work, although they typically disparage that profession. By the selective use of data, some have managed to convince themselves that they have found scientific rules for the behavior of production from oil fields and nations, when what they have done is no better than those who try to predict the stock market by looking at past trends.

Context is something that is often lacking in public policy discussions, and this is certainly true of many discussions about petroleum and energy. Numbers are given without either global or historical context: 5 million barrels a day of capacity needs to be replaced each year in the future due to depletion, but there is no mention of the fact that the industry has been doing that for decades.

The misinterpretation of short-term or transient effects as changes in fundamentals occurs again and again. The Iranian Revolution and the second Gulf War both had obvious effects on prices, but too soon analysts shifted the causality to resource scarcity. When a freeze damages the orange crop, for example, no one thinks it a permanent problem, but if a government nationalizes foreign oil companies' holdings, some act as if the resource had physically vanished. The continued willingness of some economists to argue that economic principles "proved" prices should rise, directly contrary to historical experience, is a sad commentary on the profession.

Mistaking policy decisions for physical constraints has also plagued many analysts. The big Middle Eastern oil producers cut exploration and development sharply when the oil market collapsed in the 1980s, but many think the lack of discoveries equates to a lack of undiscovered resources. And the U.S. natural gas shortage in the 1970s, which many correctly argued was due to price controls, nevertheless led to a policy of restraining production, to the detriment of the industry, consumers, and the environment. Policies might be hard to change, but they do change especially when they are obviously misguided, and industries have lost too much by refusing to consider this possibility.

Misunderstood Resources

The concept of "peak" oil needs to be retired, because the basis is a theory and mathematical calculations that are simply invalid. Most especially, the notion that at some point scarcity will lead to a peak about which nothing can be done and which will create economic and social turmoil has been repeatedly proved quite simply wrong. Resource economics as embodied by practitioners like M. A. Adelman, Richard Gordon, and others has proved valid and is disparaged only by those unfamiliar with them.

Next, the belief that resource or commodity prices must rise over the long term because of depletion or increasing demand should be completely abandoned. Obviously, oil-producing nations, oil companies, and producers of competing technologies as well as most environmentalists prefer a rising price forecast; wishful thinking is a problem hardly unique to energy. More troubling is the rapidity and conviction with which seemingly expert people will seize on arguments that a price cycle actually represents the new norm.

The production of mineral and energy resources, unlike most economic activities, occurs only where they are already located. However, the inability to predict that a given exploratory attempt will be successful should not be extrapolated to assuming no further success. Given the large resource base, global exploration can be expected to have enough success, on average, to provide needed supply assuming adequate prices and access.

Astonishingly, at the core of most neo-Malthusian, and much peak oil, research, and writings, is an assumption that technological progress will slow and cease. This is not only bizarre but counterfactual. New resources are constantly being added to humanity's larder by large numbers of government, academic, and private researchers. In the case of petroleum, George Mitchell's persistence has opened up a treasure trove of oil and gas, but many other advances that are not apparent to the general public are constantly being made. The fact that given scientific or technological advances cannot be specifically predicted does mean it should be assumed that none will occur.

Then, too, there is the assumption of no feedback effects. As Ron Bailey puts it, people are "not reproductive automatons driven by blind instinct." Nor are oil producers and consumers, but neo-Malthusian models tend to be completely static, assuming no change in response to their projected scarcities. Resource economics has long held that higher prices yield conservation, substitution, and new production technologies. This has been observed again and again, yet neo-Malthusians willfully ignore it.

Bringing us to the final mistake, namely, ignoring the impact of prices; many economists began to believe in the late 1970s that oil demand would not respond to higher prices because they hadn't done so, at least not as initially expected. The poor response proved to be more a question of lag time rather than extent. And some still think that supply doesn't respond to price increases, simply because the effect is diluted by taxes and cyclical cost increases. Looking at the closest thing to a free petroleum market, North American natural gas, shows how supply and demand will balance over the long-term, and usually at much lower prices than anticipated by the expert community.

In fact, even if the riddle of shale oil had not been unlocked, there are numerous steps that could have been taken to rebalance oil markets, from opening up new areas for drilling to ending the enormous price subsidies in many oil producing nations.

FINAL THOUGHTS

If patriotism is the last refuge of a scoundrel, perhaps hypocrisy is the first. Some Americans have been called anti-science for not accepting a particular view of climate change, but the same people who apply that misnomer have no trouble refusing to accept the scientific consensus on other issues, from oil supply, to fracking, to vaccines and genetically modified organisms. If even presidential candidates can say they disbelieve in evolution, or think vaccines cause autism, despite the huge amount of research refuting such ideas, how can the general public be expected to do better?

Of course, one answer is real science or at least real research. Analysis and open debate should provide guidance to policy-makers and the public alike, not religion or ideologies, which implies that better education is the second part of the solution. Demystifying physics, chemistry, and biology can go a long way toward helping people rely less on emotions and more on understanding.

But as cases like peak oil demonstrates so clearly, the biggest problem remains the embrace of bias by all too many people, who fervently believe things that are obviously wrong and act on those beliefs. The damage done by the peak oil movement is significant, although it pales in comparison to the opponents of vaccines and GMOs.

Education and information might not change many minds. As Max Planck noted, a scientific truth does not triumph by convincing its opponents and making them see the light, but rather because its opponents eventually die and a new generation grows up that is familiar with it. Thomas Kuhn, in his *The Structure of Scientific Revolutions*, makes a similar argument, but both essentially agree that ultimately, good theories will be validated.

However, that requires more than the superficial attention prevalent in so many debates. Many issues are complex, and frequently seemingly logical theories prove to be just as false as obviously foolish ones do. Only serious effort can resolve the uncertainties involved. Socratic debate is ancient, but rarely employed in today's society and we haven't moved much beyond the split between his successors. Plato believed that there was an inherent truth in numbers, whereas Aristotle insisted that the nature of things much be studied. In many ways, Plato could be considered a curve-fitter, much like some of the peak oil theorists, who think that mere observation of a trend is science. As for me, I will side with Aristotle.

Notes

CHAPTER 1

1. *ASPO Newsletter* 2, August 2006, published by Association for the Study of Peak Oil.

2. Jeff Rubin, *Why Your World Is about to Get a Whole Lot Smaller: Peak Oil and the End of Globalization* (New York: Random House, 2009), p. 218.

3. Daniel Yergin, *The Prize: The Epic Quest for Oil, Money and Power* (New York: Simon and Shuster, 1991), p. 780.

4. Inhttp://www.taxfoundation.org/news/show/1168.html.

5. This writing, March 2010.

6. At least one of this author's works criticizing peak oil theory is available on the Website, but a link is not displayed on the home page.

7. Jeffrey Ball, "As Prices Soar, Doomsayers Provoke Debate on Oil's Future." *Wall Street Journal,* September 21, 2004, p. A1.

8. The preceding three quotes are from Ed Porter, "Are We Running Out of Oil?" American Petroleum Institute, 1995.

9. *The Prize,* p. 52.

10. Leonardo Maugeri, *The Age of Oil:The Mythology, History, and Future of the World's Most Controversial Resource* (Westport, CT: Praeger, 2006).

11. M. King Hubbert, "Nuclear Energy and the Fossil Fuels" (Houston, TX: Shell Development Company Publication 95, June 1956).

12. Petroconsultants has since been acquired by IHS Energy.

13. The presentation can be seen at http://www.opec.org/opec_web/en/ multimedia/videoDetail.htm#1710171805001,1673236007001. The moderator's comment, and criticism from other panel members, can be seen starting at 1:03.

CHAPTER 2

1. An in-depth discussion of the evolution of thinking about long-term oil markets in the 1970s can be found in Michael C. Lynch, "Bias and Theoretical Error in Long-Term Oil Market Forecasting," in *Advances in the Economics of Energy and Natural Resources,* ed. John R. Moroney (Greenwich, CT: JAI Press, 1994).

2. George P. Shultz, *The Oil Import Question: A Report on the Relationship of Oil Imports to the National Security* (Washington, DC: Government Printing Office, February 1970).

3. M. A. Adelman, "A Long-term Oil Price Forecast," *Journal of Petroleum Technology* (1969) vol. 12, no. 22, 15151520.

4. The first Arab oil embargo occurred in 1967 and was so unsuccessful that it went largely unnoticed by the public at large.

5. John Maynard Keynes, *The General Theory of Employment, Interest and Money* (San Diego: Harcourt, Brace and World, 1965).

6. M. A. Adelman, *The Genie out of the Bottle* (Cambridge: MIT Press, 2008).

7. Robert Solow, "The Economics of Resources or the Resources of Economics?," *American Economic Review Proceedings* (1974), vol. 64, no. 2, pp. 114.

8. Harold Hotelling, "The Economics of Exhaustible Resources," *Journal of Political Economy* vol. 39 no. 2 (1931) 137–175.

9. See especially M. A. Adelman, "Mineral Depletion Theory with Special Reference to Petroleum," *Journal of Economics and Statistics* vol. 72, no. 1 (1990); 1–10 also Campbell Watkins, "The Hotelling Principle: Autobahn or Cul de Sac," *Energy Journal* vol. 13, no. 1 (1992) 1–24; Richard L.Gordon, "Hicks, Hayek, Hotelling, Hubbert, and Hysteria or Energy, Exhaustion, Environmentalism, and Etatism in the 21st Century," *Energy Journal* vol. 30, no. 1(2009) 1–30.

10. Partha Dasgupta and Heal Geoffrey, *Economic Theory and Exhaustible Resources* (Cambridge: Cambridge University Press, 1979); Michael Boskin, Marc S. Robinson, Terrance O'Reilly, and Praveen Kumar, "New Estimates of the Value of Federal Mineral Rights and Land," *American Economic Review* vol. 75, no. 3 (December 1985) 923–936.

11. Margaret E. Slade and Henry Thille, "Whither Hotelling: Tests of the Theory of Exhaustible Resources," ed. Gordon Rausser, Kerry Smith, and David Zilberman, *Annual Review of Resource Economics* 1 (2009): 252.

12. Ibid.

13. A review can be found in N. Choucri, "Analytical Specifications of the World Oil Market: A Review and Comparison of 12 models," *Journal of Conflict Resolution* vol. 23 no. 2 (1979) 346–372.

14. OECD, *Energy: Prospects to 1985* (Paris: OECD, 1974).

15. U.S. Central Intelligence Agency, "The World Oil Market in the Years Ahead" (Washington, DC: Government Printing Office, August 1979).

16. The Iranian oil crisis was more complex than just the direct loss of supply, but that is the subject of another book.

17. Edgar Allan Poe, "Maelzel's Chess Player," in *The Complete Tales and Poems of Edgar Allan Poe,* ed. Mladinska Knija (New York: Vintage Books, 1975), 421, 430.

18. This version of the figure is from the Department of Energy's *Annual Energy Outlook,* where it was updated every year. However, after 1985, it became obvious that the curve no longer worked and, by 1988, it ceased to publish the figure. The original theory was propounded by Dermot Gately in 1975, long before there were much data to validate or refute it.

19. Ronald Bailey, *The End of Doom: Environmental Renewal in the Twenty-First Century* (New York: St. Martin's Press, 2015).

20. Among peak oil advocates, Albert Bartlett almost exclusively focuses on the threat of exponential growth. Albert A. Bartlett, *The Essential Exponential! (For the Future of Our Planet)* (Example Product Manufacturer, 2004).

21. Hermann Franssen popularized this concept in the 1970s.

22. This is called "the price elasticity of supply" by economists, and, put simply, an elasticity of 0.5 means that for every 1% increase in prices, there would be a 0.5% increase in supply, after an appropriate period.

23. Central Intelligence Agency, "The International Energy Situation: Outlook to 1985" (Washington, DC: Government Printing Office, April 1977).

24. Mexico's exploratory success and tendency to overstate its reserves caused many to be too optimistic about its potential. This is a rare exception to the tendency toward resource pessimism at the time.

25. http://dieoff.org/synopsis.htm.

26. In a 1989 paper, this author noted that nearly all forecasts projected that Third World, non-OPEC oil production would cease growing and decline almost immediately, even though the producers' resource was

immature and their costs low. They have continued to grow almost continuously since. See "The Under projection of Non-OPEC Third World Oil Production," delivered at the Ninth International Conference of the International Association of Energy Economists, Calgary, Alberta, July 1987, in Proceedings.

27. C. D. Masters, D. H. Root, and E. D. Attanasi, "Resource Constraints in Petroleum Production Potential," *Science* 253, no. 5016 (July 1991): 146–152; Lester R. Brown, Christopher Flavin, and Colin Norman, "The Future of the Automobile in an Oil-Short World" (Washington, DC: Worldwatch Paper no. 32, 1979).

28. Rawleigh Warner, "Petroleum Faces Transition Period," *Petroleum 2000: Oil & Gas Journal,* August 1977.

29. Robert Stobaugh, "After the Peak: The Threat of Hostile Oil," in *Energy Future: The Report of the Energy Project at the Harvard Business School,* ed. Robert Stobaugh and Daniel Yergin (New York: Random House, 1979), 40. Note the peak referenced is U.S. oil production, not global.

30. "The Under Projection of Non-OPEC Third World Oil Production," delivered at the Ninth International Conference of the International Association of Energy Economists, Calgary, Alberta, July 1987, in Proceedings.

31. Colin J. Campbell, *The Coming Oil Crisis.* (Brentwood, Essex, England: Multi-Science Publiching Company 1997), 144 mentions this in an approving manner.

32. http://www.bloomberg.com/news/2010-08-30/pemex-plans-to-invest -269-billion-in-next-10-years-to-increase-oil-output.html.

33. Joseph Romm and Charles Curtis, "Mideast Oil Forever," *Atlantic Monthly,* April 1996.

34. Kelly M. Greenhill, *Weapons of Mass Migration: Forced Displacement, Coercion, and Foreign Policy* (Ithaca, NY: Cornell University Press, 2010), 104.

35. See Adrienne Mayor, *The Poison King, the Life and Legend of Mithradates, Rome's Deadliest Enemy* (Princeton: Princeton University Press, 2009), 315.

CHAPTER 3

1. Vito Stagliano, *A Policy of Discontent: The Making of National Energy Strategy* (Oklahoma City, OK: Pennwell Corp., 2001).

2. Robert Dodge, *Schelling's Game Theory: How to Make Decisions* (Oxford: Oxford University Press, 2012), 220.

3. BP, Chevron, Exxon, Gulf, Mobil, Shell, and Texaco.

4. See Daniel Badger and Robert Belgrave, *Oil Supply and Price: What Went Right in 1980?* (British Institute's Joint Energy Policy Programme, May 1982).

5. David Deese and Joseph Nye, eds., *Energy Security* (Cambridge: Ballinger, 1981).

6. World Nuclear Association, "Supply of Uranium" (September 2015), http://www.world-nuclear.org/info/Nuclear-Fuel-Cycle/Uranium-Resources/Supply-of-Uranium/

7. Nuclear Energy Agency, Forty Years of Uranium Resources, Production and Demand in Perspective: "The Red Book Retrospective." (Paris: OECD, 2006), 228.

8. World Nuclear Association, "Supply of Uranium."

9. Executive Office of the President, Energy Policy and Planning, Decision and Report to Congress on the Alaska Natural Gas Transportation System (Washington D.C.: Government Printing Office, September 1977). p. 95.

10. "Nader Leads Protest against a Coast Plant," *New York Times*, April 8, 1979.

11. http://nmenvirolaw.org/images/pdf/Ex._E_.pdf

12. Robin Herman, *Fusion: The Search for Endless Energy* (Cambridge: Cambridge University Press, 1990), 203.

13. Ilias Targas, "Spanish Solar Resurgence? Not on the Horizon," *PV Magazine*, February 11, 2015, http://www.pv-magazine.com/news/details/beitrag/spanish-pv-resurgence-not-on-the-horizon_100018184/#axzz3SyE VWO88.

14. Anthony Parisi, "ARCO Pays 25 Million to Speed Solar Plan," *New York Times,* January 16, 1980.

15. Thomas L. Friedman, "Industry Starts Smiling on Photovoltaics," *New York Times*, September 20, 1981.

16. Michael Parrish, "Siemens Unit Sues ARCO for $150 Million," *Los Angeles Times*, March 2, 1993.

17. Solar Facts Website 2014, http://www.solar-facts.com/panels/thin-film.php, accessed August 12, 2014.

18. Rawleigh Warner, "Petroleum Faces Transition Period," in *Petroleum 2000: Oil & Gas Journal* (Oklahoma City, OK: Pennwell Publishing, August 1977).

19. John C. Given, "Exxon Corp. Selling Reliance Electric for $1.35 Billion." *AP News,* December 12, 1986.

20. Ibid.

21. Richard B. Schmitt and Laurie P. Cohen, "Exxon's Flop in Field of Office Gear Shows Diversification Perils," *Wall Street Journal*, September 3, 1985.

22. Raymond Vernon, *Two Hungry Giants: The United States and Japan in the Quest for Oil and Ore* (Cambridge: Harvard University Press, 1983); Kent E. Calder, *Pacific Defense: Arms, Energy and America's Future in Asia* (New York: William Morrow, 1996).

23. *Business Week*, October 30, 1995, http://www.businessweek.com/1995/44/b34489.htm.

24. Standard and Poor's Industry Outlook, August 7, 1980, Section 2, 63.

25. Standard and Poor's Industry Surveys, November 15, 1984, p. 58.

26. Chalmers Johnson, *MITI and the Japanese Miracle: The Growth of Industrial Policy 1925–1975* (Stanford: Stanford University Press, 1982).

27. Richard J. Samuels, *The Business of the Japanese State: Energy Markets in Comparative and Historical Perspective* (Ithaca, NY: Cornell University Press, 1987); David D. Friedman, *The Misunderstood Miracle: Industrial Development and Political Change in Japan* (Ithaca, NY: Cornell University Press, 1988).

28. See, for example, General Accounting Office, "Uncertainty about Future Oil Supply Makes It Important to Develop a Strategy for Addressing a Peak and Decline in Oil Production," Washington, D.C., 2007. Also known as the "Hirsch report."

CHAPTER 4

1. Tertullian, "On the Soul," in *Apologetical Works and Minucius Felix Octavius* (Washington Catholic University Press, 1950), 250.

2. Robert J. Mayhew, *Malthus: The Life and Legacies of an Untimely Prophet* (Cambridge, MA: Belknap Press, 2014).

3. Barbara Tuchman, *The March of Folly: From Troy to Vietnam* (Knopf, 1985).

4. Christian Wolmar, *Blood, Iron and Gold* (New York: Public Affairs, 2010), 224.

5. Lee Edwards, "The Legacy of Mao Is Mass Murder," February 2, 2010, http://www.heritage.org/research/commentary/2010/02/the-legacy-of-mao-zedong-is-mass-murder.

6. Amartya Sen, *Poverty and Famine: An Essay on Entitlement and Deprivation* (Oxford: Oxford University Press, 1983).

7. Paul Ehrlich, *The Population Bomb* (New York: Ballantine, 1968).

8. Paul R. Ehrlich and Anne H. Ehrlich, *The End of Affluence: A Blueprint for your Future* (New York: Ballantine, 1974), 5.

9. I am indebted to Phillip Verleger for this.

10. Donella H. Meadows, Dennis L. Meadows, Jorgen Randers, and William W. Behrens III, *The Limits to Growth* (New York: New American Library, 1972).

11. Ibid., 66.

12. Dan Gardner, *Future Babble: Why Expert Predictions Are Next to Worthless, and You Can Do Better* (New York: Dutton Adult, 2011), 205.

13. Graham Turner and Cathy Alexander, "Limits to Growth Was Right. New Research Shows We're Nearing Collapse," *The Guardian,* February 2, 2014; Charles A. S. Hall and John W. Day, "Revisiting the Limits to Growth after Peak Oil," *American Scientist,* May–June 1999.

14. Charles MacKay, *Extraordinary Popular Delusions and the Madness of Crowds* (Original in 1841, republished Wordsworth Reference 1999).

15. C. J. Campbell, "Evolution of Oil Assessments," March 23, 2000, http://www.oilcrisis.com/campbell/assessments.htm.

16. Matthew Simmons, "Revisiting the Limits to Growth: Could the Club of Rome Have Been Correct, After All?" Energy White Paper, October 2000, www.greatchange.org/ov-simmons, club_of_rome_revisted.pdf.

17. Ibid., 5.

18. Meadows et al., *The Limits to Growth,* 166.

19. Donella H. Meadows, Dennis L. Meadows, and Jorgen Randers, *Beyond the Limits: Confronting Global Collapse, Envisioning a Sustainable Future* (White River Junction, VT: Chelsea Green Publishing, 1992), 68.

20. Simmons, "Revisiting *The Limits to Growth*."

21. http://www.theoildrum.com/node/6792#more.

22. Robert L. Hirsch, Roger H. Bezdek, and Robert M. Wendling, *The Impending World Energy Mess: What It Is and What It Means to You!* (Apogee Prime, 2010).

23. Campbell, "Evolution of Oil Assessments."

24. Initially, I thought the term referred to Campbell's accusation of economists as being like the conservative Spanish court fearing to sail off the edge of the earth, while he, like Columbus, was a visionary. However, the origin appears to be with Bartlett.

25. Albert A. Bartlett, "Forgotten Fundamentals of the Energy Crisis," http://www.npg.org/specialreports/bartlett_index.htm.

26. Joseph Tainter, *The Collapse of Complex Societies* (Cambridge, UK: Cambridge University Press, 1990).

27. Eric Jay Dolin, *Leviathan: The History of Whaling in America* (New York: W.W. Norton, 2007).

28. Sharon Begley, *Newsweek,* June 6, 2011, 42.

29. Timothy Egan, *The Worst Hard Time: The Untold Story of Those Who Survived the Great American Dust Bowl.* Boston: Houghton Mifflin, 2005.

30. http://www.ncdc.noaa.gov/climate-information/extreme-events/us-tornado-climatology/deadliest.

31. Michael Shermer, *The Believing Brain: From Ghosts and Gods to Politics and Conspiracies* (New York: Times Books, 2011).

32. Gardner, *Future Babble*, 217.

33. Ehrlich, *The Population Bomb*, 11 and 19.

34. Gardner, *Future Babble*, Chapter 6.

35. Ezra Vogel, *Japan as Number One: Lessons for America* (Bloomington, IN: iUniverse, 1999).

36. Paul Kennedy, *The Rise and Fall of the Great Powers: Economic Change and Military Conflict from 1500 to 2000* (New York: Random House, 1987).

37. Michael Crichton, *Rising Sun* (New York: Alfred A. Knopf, 1992).

38. Lester R. Brown and Linda Starke, *Who Will Feed China? Wake-Up Call for a Small Planet* (New York: W.W. Norton, 1995).

39. Donella Meadows and Jorgen Randers, *Limits to Growth: The 30 Year Update* (White River Junction, VT: Chelsea Green, 2004).

CHAPTER 5

1. David Brown, "Bulls and Bears Duel over Supply," *AAPG Explorer,* May 2000.

2. http://news.yahoo.com/s/livescience/20100312/sc_livescience/oilproductiontopeakin2014scientistspredict.

3. This is related to the time value of money and net present value (NPV) calculations, which take into account the fact that interest payments on borrowed money mean that money now is worth more than money later.

4. Jean Laherrere, "Oil and Natural Gas Resource Assessment: Production Growth Cycle Models," in Cutler J. Cleveland, editor-in-chief, *Encyclopedia of Energy.* New York: Elsevier, 2004. pp. 617–631.

5. R. W. Bentley, S. A. Mannan, and S. J. Wheeler, "Assessing the Date of the Global Oil Peak: The Need to Use 2P Reserves," *Energy Policy* 35, no. 12 (2007): 6364–82.

6. Colin J. Campbell, "Coming to Grips with Oil Depletion," *EV World,* October 2, 2002.

7. Jean Laherrere, "Forecasting Future Production from Past Oil Discovery," *OPEC Seminar,* September 2001.

8. Letters, *Scientific American,* October 2009.

9. Ugo Bardo, "Peak Oil: Has It Arrived?" June 2012. http://www.financialsense.com/contributors/ugo-bardi/peak-oil-has-it-arrived.

10. Colin J. Campbell and Jean Laherrere, "The End of Cheap Oil," *Scientific American,* March 1998.

11. Jean L. Laherrere, "World oil supply-what goes up must come down, but when will it peak?" *Oil &Gas Journal,* February 1, 1999.

12. Peter Baker and Dan Bilesky "Russia and U.S. Sign Nuclear Arms Reduction Pact," *New York Times,* April 9, 2010.

13. Betsy McKay, "The Flu Season That Fizzled," *Wall Street Journal,* March 2, 2010.

14. Robert Arp, *1001 Ideas That Changed the Way We Think* (New York: Simon and Schuster, 2013), 643.

15. Jonathan Swift, *Gulliver's Travels,* New York: Barnes and Noble's Classics, 2003, 197.

16. Mary Roach, *Spook: Science Tackles the Afterlife* (New York: W. W. Norton, 2005), 186.

17. Kenneth Deffeyes, *Hubbert's Peak: The Impending World Oil Shortage* (Princeton: Princeton University Press, 2001), 116.

18. C. J. Campbell, "Oil Depletion: Updated through 2001," February 2002, www.oilcrisis.com.

19. Michael C. Lynch, "The Economics of Petroleum in the Former Soviet Union," in *Gulf Energy and the World: Challenges and Threats, The Emirates Center for Strategic Studies and Research,* Abu Dhabi, 1997.

CHAPTER 6

1. *Fred Guterl*, "When Wells Go Dry," article on Ken Deffeyes, *Newsweek,* April 14, 2002, http://www.newsweek.com/2002/04/14/when-wells-go-dry.html.

2. http://easycalculation.com/funny/tricks/trick1.php.

3. Deffeyes, Kenneth. *Hubbert's Peak: The Impending World Oil Shortage*. Princeton: Princeton University Press, 2001.

4. Michael C. Lynch, "The Analysis and Forecasting of Petroleum Supply: Sources of Error and Bias," in *Energy Watchers VII,* ed. by Dorothea H. El Mallakh (Boulder, CO: International Research Center for Energy and Economic Development, 1996).

5. On Oil Drum.com, a search of Hubbert linearization yields a large number of people applying the method in various instances. http://www .theoildrum.com/story/2005/9/29/3234/46878.

6. http://www.theoildrum.com/story/2005/9/29/3234/46878.

7. http://archiver.rootsweb.ancestry.com/th/read/GENBRIT/2002-10/ 1035601064.

8. James L. Smith and James L. Paddock, "Regional Modelling of Oil Discovery and Production," *Energy Economics,* January 1984.

9. Thomas S. Ahlbrandt and T. R. Klett, "Comparison of Methods Used to Assess Conventional Undiscovered Petroleum Resources: World Examples," *Natural Resources Research,* September 2005.

10. James W. Schmoker and T. R. Klett, "Estimating Potential Reserve Growth of (Known) Discovered Fields: A Component of the USGS World Petroleum Assessment 2000," *US Geological Survey Digital Data Series, 60.*

11. USGS, "Region Assessment 2: Middle East and North Africa," *World Petroleum Assessment 2000,* R2–40.

12. Michael R. Smith, "Resource Depletion: Modeling and Forecasting Oil Production," April 2006. Presented to Oak Ridge National Laboratory.

13. Gordon M. Kaufmann, "Statistical Issues in the Assessment of Undiscovered Oil and Gas Resources," *Energy Journal,* 1993.

14. Kjell Aleklett, *Peeking at Peak Oil* (New York: Springer, 2012).

CHAPTER 7

1. Lawrence J. Drew, *Undiscovered Petroleum and Mineral Resources: Assessment and Controversy* (New York: Springer, 1996).

2. "The World's Oil Supply—1930–2050," Petroconsultants Report, October, 650p, CD-ROM.

3. M. A. Adelman, John C. Houghton, Gordon Kaufman, and Martin B. Zimmerman, *Energy Resources in an Uncertain Future: Coal, Gas, Oil and Uranium Supply Forecasting* (Cambridge, MA: Ballinger, 1983); see also James L. Smith and James L. Paddock, "Regional Modelling of Oil Discovery and Production," *Energy Economics,* January 1984.

4. Colin J. Campbell, *The Coming Oil Crisis* (Essex, UK.: Multi-Science) 1997, 74.

5. Jean Laherrere, "International conference on Oil Demand, Production and Costs—Prospects for the Future." The Danish Technology Council and the Danish Society of Engineers, Copenhagen, December 10, 2003, 18.

6. C. J. Campbell, *The Essence of Oil and Gas Depletion* (Essex, UK: Multi-Science Publishing Co., 2003), 149.

7. Kjell Aleklett, *Peeking at Peak Oil* (New York: Springer, 2012), 48.

8. UK Energy Research Centre, Global Oil Depletion: An Assessment of the Evidence for a Near-Term Peak in Global Oil Production (London: UKERC, 2009), 50.

9. Tone Sem and Denny Ellerman, "North Sea Reserve Appreciation, Production and Depletion" (Cambridge: M.I.T. Center for Energy and Environmental Policy Research, 1999), 35.

10. Jean Laherrere, "Fossil Fuels Future Production," March 2005, Bucharest, Romania.

11. Jean Laherrere, "Uncertainty on Data and Forecasts."ASPO 5 San Rossore Italy 18–19, July 2006.

12. Laherrere, Fossil Fuels Future Production.

13. James Schmoker and T. R. Klett, "Estimating Potential Reserve Growth of Known (Discovered) Fields, Chapter RG. http://certmapper.cr.usgs.gov/data/PubArchives/WEcont/chaps/RG.pdf.

14. Richard Heinberg, *The Party's Over: Oil War and the Fate of Industrial Societies.* (Gabriola Island, Canada: New Society Publishers, 2005), 114.

15. Schmoker and Klett, "Estimating Potential Reserve Growth of Known (Discovered) Fields." http://certmapper.cr.usgs.gov/data/PubArchives/WEcont/chaps/RG.pdf.

16. Matthew Simmons, "The Only Way Is Down: The High Priest of "Peak Oil" Thinks World Oil Output Can Now Only Decline." *Economist,* July 10, 2008.

17. *New York Times,* September 27, 2011.

18. https://groups.yahoo.com/neo/groups/energyresources/conversations/topics/127820.

19. William Pike, "Through the Rear-View Mirror," *World Oil,* August 2011.

20. Campbell, 2002, http://www.mbendi.com/indy/oilg/p0070.htm.

21. Matthew Simmons, *Twilight in the Desert: The Coming Saudi Oil Shock and the World* Economy (Hoboken, N.J.: Wiley, 2005) 116.

22. Matthew Simmons, *Twilight in the Desert: The Coming Saudi Oil Shock and the World* Economy (Hoboken, N.J.: Wiley, 2005) 288.

23. G. C. Watkins, "Characteristics of North Sea Oil Reserve Appreciation" (Cambridge, MA: M.I.T. Center for Energy and Environmental Policy Working Paper, 00-008WP, December 2000).

24. M. A. Adelman, Michael C. Lynch, and Kenichi Ohashi, "Supply Aspects of North American Gas Trade," in *Canadian-U.S. Natural Gas Trade,* Final Report of the International Natural Gas Trade Project, MIT Energy Laboratory Report 85-013, October 1985.

25. The following Website has a nice graphic http://www.chinaoil fieldtech.com/oilrecovery.html.

26. Rob Watts, "EnQuest to Move in on Alma and Galia Fields," *Upstream,* December 2, 2011, 27.

27. "Shell hopes to extend Life of Mars B with New Olympus TLP," *World Oil,* July 2013.

28. *Upstream,* November 25, 2011, 4.

29. Beate Schjolberg, "Subsea Plan for Gullfaks Aims to Boost Recovery," *Upstream,* May 25, 2012, 21.

30. http://www.rigzone.com/news/oil_gas/a/78707/Saudi_Aramcos _Smart_Water_May_Aid_Oil_Production#sthash.flxLqBAv.dpuf.

31. Lijun You, Yili Kang, Zhangxin Chen, Xiao Niu, and Faqiang Luo, "Optimized Fluids Improve Production in Horizontal Tarim Wells," *Oil & Gas Journal,* May 5, 2014.

32. Tan Hee Hwee, "Exxon Mobil Kicks off Studies into Guntong," *Upstream,* May 18, 2012, 21.

33. Clare Temple-Heald, Craig Davies, Natalie Wilson, and Nicola Readman, "Developing New Surfactant Chemistry for Breaking Emulsions in Heavy Oil," *Journal of Petroleum Technology* 66, no. 1 (January 2014): 30–36.

34. "New LWD Service Integrates Formation Fluid Analysis and Sampling," *Journal of Petroleum Technology* (January 2014).

35. http://www.rigzone.com/training/insight.asp?insight_id=324&c _id=22.

36. Melanie Cruthids, "Innovative Thinkers," *World Oil,* June 2014.

37. Ernest Scheyder, "Walking, 'Talking' Drilling Rigs Aim to Modernize Fracking," *Reuters,* April 15, 2014, http://www.reuters.com/ article/2014/04/15/us-schramm-shale-idUSBREA3E0NA20140415.

38. Suleiman et al., "Openhole gravel packing of fishhook wells with zonal isolation and uphill heel-to-toe packing," *World Oil,* July 2013.

39. Don Hannegan, "MPD Widens Offshore Drilling Capabilities," *Journal of Petroleum Technology* (February 2015).

40. Rifat Kayumov, Artem Klubin, Alexey Yudin, Phillippe Enkaba-bian, Fedor Leskin, Igor Davidenko, and Zdenko Kaluder, "First Channel Fracturing in Mature Well Increases Production from Talinskoe Oil Field in Western Siberia," SPE 159347, October 2012, summarized in *Journal of Petroleum Technology,* January 2014.

41. http://scitizen.com/future-energies/will-toe-to-heel-air-injection -extend-the-oil-age-_a-14-3449.html.

42. Wieqiang Feng, Qi Li, and Jian Wang, "Model Optimizes Sandstone-Carbonate Fracturing in China," *Oil & Gas Journal,* January 9, 2012.

43. Shell Alaska Vice President Pete Slaiby, *Upstream,* April 27, 2012, 36.

44. "Apache believes Stag Field Still Has Plenty More Oil to Offer." *Upstream,* May 25, 2012, 43.

45. Stephen Rassenfuss, "From Bacteria to Barrels: Microbiology Having an Impact on Oil Fields," *Journal of Petroleum Technology* (November 2011).

46. Melanie Cruthirds, "Innovative Thinking," *World Oil,* October 2013.

CHAPTER 8

1. Heinberg, 81.

2. "Current Events: Join Us As We Watch the Crisis Unfolding," January 20, 2005.http://www.princeton.edu/hubbert/current-events-05 -01.html.

3. Sam Foucher, "The Hubbert Linearization Applied on Ghawar," October 10, 2007. http://www.theoildrum.com/node/3050.

4. M. A. Adelman, *The Genie Out of the Bottle: World Oil Since 1970* (Cambridge, MA: MIT Press, 2005), 166.

5. Chip Haynes, "Ghawar Is Dying," August 2001, http://www.newcolonist .com/ghawar.html.

6. Eastex on TheOilDrum.com, April 18, 2011, http://www.theoildrum .com/node/7817.

7. The description of U.S. military action in Iraq in 2003 is semanti-cally sensitive. The Shi-ites might describe it as a liberation, the Sunnis as an occupation, the U.S. liberals as an occupation, and conservatives as a liberation. I will not take sides.

8. Mephistopheles, *Goethe's Faust,* trans. and with an introduction Walter Kaufmann (New York: Anchor Books, 1990), 32.

9. Simmons, 329–330.

10. Simmons, http://csis.org/files/attachments/040224_simmons.pdf. The Aramco slides are available in http://csis.org/files/attachments/040224_baqiandsaleri.pdf.

11. Simmons, *Twilight,* 51.

12. Louis W. Powers, *The World Energy Dilemma* (Tulsa, OK: Pennwell, 2012).

13. Simmons, *Twilight,* 178.

14. Halliburton, Oilfield Water Management, http://www.halliburton.com/en-US/ps/solutions/clean-energy/oilfield-water-management/default.page?node-id=hgjyd44q.

15. Simmons, *Twilight,* 213.

16. Ibid.

17. Carola Hoyos, "What Is Happening to Saudi Oil Production as Khurais Comes On Line," *Financial Times* (blog), June 11, 2009, http://blogs.ft.com/energy-source/2009/06/11/what-is-happening-to-saudi-oil-production-as-khurais-comes-on-stream/#axzz362turKOu.

18. Jim Jarell, "Another Day in the Desert: A Response to the Book Twilight in the Desert," *Geopolitics of Energy,* October 2005.

19. Saddad al Husseni, "Rebutting the Critics: Saudi Arabia's Oil Reserves, Production Practices Ensure Its Cornerstone Role in Future Oil Supply," *Oil & Gas Journal,* May 17, 2004.

20. Louis W. Powers, *The World Energy Dilemma* (Tulsa, OK: Pennwell, 2012).

21. Simmons, *Twilight,* 72.

22. The amount of sedimentary basin is used, which is roughly equivalent to petroleum prospective areas, taken from Bernardo Grossling, "A Critical Survey of World Petroleum Opportunities," in Project Interdependence, Library of Congress, 1977, Committee Print 95–33, 645–658.

23. Michael C. Lynch, "Crop Circles in the Desert: The Strange Controversy Over Saudi Petroleum," Occasional Paper 40, The International Research Center for Energy and Economic Development, Boulder, Colorado, 2006, 20.

24. Ibid.

25. Powers, *The World Energy Dilemma.*

26. Ahmed Maneeri, "Manifa Is an Expensive Development," *Rigzone,* July 22, 2013.

27. Tam Hunt, "Guest Post: The Saudi Oil Problem," June 6, 2012, http://www.greentechmedia.com/articles/read/Guest-Post-The-Saudi-Oil-Problem.

28. Westexas, "Hubbert Linearization Analysis of the Top Three Net Oil Exporters," January 27, 2006, http://www.theoildrum.com/story/2006/1/27/14471/5832.

29. Glada Lahn and Paul Stevens, "Burning Oil to Keep Cool: The Hidden Energy Crisis in Saudi Arabia," December 2011, Chatham House.

30. Hunt, Guest Post.

CHAPTER 9

1. Jeffrey Ball, "As Prices Soar, Doomsayers Provoke Debate on Oil's Future," *Wall Street Journal,* September 21, 2004.

2. In 1983, I actually saw an academic economist denounce a computer model as being invalid because it predicted that oil prices would decline the following year (an incredibly good prediction). His argument was that models must be stable. When another economist remarked that the model seemed to be tracking historical behavior rather well, the first economist retorted that that was irrelevant.

3. http://www.aspo-usa.com/index.php?option=com_content&task=view&id=313&Itemid=146.

4. Jean Laherrere, May 13, 2004, https://groups.yahoo.com/neo/groups/energyresources/conversations/messages/56991.

5. Robert Stobaugh and Daniel Yergin, eds., *Energy Future: Report of the Energy Project at the Harvard Business School* (New York: Ballantine, 1979), 38.

6. Ilan Berman, "Saudi Arabia's House of Cards," 2010, http://www.forbes.com/2010/07/12/saudi-arabia-royalty-oil-opinions-columnists-ilan-berman.html.

7. http://www.hubbertpeak.com/duncan/Hoovervilles&BushCamps.htm.

8. M. King Hubbert, "Nuclear Energy and the Fossil Fuels" (Houston, TX: Shell Development Company Publication 95, June 1956).

9. Office of Technology Assessment, "US Natural Gas Availability: Gas Supply through 2000" (Washington, DC: Government Printing Office, February 1985), 49.

10. Michael C. Lynch, "The Analysis and Forecasting of Petroleum Supply: Sources of Error and Bias," in *Energy Watchers VII,* ed. Dorothea H. El Mallakh (Boulder, CO: International Research Center for Energy and Economic Development, 1996).

11. L. Michael White, http://www.pbs.org/wgbh/pages/frontline/shows/apocalypse/explanation/amprophesy.html.

12. Michael C. Lynch, "The Analysis and Forecasting of Petroleum Supply: Sources of Error and Bias," in *Energy Watchers VII,* ed. Dorothea H. El Mallakh (Boulder, CO: International Research Center for Energy and Economic Development, 1996).

13. Steven Scharfman, "Petroleum Experts Debate Impending Oil Shortage," October 2002, http://www.freeinquiry.com/skeptic/ badgeology/energy/debate-review.htm.

14. http://www.economist.com/node/10496503.

15. http://www.time.com/time/specials/packages/article/0,28804,1954176 _1954175_1954172,00.html#ixzz0dFlJX8EJ.

16. http://www.rigzone.com/news/oil_gas/a/122712/Total_Oil_ Production_to_Peak_at_98M_Barrels_per_Day.

17. http://www.oilcrisis.com/campbell/assessments.htm.

18. Colin J. Campbell, "Oil Price Leap in the 90s," *Noroil,* December 1989.

19. Projecting OPEC production is not a valid test of a geological model, since market conditions affect their production decisions. Production dropped in the FSU in a manner similar to his forecast, but due to the economic collapse in the FSU rather than industry conditions. His forecast of U.S. production was very pessimistic, but the inclusion of deepwater production explains part of the error.

20. Nate Hagens, "Michael Lynch and the 'False Threat of Disappearing Oil'," August 27, 2009, http://www.theoildrum.com/node/5716.

21. Kjell Aleklett, *Peeking at Peak Oil* (New York: Springer, 2012) 138.

22. Harry R. Johnson, Peter M. Crawford, and James W. Bunger, "Strategic Significant of America's Shale Oil Resource," vol. I, *Assessment of Strategic Issues* (U.S. Department of Energy, 2004).

23. Richard Duncan and Walter Youngquist, "The World Petroleum Life-Cycle," University of Southern California, October 22, 1998, http:// dieoff.com/page133.htm.

24. M. A. Adelman and Michael C. Lynch, "Fixed View of Resource Limits Creates Undue Pessimism," *Oil & Gas Journal,* April 7, 1997.

25. Jean Laherrere, "Future Sources of Crude Oil Supply and Quality Considerations," French Petroleum Institute, June 1997.

26. Roger Blanchard, "A Look Back at North Sea Projections," April 11, 2011, http://peak-oil.org/2011/04/a-look-back-at-north-sea-oil -production-projections/.

27. This author, in a 1997 paper, examined the data available for the FSU and correctly concluded that it had the potential for a significant

production increase. The Economics of Petroleum in the Former Soviet Union," *Gulf Energy and the World: Challenges and Threats* (Abu Dhabi: The Emirates Center for Strategic Studies and Research, 1997).

28. http://www.energybulletin.net/node/38948.

29. Robert A. Wattenbarger, "Oil Production Trends in the Soviet Union," in *Advances in the Economics of Energy and the Environment,* ed. John R. Moroney (Greenwich, CT: JAI Press, 1997).

30. Sam Foucher, "Russia's Production Is About to Peak," April 24, 2008, http://www.theoildrum.com/node/3626.

31. M. Payne, "Mexican Oil Exports. Start Saying Adios!," 2008, http://peakopps.blogspot.com/2008/03/mexican-oil-exports-start-saying -adios.html.

32. Kurt Cobb, *Resource Insights,* 2013, http://resourceinsights.blog spot.com/search?q=cantarell.

33. Sam Foucher, "Potential Impact of Cantarell's Decline in Mexico's Oil Production," July 14, 2006, http://www.theoildrum.com/story/2006/7/ 12/10421/4972.

34. Michael R. Smith, "Resource Depletion: Modeling and Forecasting Oil Production," April 2006. Presented to Oak Ridge National Laboratory.

35. Werner Zittel and Jorg Schindler, "Future World Oil Supply." http://www.peakoil.net/Publications/International-Summer-School _Salzburg_2002.pdf.

36. Kjell Aleklett, "A World Addicted to Oil," Pisa, July 2006, slides.

37. C. Hall and C. J. Cleveland, "Petroleum Drilling and Production in the United States: Yield per Effort and Net Energy Analysis," *Science* 211, no. 4482 (1981): 576–79.

38. M. King Hubbert, "Techniques of Production as Applied to the Production of Oil and Gas," in S. I. Gass, Oil and Gas Supply Modelling (Washington, D.C. National Bureau of Standards 1980). Arthur L. Smith and Bryan Lidsky, "M. King Hubbert's Analysis Revisited: An Update of Lower 48 Oil and Gas Resource Base," New Orleans USAEE, 1992.

39. "Energy Prices and Energy Fundamentals: Is There a link," CGES, July 1, 2004.

40. Matthew Simmons, "Unlocking the US Natural Gas Riddle," 2002, 30.

41. Jean Laherrere," Forecast of Oil and Gas Supply to 2050." Delivered to Petrotech 2003, New Delhi, 9 and Ibid., 8.

42. Heinberg, 117.

43. http://www.theoildrum.com/node/6519#more.

44. Chris Nelder, "Reading Peak Oil Deniers Is a Waste of Time," April 28, 2009, http://www.greenchipstocks.com/articles/reading-peak -oil-deniers/486.

45. http://www.theoildrum.com/node/5716.

46. M. A. Adelman, "Natural Gas Supply in Western Europe," *Western Europe Natural Gas Trade,* Final Report of the International Natural Gas Trade Project, MIT Energy Laboratory, December 1986; M. A. Adelman "Natural Gas Supply in Asia-Pacific Region," *East Asia/Pacific Natural Gas Trade,* Final Report of the International Natural Gas Trade Project, MIT Energy Laboratory Report MIT EL86-005, March 1986; M. A. Adelman and Kenichi Ohashi, "Supply Aspects of North American Gas Trade," *Canadian-U.S. Natural Gas Trade,* Final Report of the International Natural Gas Trade Project, MIT Energy Laboratory Report 85-013, October 1985.

47. "Structural Changes in World Oil Markets and Their Impact on Market Behavior," MIT Energy Laboratory Working Paper MIT-EL 86-009WP, March 1986.

48. "The Price of Crude Oil to 2000: The Economics of the Oil Market," *Economist Intelligence Unit,* May 1989.

49. *International Petroleum Price, Supply and Demand: Projections through 2020,* Gas Research Institute, January 1996.

50. "The Economics of Petroleum in the Former Soviet Union," *Gulf Energy and the World: Challenges and Threats* (Abu Dhabi: The Emirates Center for Strategic Studies and Research, 1997).

51. Michael C. Lynch, "A New Era of Price Volatility?" *Oil & Gas Journal,* February 12, 2001.

52. "Crop Circles in the Desert: The Strange Controversy over Saudi Petroleum," Occasional Paper 40, The International Research Center for Energy and Economic Development, Boulder, Colorado, 2006.

53. http://www.opec.org/opec_web/en/multimedia/videoDetail.htm# 1710171805001,1673236007001.

54. Michael C. Lynch, "Uncertainties Threaten Natgas Development," *Oil & Gas Journal,* March 4, 2013.

55. Energy Information Administration, *Annual Energy Outlook 2014 with projections to 2040* (Washington, DC: Department of Energy, April 2014), CP-3.

CHAPTER 10

1. C. J. Campbell, *The Essence of Oil and Gas Depletion* (Essex, England Multi-Science Publishing, 2003).

2. Simon Singh, *The Code Book*: *The Science of Secrecy from Ancient Egypt to Quantum Cryptography* (New York: Anchor, 2000).

3. http://blogs.hindustantimes.com/foreign-hand/2012/07/06/fall-from -peak-oil/.

4. Questions for Harry Markopolos, Deborah Solomon, *New York Times,* February 28, 2010, 14.

5. http://www.peakoil.net/about-aspo/aspo-president.

6. As an example, see http://www.earleyenvironmentalgroup.co .uk/Environment/PeakOilTalk.asp. http://peakoil.com/generalideas/ interview-professor-kjell-aleklett-the-great-energy-journey.

7. In http://www.moneynews.com/MKTnews/Market-Collapse-Predicted -By-Scientist/2013/03/13/id/494569?promo_code=12C9B-1&utm_source =taboola. http://www.energybulletin.net/stories/2010-11-16/interview-chris -martenson-prepare-peak-oil-while-there-time.

8. Joe Romm, 2009, http://climateprogress.org/2009/12/29/michael -lynch-wager-peak-oil/#comment-248185.

9. Gilgamesh, 97.

10. Yes, I know, they didn't actually wind up fighting.

11. https://groups.yahoo.com/neo/groups/energyresources/ conversations/messages/74679.

12. Letters, *National Journal*, Fall 1998.

13. Matthew Simmons, "Peak Oil: The Reality of Depletion," May 3, 2004, Houston, Texas.

14. General Accounting Office, "Crude Oil: Uncertainty about Future Supply Makes it Important to Develop a Strategy for Addressing a Peak and Decline in Oil Production (Washington: General Printing Office: February 2007).

15. Kjell Aleklett, http://www.theoildrum.com/node/9697#more.

16. http://www.usnews.com/news/articles/2012/06/28/most-americans -believe-government-keeps-ufo-secrets-survey-finds.

17. http://www.postcarbon.org/blog-post/985668-peak-denial.

18. James Howard Kunstler, "Get Real," May 23, 2011, http://Kunstler .com/blog/2011/05/get-real.html.

19. http://www.energybulletin.net/node/52493.

20. Michael Schermer, *The Believing Brain* (New York: Times Books), 2011.

21. "Petroleum Experts Debate Impending World Oil Shortage: Round Table Discussion on World Oil Supply and the Consequences for America," Review by Steven Schafersman, October 10, 2002, http://llanoestacado.org/ freeinquiry/skeptic/badgeology/energy/debate-review.htm.

22. http://www.raisethehammer.org/article/1581/home_run_for_peak_oil.

23. http://www.businessgreen.com/business-green/news/2264554/former-chief-scientist-lends.

24. http://peakoil.com/geology/jean-laherrere-comments-on-bp-statistical-review-2012.

25. http://www.theoildrum.com/node/8395#more.

26. Lionel Badal, "How the Global Oil Watchdog Failed Its Mission," May 26, 2010, http://www.informationclearinghouse.info/article25546.htm.

27. James Howard Kunstler, "Get Real," May 25, 2011, http://Kunstler.com/blog/2011/05/get-real.html.

28. Richard Heinberg, "Peak Denial," July 2, 2012, http://www.postcarbon.org/blog-post/985668-peak-denial.

29. http://blogdredd.blogspot.com/2012/05/momcom-private-parts.html.

30. Emily Willingham, "10 Questions to Distinguish Real from Fake Science," http://www.forbes.com/sites/emilywillingham/2012/11/08/10-questions-to-distinguish-real-from-fake-science/.

31. Robert L. Park, *Voodoo Science: The Road from Foolishness to Fraud* (Oxford: Oxford University Press, 2000).

CHAPTER 11

1. Lawrence J. Drew, *Undiscovered Petroleum and Mineral Resources: Assessment and Controversy* (New York: Springer, 1996). Thomas S. Ahlbrandt and T. R. Klett, "Comparison of Methods Used to Assess Conventional Undiscovered Petroleum Resources: World Examples," *Natural Resources Research,* September 2005.

2. The primary work by this author: "The Under-Projection of Non-OPEC Third World Oil Production," delivered at the Ninth International Conference of the International Association of Energy Economists, Calgary, Alberta, July 1987; "An Omitted Variable in OECD Supply Forecasting," Proceedings delivered to the 12th Annual North American Conference, International Association of Energy Economics, Ottawa Canada, October 1990; "The Supply of Oil: The Wolf at the Door or Crying Wolf," *Forum for Applied Research and Public Policy,* Spring 1992; "The Analysis and Forecasting of Petroleum Supply: Sources of Error and Bias," in *Energy Watchers VII,* ed. by Dorothea H. El Mallakh (Boulder, CO: International Research Center for Energy and Economic Development, 1996).

3. General Accounting Office, "Uncertainty about Future Oil Supply Makes It Important to Develop a Strategy for Addressing a Peak and Decline in Oil Production," February 2007.

4. Maass, Peter, "The Breaking Point," *New York Times,* August 21, 2005; Maass, Peter, *Crude World: The Violent Twilight of Oil* (Vintage 2010).

5. Ann Muggeridge, Andrew Cockin, Kevin Webb, Harry Frampton, Ian Collins, Tim Moulds, and Peter Salino, "Recovery Rates, Enhanced Oil Recovery and Technological Limits," http://www.ncbi.nlm.nih.gov/pmc/articles/PMC3866386/.

6. David Goodstein, *Out of Gas: The End of the Age of Oil* (Norton Books, 2005).

7. ASPO NL 9/2004.

8. Jean Laherrere, "Uncertainty on Data in Forecasts," ASPO 5, Italy, July 2006.

9. For instance, "Pemex Lists E&P Projects to Counter Cantarell's Decline," *Oil & Gas Journal,* April 17, 2006.

10. http://csis.org/files/attachments/040224_simmons.pdf.

11. Matthew Simmons, "The Peak Oil Debate: Crisis or Comedy?" SPE Annual Technical Conference, Houston, September 27, 2004. http://peakoil.com/consumption/the-peak-oil-debate-crisis-or-comedy.

12. Jean Laherrere, "Reserve Growth: Technological Progress or Bad Reporting and Bad Arithmetic?," *Geopolitics of Energy,* April 1999.

13. M. Höök and K. Aleklett, "A Decline Rate Study of Norwegian Oil Production." *Energy Policy,* 2008.

14. Robert L. Hirsch, Robert Bedzek, and Robert Wendling, "Peaking of World Oil Production: Impacts, Mitigation, and Risk Management," February 2005.

15. Robert L. Hirsch, "The Inevitable Peaking of World Oil Production," Bulletin, The Atlantic Council of the United States, 2005.

16. C. J. Campbell, *The Essence of Oil & Gas Depletion: Collected Papers and Excerpts* (Essex, U.K.: Multi-Science Publishing, 2003).

17. Kenneth Deffeyes, *Hubbert's Peak: The Impending World Oil Shortage* (Princeton: Princeton University Press, 2001).

18. Jean H. Laherrere, "Future Sources of Crude Oil Supply and Quality Considerations," French Petroleum Institute Conference, Rueil-Malmaison, France, 12–13 June 1997, 12.

19. Two cases are worthy of mention, both caught by Laherrere. A graph I produced of forecasts for non-OPEC, non-FSU, non-US

production was inadvertently labeled simply non-OPEC. Laherrere insisted that this meant all of my work was invalid, an interesting stance given his many mistakes. Also, in a presentation to the European Geophysical Union, I referred to Apache's announcement that it had increased reserves at the Forties field, but I mistakenly used the number for oil in place as proved reserves (which would inflate it by a factor of 2 or 3). The point, however, was that Laherrere had insisted that he "proved" that the reserves could not be increased at all, discussed in more detail in Chapter 7.

20. Michael R. Smith, "Resource Depletion: Modeling and Forecasting Oil Production," April 2006. Presented to Oak Ridge National Laboratory; Michael R. Rodgers, "Recent Trends in Exploration Results and the Implication for Future Petroleum Liquids Supply," October 2006.

21. Bushan Bahree and Jeffery Ball, "Producers Move to Debunk Gloomy 'Peak Oil' Forecasts" *Wall Street Journal,* September 14, 2006.

22. National Petroleum Council, *Hard Truths: Facing the Hard Truths about Energy* (Washington, DC: National Petroleum Council, 2007).

23. http://energyandourfuture.org/node/5947.

24. ASPO NL 12/05.

25. Kjell Aleklett, *Peeking at Peak Oil* (New York: Springer, 2012) 50. Leif Magne Meling, "How and for How Long It Is Possible to Secure a Sustainable Growth of Oil Supply," 2003; http://www.gasandoil.com/news/europe/ece6b4f43c323611c90a9ad71b73e663.

26. Kjell Aleklett, *Peeking at Peak Oil* (New York: Springer, 2012) 261.

27. Richard Heinberg, *The Party's Over: Oil War and the Fate of Industrial Societies* (Gabriola Island, Canada: New Society Publishers, 2005) 85.

28. AndrewTopf, "The Easy Oil Is Gone, so Where Do We Look Now?," February 15, 2015, http://www.nasdaq.com/article/the-easy-oil-is-gone-so-where-do-we-look-now-cm447457.

29. M. A. Adelman, "Finding and Developing Costs in the USA 1918–1986," in *Advances in the Economics of Energy & Resources,* ed. John Moroney (JAI Press, 1990); Adelman, M. A., and Manoj Shahi, "Development Cost Estimates for Producing Countries 1958–1986," *Energy Economics*, 1989.

30. http://www.sustainablecitynews.com/ghawar-html/.

31. Michael Ruppert, *Confronting Collapse: The Crisis of Energy and Money in a Post Peak Oil World* (White River Junction, VT: Chelsea Green Publishing, 2009), 41.

32. http://tech.groups.yahoo.com/group/energyresources/message/ 56178.

33. Mahmoud Abdul Baqi and Nansen Saleri, "Fifty Year Crude Oil Supply Scenarios: Saudi Aramco's Perspective," February 24, 2004, CSIS, Washington, DC.

34. W. Zittel, J. Schindler, and L. –B. Systemtechn, "The Countdown for the Peak of Oil Production Has Begun—but What Are the Views of the Most Important International Energy Agency," http://www.resilience.org/ stories/2004-10-14/countdown-peak-oil-production-has-begun-%E2%80% 93-what-are-views-most-important-internati, October 14, 2004.

35. Sadad al-Husseini, "Rebutting the Critics: Saudi Arabia's Oil Reserves, Production Practices Ensure Its Cornerstone Role in Future Oil Supply," *Oil & Gas Journal,* May 17, 2004.

36. Sadad al-Husseini, "Why Higher Oil Prices Are Inevitable This Year, Rest of Decade," *Oil & Gas Journal,* August 2, 2004.

37. At a conference in Sweden, Kjell Aleklett remarked that he had recently been to Abu Dhabi, where those in the industry assured him it was getting very hard to increase production. He did not provide further evidence, beyond the typical general statements, nor respond to the substantive criticisms made by this author during the presentations. At a conference in Abu Dhabi nearly a decade ago, Jean Laherrere showed his evidence that nearly all of the oil in the Middle East had been found, and the audience was distinctly unimpressed.

38. Drew, *Undiscovered Petroleum and Mineral Resources.*

39. Thomas S. Ahlbrandt and T. R. Klett, "Comparison of Methods Used to Assess Conventional Undiscovered Petroleum Resources: World Examples," *Natural Resources Research,* September 2005.

40. Interview in *Playboy,* January 2007.

41. Richard Heinberg, *The Party's Over: Oil War and the Fate of Industrial Societies.* (Gabriola Island, Canada: New Society Publishers, 2005). 100; Roger Blanchard, *The Future of Global Oil Production: Facts, Figures, Trends and Projections, by Region.* (London: McFarland and Company, 2005). op. cit. p. 23; Kjell Aleklett, *Peeking at Peak Oil* (New York: Springer, 2012), 48.

42. Michael Ruppert, *Confronting Collapse: The Crisis of Energy and Money in a Post Peak Oil World* (White River Junction, VT: Chelsea Green Publishing, 2009).

43. Ibid., 91.

44. Colin J. Campbell, *The Coming Oil Crisis.* (Brentwood, Essex, England: Multi-Science Publishing Company 1997), 121.

45. Ibid., 165.

46. http://www.theoildrum.com/node/9869#more.

47. David Goodstein, *Out of Gas: The End of the Age of Oil* (Norton, 2005).

48. Jean Laherrere, "Is the USGS Assessment Reliable?" Published on the cyber conference of the World Energy Council on May 19, 2000. http://energyresource2000.com/.

49. C. D. Masters D. H. Root, E. D. Attanasi, "World Oil and Gas Resources: Future Production Realities," in *Annual Review of Energy* (Greenwich, CT: JAI Press 1990), 26.

50. C. J. Campbell, *The Coming Oil Crisis* (Brentwood, Essex, England: Multi-Science Publishing Company 1997).

51. Richard Heinberg, *The Party's Over: Oil War and the Fate of Industrial Societies* (Gabriola Island, Canada: New Society Publishers, 2005), 109.

52. Paul Roberts, *The End of Oil: On the Edge of a Perilous New World* (Boston: Houghton Mifflin 2004), 2.

53. Matthew R. Simmons, "Depletion and US Energy Policy," International Workshop on Oil Depletion, Uppsala University, May 23, 2002, 5.

54. Jaromar Benes, Marcelle Chauvet, Ondra Kamenik, Michael Kumof, Douglas Laxton, Susanna Mursula, and Jack Selody, "The Future of Oil: Technology versus Geology," IMF Working Paper WP 12/109, May 2012.

55. Roland Watson, "Peak Oil 2005?," http://www.resilience.org/stories/2005-07-15/peak-oil-2005.

56. Durham Louise, "The Elephant of All Elephants," *AAPG Explorer,* January 2005.

57. http://dieoff.org/synopsis.htm.

58. Ibid.

59. https://groups.yahoo.com/neo/groups/energyresources/conversations/messages/2846.

60. Matthew Simmons, *Twilight in the Desert: The Coming Saudi Oil Shock and the World* Economy (Hoboken, NJ: Wiley, 2005), 51.

61. Michael Ruppert, *Confronting Collapse: The Crisis of Energy and Money in a Post Peak Oil World* (White River Junction, VT: Chelsea Green Publishing, 2009).

62. "Mexico Discovers Huge Oil Field," *BBC,* March 15, 2006. http://news.bbc.co.uk/2/hi/americas/4808466.stm.

63. http://www.cbsnews.com/news/huge-oil-reserve-found-in-the-gulf/.

64. http://www.offshore-technology.com/projects/lapa-oilfield-santos-basin/.

65. Peter Schweizer, *Victory: The Reagan Administration's Secret Strategy to Hasten the Collapse of the Soviet Union* (Atlantic Monthly Press, 1994).

66. Ibid., C. J. Campbell, *The Coming Oil Crisis* (Brentwood, Essex, England: Multi-Science Publishing Company 1997), 55–57.

67. Michael Ruppert, *Confronting Collapse: The Crisis of Energy and Money in a Post Peak Oil World* (White River Junction, VT: Chelsea Green Publishing, 2009)*,* 46.

68. General Accounting Office, "Crude Oil: Uncertainty about Future Supply Makes it Important to Develop a Strategy for Addressing a Peak and Decline in Oil Production." (Washington: General Printing Office: February 2007).

69. http://www.washingtonpost.com/wp-dyn/content/article/2008/07/28/AR2008072802905_pf.html.

70. Actually, Jamestown was not settled until 1607.The first Africans were brought in 1621 to the colony. http://www-cgi.cnn.com/US/9510/megamarch/10-16/transcript/index.html.The two monuments are obviously not 19 feet high, but perhaps he was referring to the statues inside them.

71. *ASPO Newsletter,* April 2004, 3.

CHAPTER 12

1. Chris Nelder, "Reading Peak Oil Deniers Is a Waste of Time," August 28, 2009, http://www.greenchipstocks.com/articles/reading-peak-oil-deniers/486.

2. See Michael Shermer, *The Believing Brain*: From Ghosts and Gods to Politics and Conspiracies—How We Construct Beliefs and Reinforce Them as Truths (New York: Times Books 1994).

3. http://www.chevron.com/chevron/pressreleases/article/07052005_newchevronadvertisingtargetsdialogueaboutglobalenergyissues.news.

4. David Goodstein, *Out of Gas: The End of the Age of Oil* (New York: Norton, 2004).

5. This appears to have been one of the funders of Lovins et al., *Winning the Oil Endgame: Innovation for Profits, Jobs, and Security* (Snowmass, Colorado: Rocky Mountain Institute, 2005); see Ibid., 278.

6. Michael J. Lynch, *Big Prisons, Big Dreams: Crime and the Failure of America's Penal System (Critical Issues in Crime and Society)* (New Brunswick: Rutgers University Press, 2007).

7. Lester R. Brown, Christopher Flavin, and Colin Norman, "Running on Empty: The Future of the Automobile in an Oil-Short World" (New York: W.W. Norton, 1979).

8. http://www.amazon.com/James-Howard-Kunstler/e/B000APLGD0/ref=la_B000APLGD0_pg_1?rh=n%3A283155%2Cp_82%3AB000APLGD0&ie=UTF8&qid=1441658602.

9. Stan Cox, *Losing Our Cool: Uncomfortable Truths about Our Air-Conditioned World (and Finding New Ways to Get through the Summer)* (New York: New Press, 2010).

10. Kirkpatrick Sale, *Rebels against the Future: The Luddites and Their War on the Industrial Revolution: Lessons for the Computer Age* (Boston, MA: Addison-Wesley, 1995).

11. Joel Mokyr, *The Lever of Riches: Technological Creativity and Economic Progress* (Oxford: Oxford University Press, 1992), 259.

12. Robert Ferguson, *The Vikings: A History* (New York: Viking Adult, 2009), 334.

13. Freidlander et al. Saul Friedlander, ed., *Visions of Apocalypse: End or Rebirth?* (Teaneck, NJ: Holmes and Meier, 1985), 71.

14. King James Bible, "Daniel," chapter 8, verse 19.

15. David Abulafia, *The Great Sea: A Human History of the Mediterranean* (Oxford: Oxford University Press, 2011), 471.

16. Saul Friedlander, ed., *Visions of Apocalypse: End or Rebirth?* (Teaneck, NJ: Holmes and Meier, 1985), 71.

17. *http://www.youtube.com/watch?v=0Hrpv21Ubf8&list=PLB81766ED6AA72EB9&index=2.*

18. ASPO NL 40.

19. Sam Foucher, "Russia's Oil Production Is about to Peak," April 24, 2008. http://www.theoildrum.com/node/3626.

20. Martin Payne, "2nd Largest Oil Field in the World, Mexico's Cantarell, Declining Rapidly!," http://energyskeptic.com/2014/2nd-largest-oil-field-in-the-world-mexics-cantarelle-declining-rapidly/.

21. Ed Blanche, "Shell Scandal Points to Exaggerated Estimates of Oil Reserves," April 17, 2004. http://www.countercurrents.org/peakoil-blance170404.htm.

CHAPTER 13

1. National Petroleum Council, *Hard Truths: Facing the Hard Truths about Energy* (Washington, DC: National Petroleum Council, 2007); Donald Paul Hodel and Ronald Dietz, *Crisis in the Oil Patch: How*

America's Oil Industry Is Being Destroyed and What Must Be Done to Save It (Washington, DC: Regnery Publishing, 1994); John Hofmeister, *Why We Hate the Oil Companies: Straight Talk from an Industry Insider* (New York: Palgrave MacMillan, 2010).

2. "Confederates Attack Oilfield," American Oil and Gas Historical Society, http://aoghs.org/oil-amanac/confederates-attack-oilfield/.

3. Kathleen Conti, "Solar Projects Increasingly Meet Local Resistance," Kathleen Conti. In: *Boston Globe,* May 5, 2013.

4. http://globalwarming.markey.house.gov/mediacenter/pressreleases _id=0217.html.

5. http://www.nytimes.com/2013/05/23/opinion/nocera-here-comes -the-sun.html?ref=joenocera&_r=0.

6. Hofmeister, *Why We Hate the Oil Companies* (New York: Palgrave MacMillan, 2010).

7. Guy Chazan, " 'Terrifying' Oil Skills Shortage Delays Projects and Raises Risks," *Financial Times,* July 16, 2014.

8. Number of New Graduates in Petroleum Engineering Increases, Society of Petroleum Engineers White Paper, January 27, 2010. As of May 2013, Bureau of Labor Statistics, http://www.bls.gov/oes/current/ oes172171.htm.

9. E. W. Hough and H. C. Simrall, "The Role of the Small Petroleum Engineering Department." *Journal of Petroleum Technology*, 1962.

10. Ibid.

11. Roman Kilisek, "Data Avalanche and the Great Crew Change in the Oil and Gas Industry," September 9, 2014. http://breakingenergy .com/2014/09/09/data-avalanche-and-the-great-crew-change-in-the-oil -and-gas-industry/?utm_source=breakingenergy&utm_medium=module &utm_campaign=Career%20Insights.

12. Robert L. Bradley, ed., *Oil, Gas and Government: The U.S. Experience*, vols. 1 & 2 (Washington, DC: Rowan & Littlefield, 1995).

13. American Petroleum Institute, "Environmental Expenditures by the U.S. Oil and Gas Industry: 1990–2014," 2014.

14. Conglin Xu, "Capital Spending in US, Canada to Rise Led by Pipeline Investment Boom," *Oil & Gas Journal,* March 4, 2013.

15. David Wethe, "Rig Shortage Means Record $4.5 Billion Blowout Binge: Energy," *Bloomberg,* August 10, 2012, http://www.bloomberg .com/news/articles/2012-08-09/rig-shortage-means-record-4-5-billion -blowout-binge-energy.

16. Jaxon Van Derbeken, "Benicia Sees Cash in Crude Oil; Neighbors See Catastrophe," *SF Gate,* October 24, 2014, http://www.sfgate.com/

bayarea/article/Benicia-sees-cash-in-crude-oil-neighbors-see-5843785
.php.

17. Ernst & Young LLP, "Investment and Other Uses of Cash Flow by the Oil Industry 1992–2006," Report to the American Petroleum Institute, May 2007.

18. National Petroleum Council, *Hard Truths.*

19. Xu Conglin, "Barclays: Global E&P Spending to Reach Record High in 2013," *Oil & Gas Journal,* June 4, 2013, http://www.ogj.com/articles/2013/06/barclays—global-e-p-spending-to-reach-record-high-in-2013.html.

20. Luis Miguel Labardini, "Seven Misconceptions about the Oil Industry in Mexico," *Informe de Petroleo,* May 25, 2008.

21. http://www.oil-price.net/en/articles/oil-and-refineries.php.

22. http://www.eia.gov/dnav/pet/hist/LeafHandler.ashx?n=PET&s=8_NA_8O0_NUS_C&f=A.

23. http://www.asce.org/Content.aspx?id=25562.

24. Phil, Hart "The Cost of Corrosion," March 23, 2009. http://anz.theoildrum.com/node/5215.

25. http://abcnews.go.com/US/story?id=96090.

CHAPTER 14

1. Robert Hirsch, "The Inevitable Peaking of World Oil Production," *Bulletin, Atlantic Council of the United States*, October 2005.

2. Matthew R. Simmons, *Twilight in the Desert* (New York: Wiley, 2005), 302.

3. Ibid., 288.

4. M. A. Adelman, *The Economics of Petroleum Supply* (Cambridge, MA: MIT Press, 1993); see also Carol A. Dahl, *International Energy Markets: Understanding Pricing, Policies, and Profits* (Tulsa, OK: Pennwell Publishing, 2015).

5. http://www.theoildrum.com/node/9474.

6. UK Energy Research Centre, *Global Oil Depletion: An Assessment of the Evidence for a Near-Term Peak in Global Oil Production*, London, 2009, p. 1.

7. International Energy Agency, *World Energy Outlook*, Paris, 2008.

8. M. A. Adelman, "Mineral Depletion Theory with Special Reference to Petroleum," *Review of Economics and Statistics*, 1990.

9. Stuart McCarthy, "Peak Oil and the Australia's National Infrastructure," October 2008, http://www.infrastructureaustralia.gov.au/public

_submissions/published/files/266_australianassociationforthestudyofpeak oilandgas_SUB.pdf.

10. Kjell Aleklett, *Peeking at Peak Oil* (New York: Springer, 2012), 80.

11. Ibid., 81.

12. Ibid., 140.

13. "Revealing Insights into Peak Oil by a Former Bush Insider," http://www.fromthewilderness.com/members/053103_aspo.html, June 12, 2003.

14. Cambridge Energy Research Associates, "Finding the Critical Numbers: What are the Real Decline Rates for Global Oil Production." 2008.

15. Mikael Hook, R. Hirsch, and K. Aleklett, "Giant Oil Field Decline Rates and Their Influence on World Oil Production," *Energy Policy*, June 2009.

16. Russell McClulley, "Apache Finds New Opportunities at Forties Field," *Offshore,* August 8, 2013, http://www.offshore-mag.com/articles/print/volume-73/issue-8/north-sea/apache-finds-new-opportunities-at -forties-field.html.

17. Mikael Hook et al., *Giant Oil Field Decline Rates and Their Influence on World Oil Production*, 2009.

18. Data from https://www.og.decc.gov.uk/pprs/full_production.htm.

19. http://www.theguardian.com/business/2011/feb/08/saudi-oil-reserves -overstated-wikileaks.

CHAPTER 15

1. Daniel Yergin, *The Prize* (Simon and Schuster, 1991), 52.

2. James A. Clark and Michel T. Halbouty, *Spindletop: The True Story of the Oil Discovery That Changed the World* (Gulf Publishing Company, 1952), 114.

3. Cited in Daniel Yergin, *The Prize* (Simon and Schuster, 1991), 164.

4. Jean Laherrere, Letter to Philip Watts, October 13, 2001.

5. World Energy Assessment Team, "Greenland, Assessment Results Summary," U.S. Geological Survey, 2000, 1.

6. Colin J. Campbell, "Oil Price Leap in the 1990s," *Noroil*, December 1989 and *ASPO Newsletter*.

7. Charles A. S. Hall and Kent A. Klitgaard, *Energy and the Wealth of Nations: Understanding the Biophysical Economy* (New York: Springer, 2012), 324.

8. *Playboy,* February 2007.

9. Cited in Alex Epstein, *The Moral Case for Fossil Fuels* (New York: Penguin, 2014), 16.

10. The statistical workbook can be downloaded from www.bp.com.

11. H. -H. Rogner, "An Assessment of World Hydrocarbon Resources," *Annual Review of Energy and Environment* (Palo Alto: Annual Reviews, 1997).

12. Jean Laherrere, "Fossil Fuels Future Production," Bucharest Romania, March 2005.

13. M. A. Adelman John Houghton, Gordon Kaufmann, and Martin Zimmerman, *Energy Resources in an Uncertain Future: Coal, Gas, Oil and Uranium Supply Forecasting* (Cambridge, MA: Ballinger, 1983). Cited in M. A. Adelman, *The Economics of Petroleum Supply.* (Cambridge, MA: MIT Press, 1993), 127.

14. http://www.spe.org/industry/docs/GlossaryPetroleumReserves-ResourcesDefinitions_2005.pdf is a glossary produced by the Society of Petroleum Engineers, but no author or other information is given.

15. Colin J. Campbell, *The Coming Oil Crisis* (Essex, UK.: Multi-Science) 1997, 71, 74, 91.

16. Josef Chmielowski, "BP Alaska Heavy Oil Production from the Ugnu Fluvial-Deltaic Reservoir." AAPG Search and Discovery Article 80289, May 27, 2013.

17. P. R. Felder Zitha, D. Zornes, K. Brown, and K. Mohanty, "Increasing Hydrocarbon Recovery Factors," http://www.spe.org/industry/increasing-hydrocarbon-recovery-factors.php.

18. National Petroleum Council, *Hard Truths: Facing the Hard Truths about Energy* (Washington, DC: National Petroleum Council, 2007), 109.

19. Campbell and Laherrere, 1998, op. cit.

20. N. J. Smith and G. H. Robinson, Technology Pushes Reserves 'Crunch' Date Back in Time," *Oil & Gas Journal*, April 7, 1997.

21. National Petroleum Council, *Hard Truths: Facing the Hard Truths about Energy* (Washington, DC: National Petroleum Council, 2007), 109, puts it at "about one-third."

22. Mahmoud Abdul Baqi and Nansen Saleri, "Fifty Year Crude Oil Supply Scenarios: Saudi Aramco Perspective," Presented at CSIS February 24, 2004, 23.

23. Vello Kuuskra, "Undeveloped Oil Resources: a Big Target for Enhanced Oil Recovery," *World Oil,* August 2006.

24. John Noble Wilford, "Creatures of the Cambrian May Have Lived on," *New York Times,* May 17, 2010, http://www.nytimes.com/2010/05/18/science/18fossil.html?_r=0.

25. Peter Odell, *Why Carbon Fuels Will Dominate the 21st Century Energy Economy* (Essex, U.K.: Multi-Science Publishing, 2004).

26. Peter R. Odell and K. E. Rossing, *Optimal Development of North Sea Oil Fields* (London: Kogan Page Ltd, 1976), as one example.

27. H. H Rogner, "An Assessment of World Hydrocarbon Resources," *Annual Review of Energy and Environment* (Palo Alto, CA: Annual Reviews, 1997).

28. Energy Technology Systems Analysis Programme, "Unconventional Oil and Gas Production" (Paris: International Energy Agency, May 2010).

CHAPTER 16

1. Chris Nelder (2012), http://www.smartplanet.com/blog/the-energy-futurist/the-cost-of-new-oil-supply/.

2. In economic terms, the profits that accrue from mineral resources are called "rents" to distinguish them from profits created by firms. In other words, the value of the resource is not simply the result of corporate actions, as in the manufacture of a car, but to their intrinsic value.

3. M. A. Adelman, "Finding and Developing Costs in the USA, 1945–1986," in *Advances in the Economics of Energy and Resources,* ed. John R. Moroney (JAI Press, 1991).

4. Peter Eglington and Maris Uffelman, "Observed Costs of Oil and Gas Reserves in Western Canada, 1965–1979," Economic Council of Canada, Ottawa, Canada, Discussion Paper No. 235, August 1983.

5. OGJ Staff, "UK North Sea Causeway, Fionn Field Start-Ups Near," *Oil & Gas Journal,* August 21, 2012, http://www.ogj.com/articles/2012/08/uk-north-sea-causeway-fionn-field-start-ups-near.html; "BP, Partners to Invest L10 Billion in UK North Sea Projects," *Oil & Gas Journal,* October 24, 2011.

6. See Adelman, "Mineral Depletion with Special Reference to Petroleum," *Review of Economics and Statistics,* 1990.

7. Adelman and Lynch 1985, op. cit.

8. Scotia Howard Weil, "2014 F&D Cost Study," http://www.howardweil.com/docs/Reports/Conference/FDStudy.pdf.

9. His research has been summarized in his collected papers, *The Economics of Petroleum Supply* (MIT Press, 2003), and Carol Dahl has done an excellent job of summarizing the equations used to convert expenditures into costs; see Carol A. Dahl, *International Energy Markets: Understanding Prices, Policies and Profits* (Pennwell Publishing, 2004).

10. Specifically, the implicit price deflator of the U.S. GDP, published annually by the Council of Economic Advisors in the Economic Report to the President.

11. M. A. Adelman, "The Competitive Floor Price to World Oil Prices." *The Energy Journal*, October 1986.

12. John E. Tilton, *On Borrowed Time: Assessing the Threat of Mineral Depletion* (Washington, D.C.: RFF Press, 2002).

13. Douglas R. Bohi, "Changing Productivity in U.S. Petroleum Exploration and Development. Washington, D.C.: RFF Working Paper 98–38, 1998.

14. M. A .Adelman with Manoj Shahi, "Oil-Development-Operating Cost Estimates 1955–1985," *Energy Economics,* January 1989.

15. American Petroleum Institute, "Environmental Expenditures by the U.S. Oil and Gas Industry," http://www.api.org/~/media/files/publications/environmental-expenditures-2014.

16. Matt Zbrowski, "E&Y: US Firms' Capex, Revenues, Reserves up Again in 2014, but Cuts Loom," *Oil & Gas Journal,* June 15, 2015.

17. International Energy Agency, *World Energy Outlook* 2000. Paris, 79.

18. Production and reserves from Canadian Association of Petroleum Producers, *Statistical Handbook 2015.* Production time is approximate from the point of significant levels.

CHAPTER 17

1. There are rare exceptions, where a government (usually) pays for "access" to "secure" oil, but even then, the premium is marginal.

2. Mason Inman, "How to Measure the True Cost of Fossil Fuels," *Scientific American,* April 1, 2013.

3. David Pimentel, "Energy Balance, Economics and Environmental Effects Are Negative," *Natural Resources Research,* June 2003.

4. C. A. S. Hall and C. J. Cleveland, "Petroleum Drilling and Production in the United States: Yield per Effort and Net Energy Analysis." *Science* 211 (1981): 576–79; Megan C. Guilford, Charles A.S. Hall, Pete O' Connor, and Cutler J. Cleveland, "A New Long Term Assessment of Energy Return on Investment (EROI) for U.S. Oil and Gas Discovery and Production," *Sustainability,* 2011.

5. Charles A. S. Hall and Kent A Klitgaard, *Energy and the Wealth of Nations: Understanding the Biophysical Economy* (New York: Springer 2012), 315.

6. Megan C. Guilford, Charles A. S. Hall, Pete O' Connor, and Cutler J. Cleveland, "A New Long Term Assessment of Energy Return on Investment (EROI) for U.S. Oil and Gas Discovery and Production," *Sustainability,* 2011.

7. http://www.krandeegroup.com/oil_and_gas_investing.html.

8. In my 1991 paper, "A Post-OPEC World?" I described how the politics and management of the oil market had changed from the 1970s.

9. "Confederates Attack Oilfield," *American Oil and Gas Historical Society,* http://aoghs.org/oil-amanac/confederates-attack-oilfield/.

CHAPTER 18

1. http://www.bloomberg.com/news/articles/2015-07-20/oil-guru-who -called-2014-slump-sees-return-to-100-crude-by-2020.

2. https://tshaonline.org/handbook/online/articles/doe01.

3. Peter Schweizer, *Victory: The Reagan Administration's Secret Strategy that Hastened the Collapse of the Soviet Union* (Atlantic Monthly Press, 1994).

4. Colin J. Campbell, *The Coming Oil Crisis* (Brentwood, Essex, England: Multi-Science Publishing Company 1997), 56.

5. Energy Modeling Forum 6, *World Oil* (Terman Engineering Center, Stanford University, 1982).

6. Global Petroleum Division, Chase Manhattan Bank, "World Oil and Gas 1985" (New York: Chase Manhattan Bank, August 1985).

7. I'm indebted to Philip K. Verleger for this quote.

8. "Puzzled Oilmen Ask 'Where Have All the Barrels Gone?," *Petroleum Intelligence Weekly,* November 25, 1985, 3.

9. There are some consultants who claim to be able to estimate oil at sea, but it has not been as reliable as would be hoped.

10. Michael C. Lynch, "Structural Changes in World Oil Markets and Their Impact on Market Behavior," MIT Energy Laboratory Working Paper MIT-EL 86-009WP, March 1986.

11. Robert L. Hirsch, "Impending US Energy Crisis," *Science,* March 20, 1987.

12. *Petroleum Economist,* October 1989.

13. C. J. Campbell, "Oil Price Leap in the 1990s." *Noroil,* December 1989.

14. Colin J. Campbell, *The Coming Oil Crisis* (Brentwood, Essex, England: Multi-Science Publishing Company 1997), 144.

15. Donald Hodel and Robert Dietz, *Crisis in the Oil Patch: How America's Energy Industry Is Being Destroyed and What Must Be Done to Save It* (Regnery Publishing, 1994).

16. Peter Coy Gary McWilliams and John Rossant, "The New Economics of Oil." *Business Week,* November 3, 1997, http://www.businessweek.com/1998/35/z3551001.htm.

17. From Firesign Theater.

18. Vahan Janjigian, Stephen M. Horan, and Charles Trzcinka, *The Forbes Crash Course in Investing*, Forbes Media 2014.

19. *The Telegraph,* UK, September 21, 2009, http://www.telegraph.co.uk/finance/oilprices/6215333/Oil-prices-risk-rebounding-above-100-says-Total-chief-Christophe-de-Margerie.html.

20. Dmitry Zhdannikov and Claire Milhench, "Saudis, Soaring Costs May Keep Oil above $100," *Reuters,* May 16, 2012, http://www.reuters.com/article/2012/05/16/usenergysummitoilpriceidUSBRE84F0RW20120516.

CHAPTER 19

1. http://www.azimuthproject.org/azimuth/show/Biomass.

2. http://eaglefordshale.com/companies/.

3. Daniel M. Jarvie, "Unconventional Oil Petroleum Systems: Shales and Shale Hybrids," http://www.searchanddiscovery.com/documents/2011/80131jarvie/ndx_jarvie.pdf.

4. "Worldwide Review," *Oil & Gas Journal.* I have excluded the reserves of heavy oil in Canada and Venezuela from the totals.

5. Advanced Resources International, "Technically Recoverable Shale Oil and Shale Gas Resources: An Assessment of 137 Shale Formations in 41 Countries Outside the United States" (Washington, DC: Energy Information Administration, June 2013).

6. Energy Information Administration, Review of Emerging Resources: U.S. Shale Gas and Shale Oil Plays" (Washington, DC: Department of Energy, 2011).

7. Deepwater Horizon Study Group, "Final Report on the Investigation of the Macondo Well Blowout" (Berkeley, CA: Center for Catastrophic Risk Management, March 1, 2011).

8. In one notable episode, the drill stopped working and the owner had to come to the site to figure out why. The rig was out of gas and the operator hadn't checked the tank.

9. Nat gas. Berman, Arthur, "The Shale Delusion: Why the Party is Over for Tight Oil," September 14, 2015. http://oilprice.com/Energy/Crude-Oil/The-Shale-Delusion-Why-The-Partys-Over-For-US-Tight-Oil.html.

10. DaveCohen, "A Shale Gas Boom?" June 25, 2009, http://www.resilience.org/stories/2009-06-25/shale-gas-boom.

11. John Krohn, "The 'Top Ten, Mistakes Letterman Made on HF." October 12, 2010, http://energyindepth.org/marcellus/the-top-ten-mistakes-letterman-made-on-hf-3/; *New York Times*, July 9, 2015.

12. Michael C. Lynch, "Fracking Opponents Ditch Science, Embrace Hysteria," July 9, 2015. http://www.forbes.com/sites/michaellynch/2015/07/09/fracking-opponents-ditch-science-embrace-hysteria/.

13. The various studies are listed in https://www.edf.org/climate/methane-studies.

14. Environmental Protection Agency, "Assessment of the Potential Impacts of Hydraulic Fracturing for Oil and Gas on Drinking Water Resources: *Executive Summary*," June 2015, http://www2.epa.gov/sites/production/files/2015-07/documents/hf_es_erd_jun2015.pdf.

15. D. Nathan Meehan, "Challenges Remain in Exporting North America's Shale Experience," *World Oil,* December 2013.

16. Robert A., III Hefner, *The Grand Energy Transition: The Rise of Energy Gases, Sustainable Life and Growth, and the Next Great Economic Expansion* (John Wiley and Sons, 2009).

17. Vello Kuuskra, Scott Stevens, Tyler van Leeuwen and Keith Modhe, *World Shale Gas Resources: An Initial Assessment of 14 Regions Outside the United States* (Washington, D.C.: Energy Information Administration: 2011).

18. *International Oil Daily,* November 21, 2012.

19. Vello Kuuskra, Scott Stevens, Tyler van Leeuwen and Keith Modhe, *World Shale Gas Resources: An Initial Assessment of 14 Regions Outside the United States* (Washington, DC: Energy Information Administration: 2011).

20. OGJ Editors, "Continental: Bakken's Giant Scope Underappreciated," http://www.ogj.com/articles/2011/02/continental—bakken.html.

21. Advanced Resources International, "Shale Gas and Shale Oil Resource Assessment Methodology," June 2013. https://www.eia.gov/analysis/studies/worldshalegas/pdf/methodology_2013.pdf.

22. http://www.petroleumnewsbakken.com/pntruncate/138545455.shtml; http://www.bbc.com/news/uk-england-33018802.

23. http://www.ogj.com/articles/uogr/print/volume-2/issue-3/permian-operators-increasingly-target-shale-as-new-technology-rejuvenates-legacy-oil-field.html.

24. *Upstream,* May 1, 2015.

25. Jan Stuart, "Rebalancing: Oil Markets Have Passed the Low (We Think)—So Now What?," *Credit Suisse,* April 2015.

26. Russell Gold, "Easy Access to Money Keeps Oil Flowing." *Wall Street Journal,* June 1, 2015, b1.

27. Matt Zborowski, "CERAWeek: Regulatory Hurdles Still Threaten U.S. Shale, CEOs say." *Oil & Gas Journal,* April 27, 2015.

28. Ibid.

29. Ibid.

30. Ibid.

31. Jim Redden, "Shaletech: Permian Shales," *World Oil,* April 2015.

32. http://thebakken.com/articles/1101/halconundefineds-bakken -well-costs-decline-as-production-increases

33. Company Reports.

34. Lynn Doan and Dan Murtaugh, "Shale Oil Boom Could End in May after Price Collapse," April 13, 2015, http://www.bloomberg.com/ news/articles/2015-04-13/shale-oil-boom-seen-ending-in-may-after -price-collapse.

35. Citicorp, "Energy 2020: North America, the New Middle East?," March 20, 2012.

36. http://permianshale.com/news/id/121319/signs-point-to-recovery/.

37. Ernest Scheyder and Mike Stone, "Oxy to Exit North Dakota's Oil Fields in Sale to Private Equity Fund," October 15, 2015, http://www .reuters.com/article/2015/10/15/us-limerock-occidental-bakken-idUSKCN 0S92QR20151015.

38. http://www.cnbc.com/id/.

39. http://thebakken.com/articles/1101/halconundefineds-bakken -well-costs-decline-as-production-increases.

40. Troy A. Cook, "Procedure for Calculating Estimated Ultimate Recoveries of Bakken and Three Forks Formations Horizontal Wells in the Williston Basin" (Washington, DC: US Geological Survey, Open-File Report 2013-1109).

CHAPTER 20

1. Norwegian Petroleum Directorate, *Facts 2014* (Oslo, 2014), 13. Norwegian Petroleum Directorate.

2. "Hurricane Blows in Optimism," *Upstream,* September 18, 2015, p. 28.

3. Rob Watts and Eric Means, "Statoil Working on Slimmed-Down Bressay Solution," *Upstream,* September 4, 2015, p. 2; "Mariner Jacket

Put in Place," *Upstream*, September 4, 2015, p. 3. IainEsau, "Yards Wait for Captain's Orders," *Upstream*, September 4, 2015, p. 19.

4. Iain Esau, "Statoil Spuds Wildcat off Newfoundland," *Upstream*, September 18, 2015, p. 6.

5. "Pemex to Invest $6B in Cantarell Production," *Offshore*, July 3, 2014, http://www.offshore-mag.com/articles/2014/07/pemex-to-invest-6b-in-cantarell-production.html.

6. Luis Miguel Labardini, "Seven Misconceptions about the Oil Industry in Mexico," *Informe de Petroleo*, May 25, 2008.

7. Trent Jacobs, "Major Oil and Gas Discoveries Offshore Mexico," *Journal of Petroleum Technology*, August 2015, p. 41.

8. OGJ Editors, "Pemex Lets Contract in Ayatsil Development," *OGJ Online*, June 25, 2014.

9. http://elpais.com/elpais/2014/09/22/inenglish/1411395118_595417.html.

10. https://mninews.marketnews.com/content/mexicos-calderon-announces-1st-deep-water-oil-discovery.Other reports gave much lower numbers, for example, 3P reserves of 350 million barrels oil equivalent. http://www.offshore-mag.com/articles/2012/08/pemex-makes-new-deepwater.html.

11. Peter Kiernan and Roger Knight, "Assessing Mexico's Offshore Potential," *Offshore*, June 1, 2011.

12. Advanced Resources International, *Technically Recoverable Shale Oil and Gas Resources: An Assessment of 137 Shale Formations in 41 Countries outside the United States* (Washington, D.C.: U.S. Department of Energy, June 2013), 8.

13. L. R. Guardado, A. R. Spadini, J. S. L. Brandao, and M. R. Mello, *AAPG Memoir 73*, Chapter 22: Petroleum System of the Campos Basin, Brazil.

14. Sabrina Valle and Juan Pablo Spinetto, "Petrobras Foregoes Growth in 37% Spending Cut to Ease Debt," http://www.bloomberg.com/news/articles/2015-06-29/petrobras-slashes-5-year-spending-plan-to-130-3-billion, June 29, 2015.

15. http://www.upi.com/Business_News/Energy-Resources/2015/07/06/Report-Brazils-Petrobras-wading-dark-waters/2761436177632/.

16. "Tullow Cuts Exploration Budget Amid Falling Oil Prices," November 13, 2014, http://www.ogj.com/articles/2014/11/tullow-shifts-exploration-focus-amid-falling-oil-prices.html.

17. OGJ Editors, "ENI Reports Flow Test Rates for Well Offshore Congo (Brazzaville)," *Oil & Gas Journal*, December 15, 2014; Laura

Whiteand Arran Waterman, "Regional Report: West Africa," *World Oil*, August 2014.

18. Andrew Rosati and Joe Carroll, "Exxon's Guyana Oil Discovery May Be 12 Times Larger Than Economy," July 21, 2015, http://www.bloomberg.com/news/articles/2015-07-21/exxon-s-guyana-oil-find-may-be-worth-12-times-the-nation-s-gdp.

19. Michael C. Lynch, "The Economics of Petroleum in the Former Soviet Union," *Gulf Energy and the World: Challenges and Threats* (Abu Dhabi: The Emirates Center for Strategic Studies and Research, 1997).

20. "Boost for Kazakh Pipeline," *Upstream*, September 18, 20115, p. 22.

21. OPEC, *World Oil Outlook: 2014* (Vienna: OPEC), 82.

22. Hashem Kalantari, "Iran Targets 45 Oil and Gas Projects in Plan to Boost Production," August 13, 2015, http://www.bloomberg.com/news/articles/2015-08-13/iran-targets-45-oil-and-gas-projects-in-plan-to-boost-production.

23. http://www.eia.gov/todayinenergy/detail.cfm?id=23472, October 23, 2015.

24. http://mnr.krg.org/index.php/en/oil/vision.

25. "PDVSA to Boost Orinoco Oil Production," http://www.pennenergy.com/articles/pennenergy/2014/03/oil-and-gas-pdvsa-to-boost-orinoco-oil-belt-production.html, March 7, 2014.

26. http://www.eia.gov/beta/international/analysis.cfm?iso=AGO.

27. Shrikesh Laximidas, "Cobalt Makes Biggest Oil Find Yet in Kwanza Basin off Angola," May 1, 2014,http://www.reuters.com/article/2014/05/01/angola-cobalt-intl-idUSL6N0NN34S20140501. Iain Esau, "BP Runs Rule over Leda Appraisal Well Results," *Upstream*, October 2, 2015, p. 18.

CHAPTER 21

1. Papers such as Cremer and Weitzman sought to explain why higher prices in, for example, the mid-1970s seemed to lead to restricted supply using, in part, the idea of limited absorptive capacity (El Mallakh, 1981) to explain why oil exporters might seek to limit their revenue. In practice, it proved that the absorptive capacity was not the same as the ability to spend money. The latter increased rapidly such that the curve did not possess long-term predictive ability. J. Cremer and M. L. Weitzman (1976), "OPEC and the Monopoly Price of World Oil," *European Economic*

Review, vol. 8: 155–164. Ragaei El Mallakh, *The Absorptive Capacity of Kuwait: Domestic and International Perspectives* (Lexington, MA: Lexington Books, 1981).

2. Most notoriously, Colin Campbell and Jean Laherrere in "The End of Cheap Oil," *Scientific American*, March 1998, misled many into thinking there was a sound theory behind the idea of peak oil and that empirical evidence confirmed it. Neither was true then or is now.

3. The problem was worsened by the government set exchange rate of the ruble, thought to be four times the market rate. My calculations at the time were that costs were on the order of $2–4/barrel Michael C. Lynch, "The Economics of Petroleum in the Former Soviet Union," in *Gulf Energy and the World: Challenges and Threats*, (Abu Dhabi: The Emirates Center for Strategic Studies and Research, 1997).

4. Thomas R Stauffer, "Indicators of Crude Oil Production Costs: The Gulf versus non-OPEC Sources." (Boulder, Colo.: ICEED 1993). Marie N. Fagan, "Resource Depletion and Technical Change: Effects on U.S. Crude Oil Finding Costs from 1977 to 1994", *The Energy Journal*, vol. 18, No. 4, pp. 91–105.

5. M. A. Adelman, "The Competitive Floor to World Oil Prices," *The Energy Journal*, October 1986.

6. H. J. Barnett and C. Morse, *Scarcity and Growth* (Washington, DC: Resources for the Future, 1963).

7. Although he made the argument many times, probably the first published version was "Economics of Exploration and Production," *Geoexploration*, 1970, reprinted in Adelman (1993).

8. "A Post-OPEC World? The Long-Term Impact of the 1990 Oil Crisis," *Journal of Economic Democracy*, July–September 1991.

CHAPTER 22

1. Jeff Rubin, *Why Your World is about to Get a Whole Lot Smaller* (New York: Random House, 2009). cit.

2. Matthew Philips, "Cheap Oil Is *Bad* for the Economy (at Least, So Far)," July 15, 2015, http://www.bloomberg.com/news/articles/2015-07-15/cheap-oil-is-bad-for-the-economy-at-least-so-far-.

3. Ibid.

4. A. Gary Shilling, "A. Gary Shilling: Will Oil Cause the Next Recession?" September 2, 2015, http://wap.business-standard.com/article/opinion/a-gary-shilling-will-oil-cause-the-next-recession-115090201437_1.html.

5. Justin Lahart, "Cheap Oil Should Fuel Economy at Last," *Wall Street Journal*, May 11, 2015, p. c6.

6. Andrea Hopkins, "Calgary Home Prices Plunge in Biggest Monthly Drop on Record," *Financial Post*, June 12, 2015, http://business.financial post.com/personal-finance/mortgages-real-estate/calgary-home-prices -plunge-in-biggest-monthly-drop-on-record; And Penty, Jeremy Van Loon, and Robert Tuttle, "Calgary Stampede to Be More Subdued Following Oil Collapse," http://www.bloomberg.com/news/articles/2015 -06-30/calgary-stampede-is-more-subdued-following-oil-collapse, June 30, 2015.

7. Clifford Krauss, "Sinking Oil Prices Are Lowering the Boom in Texas," *New York Times*, August 15, 2015, p. a1.

8. Chris Tomlinson, "Texas Economic Growth Rate Slowing to 1.3 Percent," *Houston Chronicle*, September 18, 2015, http://www.houston chronicle.com/business/outside-the-boardroom/article/Texas-economic -growth-rate-slowing-to-1-3-percent-6513769.php.

9. Alex Nixon, "PA. Unemployment Rate Hovers at 5.4%," http:// triblive.com/business/headlines/9111333-74/jobs-month-state#axzz3 mIpQVGbA, September 18, 2015.

10. "Oil Is Not Food but Food Is Oil: The Imminent Crisis of Food Production Dependence on Oil," http://peakoil.com/consumption/oil-is -not-food-but-food-is-oil-the-imminent-crisis-of-food-production-dependence -on-oil, January 1, 2011.

11. Bill Lehane, Firat Kayakiran, and Angelina Rascouet, "Gloom-to-Boom Oil Refiners Set Up Fuel Crunch as Work Shelved," September 16, 2015, http://www.bloomberg.com/news/articles/2015-09-16/gloom-to -boom-oil-refiners-set-up-fuel-crunch-as-work-shelved.

12. Jane Martinson, "Al-Jazeera to Cut Hundreds of Jobs," September 22, 2015, http://www.theguardian.com/media/2015/sep/22/al-jazeera-expected -to-cut-hundreds-of-jobs-qatar-oil-price?CMP=share_btn_tw.

13. Chris Helman, "Oil Goes Down, Bankruptcy Goes Up—These 5 Frackers Could Be Next to Fall," http://www.forbes.com/sites/christopher helman/2015/08/17/as-oil-goes-down-bankruptcies-go-up-these-5-frackers -could-be-the-next-to-fall/; Chris Helman, "Not Dead Yet: Shares in Oil Driller Halcon Surge on Debt Swap," http://www.forbes.com/sites/ christopherhelman/2015/08/28/not-dead-yet-shares-in-oil-driller-halcon -surge-on-debt-swap/.

14. Bradley Olson, "U.S. Shale Drillers Are Drowning in Debt," http:// www.bloomberg.com/news/articles/2015-09-17/an-oklahoma-of-oil-at -risk-as-debt-shackles-u-s-shale-drillers, September 18, 2015.

15. Emily Glazer, Ryan Tracy, and Rahel Louise Ensign, "Energy Lending Caught in Squeeze," *Wall Street Journal*, September 24, 2015, p. C1.

16. Alison Sider, "Frackers Who Drove Boom Struggle to Survive," *Wall Street Journal*, September 24, 2015, p. B1.

17. Krauss, "Sinking Oil Prices."

18. "Halliburton Swings Axe," http://www.offshoreenergytoday.com/halliburton-swings-axe-again/; Sneha Banerjee, "It's a $15 Billion Oil Megamerger," http://www.businessinsider.com/its-a-15-billion-oil-mega-merger-2015-8, August 26, 2015.

19. David Wethe, "The Surprisingly Big Market for Sand Just Collapsed," September 21, 2015, http://www.bloomberg.com/news/articles/2015-09-21/oil-bust-claims-unusual-american-victim-far-away-from-shale-rigs.

20. Tim Loh and Mario Parker, "Mines in America's Coal Country Just Sold for a Total of Nothing," September 21, 2015, http://www.bloomberg.com/news/articles/2015-09-22/mines-in-america-s-coal-country-just-sold-for-a-total-of-nothing.

21. Rebecca Penty, "Shell Halts Alberta Oil-Sands Project after Leaving Arctic," http://www.bloomberg.com/news/articles/2015-10-27/shell-to-halt-carmon-creek-oil-sands-project-in-alberta, October 27, 2015.

22. Robbie Whelan, "Oil-Patch Slump Spurs Wave of Trucking Deals," *Wall Street Journal*, August 11, 2015.

23. Trafis Team, "Falling Crude Prices Could Begin to Hurt Caterpillar," Forbes, November 24, 2014, http://www.forbes.com/sites/greatspeculations/2014/11/21/falling-crude-oil-prices-could-begin-to-hurt-caterpillar/; Dan Shingler, "Plummeting Price of Oil Is Weakening Steel Industry," January 11, 2015, http://www.crainscleveland.com/article/20150111/SUB1/301119978/plummeting-price-of-oil-is-weakening-steel-industry.

24. Michael Lynch, "Oil Scarcity, Oil Crises, and Alternative Energies: Don't Be Fooled Again," *Applied Energy*, 1999.

25. Mike Ramsey, "Honda to Discontinue CNG and Hybrid Civic Vehicles, *Wall Street Journal*, June 16, 2015, p. c3.

26. Bill Vlasic, "Toyota Fashions a Bolder Image for the Practical Prius," *New York Times*, September 10, 2015, p. C3.

27. Joseph J. Romm, *The Hype about Hydrogen: Fact and Fiction in the Race to Save the Climate* (Washington, D.C.: Island Press, 2004).

28. Peter Bennett, "UK Solar Market Leader Lightsource 'Surprised and Disappointed' by FiT cuts," September 7, 2015, http://www.pv-tech

.org/news/uk_solar_market_leader_lightsource_surprised_and_disappointed _by_fit_cuts.

29. "Japan to Curb Solar Power Amid Cost Concern," *Nikkei*, April 28, 2015, http://asia.nikkei.com/Politics-Economy/Policy-Politics/Japan-to -curb-solar-power-amid-cost-concern.

30. Tim Loh, "What Cheap Gas Did to Tens of Thousands of Coal Jobs," August 20, 2015, http://www.bloomberg.com/news/articles/2015 -08-20/what-cheap-gas-did-to-tens-of-thousands-of-coal-jobs.

31. Andres Schipani, "Venezuela: Terrorized by Oil Price Drop," *Financial Times*, August 9, 2015, http://www.ft.com/cms/s/0/c9c4b05c -0b81-11e5-994d-00144feabdc0.html#axzz3lprR7enb.

32. Terry Lynn Karl, *The Paradox of Plenty: Oil Booms and Petro-States* (Oakland: University of California Press, 1997).

33. Schipani, "Venezuela."

34. Francisco Parra, *Oil Politics: A Modern History of Petroleum* (London: I. B. Tauris, 2004).

CHAPTER 23

1. John Schwartz, *New York Times*, March 16, 2015.

2. http://zeropollutionmotors.us/; "Difference Engine: End of the Electric Car?" *The Economist*, October 15, 2012, http://www.economist .com/blogs/babbage/2012/10/nitrogen-cycle.

3. New York, http://www.wsj.com/articles/SB10001424127887 3240489045783197833080709860.

4. Russell Phillips, "An Ineffective System: The M247 Sergeant York," http://www.russellphillipsbooks.co.uk/articles/an-ineffective -system-the-m247-sergeant-york/.

5. http://www.greenpeace.org/international/en/publications/reports/ green-power-for-electric-cars/.

6. *The Hydrogen Economy*. New York: Putnam, 2002.

7. Matthew Wald, "Nuclear Power Renaissance?" *Technology Review*, November/December 2009.

8. Thomas Turrentine and Kenneth Kurani, "The Household Market for Electric Vehicles: Testing the Hybrid Household Hypothesis—A reflexively Designed Survey of New-Car-Buying, Multi-Vehicle California House-holds," UC Davis Institute of Transportation Studies, May 15, 1995.

9. Aaron Kessler, "Automakers Join to Expand Network of High-Speed Charging Stations," *New York Times*, January 23, 2015, p. b3.

10. Steve Levine, *The Powerhouse*. New York: Viking, 2015.

11. Harold J. Barnett and Chandler Morse, *Scarcity and Growth: The Economics of Natural Resource Availability*. Washington, D.C.: RFF Press, 1965.

12. Joseph J. Romm, *The Hype about Hydrogen: Fact and Fiction in the Race to Save the Climate*. Washington, D.C.: Island Press, 2005; Brown et. al., op. cit.

13. http://content.time.com/time/specials/2007/article/0,28804,1658545_1658544_1658535,00.html.

14. Tony Aardvark, "Greens Admit Electric Cars Are Not That Green," http://toryaardvark.com/2013/02/07/greens-admit-electric-cars-are-not-that-green/#more-22510.

15. http://www.motherearthnews.com/green-transportation/electric-vehicle-zmaz97fmzgoe.aspx.

16. "Tesla Owners Frustrated by Recharge Waits: Complaints about Long Lines to Top-Off Batteries Has Sparked Warnings to Frequent Users," http://www.wsj.com/articles/tesla-owners-frustrated-by-recharge-waits-1435690694.

17. Jack Spring, "China to Roll Back Electric Vehicle Subsidies Faster," Reuters, April 29, 2015, http://www.reuters.com/article/2015/04/29/china-autos-environment-idUSL4N0XQ23P20150429.

18. http://www.energybulletin.net/node/53404.

19. Amory B. Lovins, "Hypercar Vehicles: Frequently Asked Questions," Rocky Mountain Institute, 1998, p. 7.

20. http://www.theguardian.com/environment/2014/feb/17/amory-lovins-renewable-energy.

21. http://www.caranddriver.com/bmw/i3/pricing/trims.

22. https://medium.com/solutions-journal-summer-2014/the-hypercar-lives-meet-vws-xl1-97603e97612f.

23. Jacqueline Leslie, "Dawn of the Hydrogen Age," *Wired*, 1997, http://archive.wired.com/wired/archive/5.10/hydrogen_pr.html.

24. "Electrifying," *The Economist*, December 18, 1997.

25. Jacqueline, "Dawn of the Hydrogen Age."

26. Kelly Hearn, "Woolsey Champions Ethanol," http://www.upi.com/Science_News/2001/11/01/Woolsey-champions-ethanol/31391004642764/, November 1, 2001.

27. Jim Lane, "EPA Slashes Biofuels Targets for 2014, 2015, 2016 under Renewable Fuel Standard," *Biofuels Digest*, May 29, 2015.

28. AF Staff, "California Embarks on 1000 Methanol Fleet," July 1982, http://www.automotive-fleet.com/article/story/1982/07/california-embarks-on-1000-vehicle-methanol-fleet.aspx.

29. Takeshi Sugiyama, "An Analysis of Alternative Fuels Promotion: The Case of Synthetic Gasoline Production in New Zealand," Thesis, Master of Science in Technology and Policy, Massachusetts Institute of Technology, 1994.

30. David Talbot, "Does Lockheed Martin Really Have a Breakthrough Fusion Machine?" *Technology Review*, October 20, 2014, http://www.technologyreview.com/news/531836/does-lockheed-martin-really-have-a-breakthrough-fusion-machine/.

31. David L. Chandler, "A Small, Modular Fusion Power Plant," MIT News Office, August 10, 2015, http://newsoffice.mit.edu/2015/small-modular-efficient-fusion-plant-0810.

32. http://spectrum.ieee.org/energy/nuclear/inside-the-dynomak-a-fusion-technology-cheaper-than-coal.

33. http://cleantechnica.com/2015/03/28/ev-demand-growing-global-market-hits-740000-units/; http://press.ihs.com/press-release/automotive/slower-not-lower-ihs-automotive-forecasting-886-million-unit-global-light-v.

34. Michael Lind, *Land of Promise: An Economic History of the United States.* Harper Collins, 2012, 93.

35. http://www.bellhelicopter.com/MungoBlobs/118/391/407GXP_WALK_150116-R01_EN_WEB.pdf; http://www.rolls-royce.com/customers/defence-aerospace/products/trainers/m250-turboshaft/engine-specifications.aspx.

36. http://www.toyota.com/camry/features/capacities/2532/2540/2546/2548.

37. Cliff Saran, "Apollo 11: The Computer That Put Man on the Moon," Computerweekly.com, July 2009, http://www.computerweekly.com/feature/Apollo-11-The-computers-that-put-man-on-the-moon.

38. Anna Simet, "Advanced Biofuels Industry: RFS Proposal Is Damaging Industry," http://www.ethanolproducer.com/articles/12630/advanced-biofuel-industry-rfs-proposal-is-damaging-industry, September 17, 2015.

39. Bram Stoker, *Dracula.* New York: Vintage Books, 2007, 132.

Index

About the Author

MICHAEL C. LYNCH is president of Strategic Energy and Economic Research, Inc. He serves as a lecturer in the MBA program at Vienna University, Austria, and blogs on energy for Forbes.com. Lynch has published numerous articles and reports on petroleum economics, such as "The Fog of Commerce: The Failure of Long-term Oil Market Forecasting" and "The Next Oil Crisis." He earned his bachelor of science and master of science degrees from Massachusetts Institute of Technology and conducted research there for nearly two decades.